Conference Proceedings of the Society for Experimental Mechanics Series

For other titles published in this series, go to
www.springer.com/series/8922

Tom Proulx
Editor

MEMS and Nanotechnology, Volume 2

Proceedings of the 2010 Annual Conference on Experimental and Applied Mechanics

Springer

Editor
Tom Proulx
Society for Experimental Mechanics, Inc.
7 School Street
Bethel, CT 06801-1405
USA
tom@sem1.com

ISSN 2191-5644 e-ISSN 2191-5652
ISBN 978-1-4419-8824-9 e-ISBN 978-1-4419-8825-6
DOI 10.1007/978-1-4419-8825-6
Springer New York Dordrecht Heidelberg London

Library of Congress Control Number: 2011923429

Printed on acid-free paper

Springer is part of Springer Science+Business Media (www.springer.com)

Preface

MEMS and Nanotechnology, collecting work comprising the **11th International Symposium on MEMS and Nanotechnology**, represents one of six tracks of technical papers presented at the Society for Experimental Mechanics Annual Conference & Exposition on Experimental and Applied Mechanics, held at Indianapolis, Indiana, June 7-10, 2010. The full proceedings also include volumes on: Dynamic Behavior of Materials, Role of Experimental Mechanics on Emerging Energy Systems and Materials, Application of Imaging Techniques to Mechanics of Materials and Structures, Experimental and Applied Mechanics, along with the Symposium on Time Dependent Constitutive Behavior and Failure/Fracture Processes.

Each collection presents early findings from experimental and computational investigations on an important area within Experimental Mechanics. The current volume on *MEMS and Nanotechnology* includes studies on:

Energy Harvesting and Scaling Effects
Metrology and Standards in MEMS and Nanotechnology
Carbon Nanotubes
Friction and Tribology
Meta-materials
Optical Methods for MEMS and Nano
Adhesion and Stiction
Sensors and Actuators

Microelectromechanical systems (MEMS) and nanotechnology are revolutionary enabling technologies (ET). These technologies merge the functions of sensing, actuation, and controls with computation and communication to affect the way people and machines interact with the physical world. This is done by integrating advances in various multidisciplinary fields to produce very small devices that use very low power and operate in many different environments. Today, developments in MEMS and nanotechnology are being made at an unprecedented rate, driven by both technology and user requirements. These developments depend on micromechanical and nanomechanical analyses, and characterization of structures comprising nanophase materials.

To provide a forum for an up-to-date account of the advances in the field of MEMS and nanotechnology and to promote an alliance of governmental, industrial, and academic practitioners of ET, SEM initiated a *Symposium Series on MEMS and Nanotechnology*.

The 2010 Symposium is the eleventh in the series and addresses pertinent issues relating to design, analysis, fabrication, testing, optimization, and applications of MEMS and nanotechnology, especially as these issues relate to experimental mechanics of microscale and nanoscale structures.

It is with deep gratitude that we thank the organizing committee, session chairs, authors, participants, and SEM staff for making the 11th-*ISMAN* a valuable and unforgettable experience.

The Society would like to thank the organizers of the track, Cosme Furlong, *Worcester Polytechnic Institute;* Gordon A. Shaw, III, *National Institute of Standards and Technology;* Barton Prorok, *Aubum University;* Ryszard J. Pryputniewicz, *Worcester Polytechnic Institute* for their efforts.

Contents

Proceedings of the SEM Annual Conference
June 7-10, 2010 Indianapolis, Indiana USA

Nanomechanical standards based on the intrinsic mechanics of molecules and atoms

Jon R. Pratt, Gordon A. Shaw, and Douglas T. Smith

National Institute of Standards and Technology (NIST)
Gaithersburg, MD, USA 20899, jon.pratt@nist.gov

ABSTRACT

For more than a decade, instruments based on local probes have allowed us to "touch" objects at the nanoscale, making it possible for scientists and engineers to probe the electrical, chemical, and physical behaviors of matter at the level of individual atoms and molecules. In principle, physical interactions on this scale are characterized by fixed, unique values that need only be reliably measured in terms of accurately realized units of force and length to serve as standards. For example, the silicon lattice spacing is often used as a convenient ruler for estimating length in atomic scale images, since this lattice spacing has been independently measured using x-ray interferometry. Recently, the force-induced failure of DNA, often referred to as the overstretch condition, has been proposed as both a standard of force and length in single-molecule bio-physics experiments. Still other nanomechanics researchers have suggested that the rupture force of a single-atom chain is unique to a given metal, and that this intrinsic force can be used to calibrate atomic break junction experiments. In both these examples, a fundamental assumption is that the irreducible nature of nanoscale experimentation, in this case tensile testing, yields consistency befitting a standard. This paper offers context and a condensed overview of recently published results from the NIST Small Force Metrology Laboratory regarding new instruments and capabilities we have developed to examine this fundamental assumption. The reviewed papers describe new test platforms, techniques, and calibration procedures that allow us to bring accurate picoscale measurements of both length and force to bear on the problems of single-molecule and single-atom tensile testing.

Keywords: small force, electrostatic force balance, atomic force microscope, colloidal probe calibration.

INTRODUCTION

An intrinsic standard is an invariant of nature that can be linked to a unit of measure. The triple point of water as a calibration reference in thermometry is a familiar example; the silicon lattice spacing, often used as a convenient ruler for estimating length in atomic scale images, is another. In the realm of piconewton to nanonewton forces, there are a handful of phenomena already referred to as "standards" in biophysical investigations of single-molecule responses to mechanical deformation and also in similar nanomechanical studies of atomic-size metal contacts. These phenomena, which have been measured in numerous experiments using a variety of techniques, are beginning to be used as convenient reference points for the calibration of the very instruments used in such experimentation. The National Institute of Standards and Technology (NIST) is seeking to securely anchor these reference points to known values by measuring intrinsic force interactions using rigorously calibrated experiments, both physical and computational, that emphasize absolute accuracy through traceability to accepted standards of force and length as defined in the International System of Units (SI). The following paper is drawn from the major published results of these efforts, using the notion of an intrinsic standard to provide a unifying backdrop for the discussion.

The paper is organized into four sections; the first two sections deal with specific intrinsic forces, while the latter two sections focus on the more general problems associated with calibrating force and length at the nanoscale. In

DOI 10.1007/978-1-4419-8825-6_1, © The Society for Experimental Mechanics, Inc. 2011

general, the sections comprise previously published source material assembled and edited here to support the keynote address delivered at the 2010 Annual Meeting of the Society for Experimental Mechanics. For example, the first section, **Au Break Junctions**, provides a brief review of Reference [1] where we describe new NIST capabilities developed to study the mechanics and electrical properties of gold nanowires. Gold nanowires apparently neck down under tension to form a single chain of atoms, a physical constriction that leads to the coincidence of quantized force and quantized electrical impedance. It is believed that gold nanowires form in at least three different atomic arrangements. It is not known if all such structures form single-atom chains when stretched to failure and, if so, whether or not the strength of the final bond depends on the initial structure. NIST has been seeking to quantify, to the extent possible, these potential systematic variations. First principles quantum mechanics models along with a new Feedback Stabilized Break Junction (FSBJ) experimental platform have been developed and used at NIST to examine how structure and conductance evolve during elongation, with the ultimate aim to reveal their effect on force at rupture; here, we offer a brief snapshot of this exciting new work. Obviously, the interested reader is encouraged to read the source material for more detail.

The intrinsic force associated with the overstretching-induced conformation change of DNA is reported to occur under a tensile force of approximately 65 pN. This force has been measured using both scanning probe microscope (SPM) and optical trapping instruments and has been proposed as both a length and force standard. The quoted force value has known dependencies on environmental factors such as pH, salinity, and temperature. NIST is working to determine how to accurately control these variables to yield reliable data, and we present a small indication of our new single-molecule expertise in the section headed **DNA Overstretch**.

The experimental platforms necessary to investigate these disparate physical phenomena share two things in common. First, both require the accurate measurement of forces at levels at or below a few nanonewtons. Second, because the primary experiments are analogous to macroscale tensile testing, they each require a means of monitoring the elongation of a mechanical specimen with resolution of a few picometers or less. Consequently, **Picometer Fiber Interferometry** reviews recent developments at NIST regarding the use of fiber interferometry to measure displacements of a nanometer or less. This section provides a highly condensed introduction to the topic; for details the reader should see Reference [2]. Finally, **Piconewton Force Calibration** summarizes our attempts to improve the calibration of atomic force microscope (AFM) cantilever stiffness and to create known forces at or below the level of a nanonewton. It provides a succinct review of our previous research, and then communicates the key findings from Reference [3] regarding the influence of friction in cantilever-on-cantilever calibration, and from References [4-7] regarding our new capabilities to apply electrostatic forces of known magnitudes directly to electrically conducting instrumented indentation and AFM probes.

Au Break Junctions

We have developed an experimental platform, which we refer to as a feedback-stabilized break junction (FSBJ), to create and deform stable atomic-scale contacts, and have used that platform to probe the phenomenon of quantized electrical conductance in Au nanowires (NWs) and single-atom chains (SACs) [1]. In such a system, electrical conductivity, σ, is known to be quantized in units of $G_0 = 2e^2/h$; that is, $\sigma = nG_0$ for integer n, with e the charge of the electron and h Plank's constant [8-10]. The conductivity for the $n = 1$ state corresponds to a contact resistance of 12.9 kΩ. Quantized conductance has been observed many times, but experimental instabilities typically limit the time a given contact stays in a low-n state to milliseconds [11], and often the presence of quantized states must be inferred from histograms compiled from hundreds or thousands of junction breaks [12].

Because the $n = 1$ conduction state is believed to occur when there is only one electron conduction channel through the contact [13,14], the ability to maintain that state indefinitely would clearly demonstrate exceptional experimental stability. The NIST FSBJ described in [1] and shown in the photos of Figure 1 is proving to be just such an exceptionally stable experimental platform. By positioning an interferometer cavity directly between an Au surface and probe mount, we have significantly tightened the displacement measurement frame relative to that achieved in prior work. This has allowed us to close a servo loop around the junction separation with long-term, picometer stability, so as to remove thermal drift and low-frequency vibration artifacts and thereby reduce the need for rigorous environmental isolation that is often encountered in these types of experiments. In the near future, the NIST FSBJ will incorporate a stiff elastic force sensor (e.g., an AFM style colloidal probe) coupled with a high resolution fiber interferometer (see Figure 1), so that direct measurements can be made of bond stiffness and breaking force in SACs.

Figure 1. Feedback-stabilized break junction platform (A) high-vacuum cryogenic chamber with long standoff microscope (B) chamber lid removed (C) mechanics including probe positioning stage (D) view of tip approaching an AFM colloidal probe as seen using the long standoff microscope. The fiber labeled force records the deflection of the colloidal probe cantilever, while the fiber labeled probe monitors the displacement of the probe stage with respect to their common plane (both fibers are in a single, double bore ferrule that is rigidly attached to the cantilever holder base) for the feedback stabilization of the probe sensor separation.

DNA Overstretch

Another example of an intrinsic force is the overstretch transition of DNA [15, 16]. During single-molecule force measurements, a transition occurs at approximately 65 piconewtons in which the DNA molecule elongates significantly with very little extra applied force. The force-displacement curve of a single molecule of DNA measured with a force-calibrated AFM is shown in Figure 2, as described previously [17]. This particular molecule was amplified from a segment of the plasmid vector pBR322, and was measured in Tris/sodium chloride buffer. The transformation can be seen as a plateau in the curve. The methods developed for small force measurement at NIST are being used to calibrate the force at which this plateau occurs. The DNA itself then will become a force reference that will allow the calibration of a wide variety of force measuring instruments such as optical, magnetic, and dielectrophoretic tweezers.

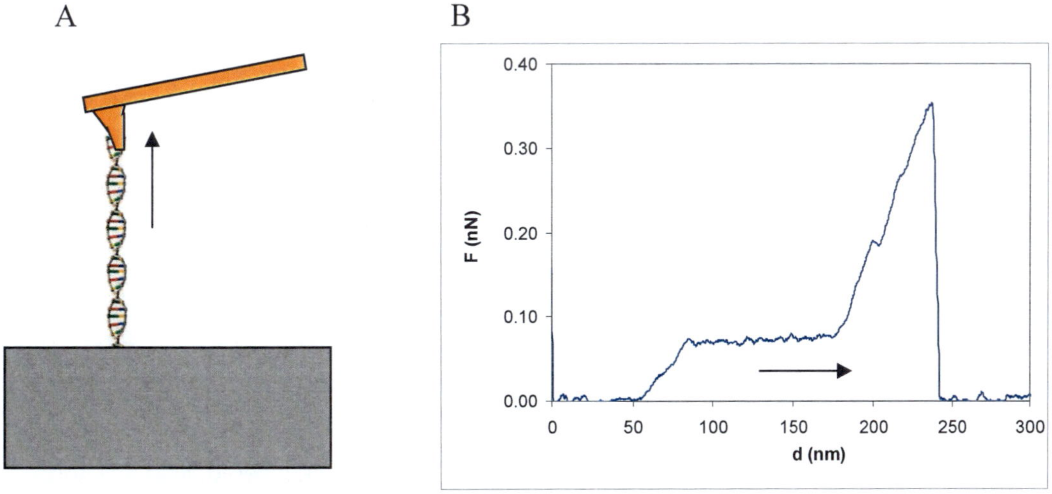

Figure 2. DNA as an intrinsic force standard. (A) Schematic of a single molecule of DNA being stretched by an AFM cantilever. (B) Force displacement curve showing the overstretch transformation of a single DNA molecule measured using an AFM. Arrow indicates the direction in which the DNA was stretched.

Picometer Fiber Interferometery

Clearly, when dealing with single-atom contact mechanics, such as that realized by the interaction of a probe tip with a flat surface, relative displacement of the tip with respect to the substrate must be monitored with resolution well below the lattice spacing of the atoms (< 0.1 nm). Similarly, the interaction forces must be resolved to within at least a few percent of the rupture force of an atomic bond, which has been observed to be between 1 nN and 2 nN for a pair of gold atoms [18]. Assuming that this force is measured using a spring sufficiently stiff to avoid snap in, a stiffness anticipated to be around 50 N m^{-1}, the detector on the force sensor must resolve relative displacement of the spring to 1.5 pm to achieve even a 5 % uncertainty in determination of the rupture force. The requirements for atomic-scale length metrology in these experiments are clear, and point to the need for an accurate sensor capable of recording full-scale relative displacements on the order of 10 nm with percent-level linearity and a noise floor of picometers, in a bandwidth from DC to hundreds of Hertz.

Accurate measurement of displacement requires calibration via comparison to an absolute standard. The unit of length in the International System of Units (SI) is most accurately realized in terms of the wavelength of light; thus, the most accurate determinations of displacements are typically achieved by incorporating measurement tools developed around optical interference, where the displacement is directly compared to the absolute SI standard, minimizing the accrual of uncertainty due to successive calibrations. Simple homodyne interferometers are often employed when the displacement of interest is expected never to exceed the fringe spacing. Robust interferometers based on the low-finesse Fabry-Perot (FP) cavity formed between the end of an optical fiber and the reflective surface of the cantilever spring have been demonstrated for scanning probe microscopes, and we have developed such a fiber-optic interferometer (see schematic in Figure 3) optimized for best performance in the frequency range from DC to 1 kHz, with displacement linearity of 1 % over a range of ± 25 nm, and noise-limited resolution of 2 pm [2]. The interferometer uses a tunable infrared laser source (nominal 1550 nm wavelength) with high amplitude and wavelength stability, low spontaneous self-emission noise, high sideband suppression and a coherence control feature that broadens the laser linewidth and dramatically lowers the low-frequency noise in the system. The amplitude stability of the source, combined with the use of specially manufactured "bend-insensitive" fiber and all-spliced fiber construction, result in a robust homodyne interferometer system that achieves resolution of 40 fm·$Hz^{-1/2}$ above 20 Hz and approaches the shot-noise-limit of 20 fm·$Hz^{-1/2}$ at 1 kHz for an optical power of 10 μW, without the need for differential detection.

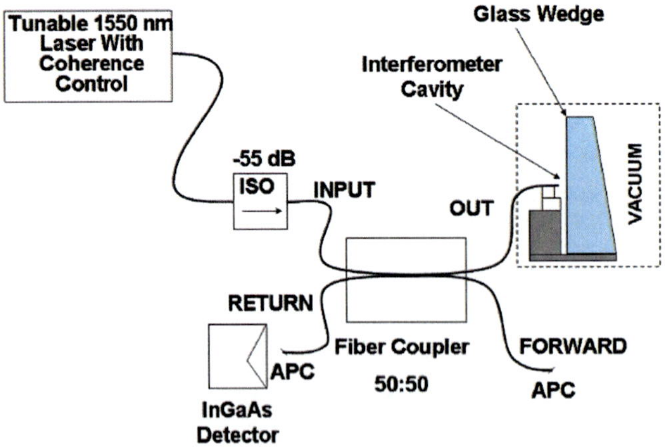

Figure 3. Schematic diagram of the fiber-optic interferometer components showing the laser, optical isolator (ISO), evanescent wave coupler, single InGaAs detector, and a test interferometer cavity. Note that the forward and return ports of the coupler are terminated using angle polished connectors (APC) to minimize interference in the parasitic cavities along these fiber runs.

The light source for our interferometer can be tuned stably and accurately over the wavelength range 1440 nm to 1640 nm. This tuning range enables two very convenient operational features. First, when we either have no control over the macroscopic cavity spacing, as for example in the case of measuring the deflection of a cantilever force sensor, or we choose not to change the cavity out of concern for dimensional stability, we can still operate the interferometer at its maximum sensitivity by tuning λ, so long as the cavity is large enough to contain a quadrature point within the tuning range of the laser. Second, we can make an absolute determination of a fixed cavity length h by sweeping wavelength. Since maxima in reflected intensity occur when $4h/\lambda = 2m+1$, assuming no phase shift at the interfaces (other than standard phase inversion), and minima occur when $4h/\lambda = 2m$, it follows directly that h can be calculated from the measured wavelength values λ_m and λ_{m+1} of two consecutive fringe maxima or minima:

$$h = \frac{1}{2} \frac{\lambda_m \lambda_{m+1}}{(\lambda_{m+1} - \lambda_m)}.$$

For the tuning range of our laser, this means that we can make an absolute determination of cavities as small as h = 6 μm. If a consecutive maximum and minimum are used to calculate h, 3 μm cavities can be measured. Accuracy is limited by our ability to measure λ_m and λ_{m+1}. With curve fitting, wavelengths of maxima and minima can typically be determined within a range of ± 0.1 nm, permitting cavity length determination to better than one part in 1000.

Piconewton Force Calibration

We have been exploring methods to calibrate microcantilever force sensors, such as those used in atomic force microscope experiments. Our initial work in this field focused on creating a system for accurately realizing or creating known forces of appropriate magnitudes. The NIST Electrostatic Force Balance (EFB), which we created for this purpose [19], is an electromechanical balance system that we have used to measure the force versus displacement response of so-called reference cantilevers [20] for use in cantilever-on-cantilever calibrations, and the force sensitivity of piezoresistive cantilever force sensors that can be similarly used to calibrate the force sensitivity of AFMs [21,22]. This work has been extended recently in two important directions.

First, we have found that there are systematic differences between loading and unloading force versus displacement curves obtained using an AFM to probe another cantilever. In Reference [3], the appearance of hysteresis in the slopes of loading curves when probing compliant surfaces with AFM has been explained with an analytical model describing the bending of the test cantilever under the combined influence of a normal contact force and a tangential sliding frictional force. Expressions derived in the paper allowed us to determine coefficients of sliding friction, as well as the stiffness of a sample, provided the spring constant of the AFM cantilever was known. More importantly within the present context, the results were also applied to determine the stiffness of an AFM cantilever using a cantilever-on-cantilever spring constant calibration. A unique aspect of this investigation was that the stiffness of both the AFM cantilever and reference cantilevers was known *a priori*, based on absolute calibrations performed using the NIST EFB. Making use of corrections described in the paper, it was possible to measure the unknown cantilever stiffness using an accurately calibrated cantilever reference within a relative standard uncertainty of approximately 3 %. Remarkably, similar precision was achieved in the determination of the coefficient of sliding friction. From a practical standpoint, a very useful observation made in the paper is that the accuracy of cantilever-on-cantilever calibration can be markedly improved by simply taking the mean of the slopes recorded during loading and unloading against the reference cantilever.

The second important direction of our recent work has been extending the use of calculable electrostatic force to the direct calibration of indentation and colloidal probe force sensors [4-6]. In Reference [4], the capacitance gradient between a spherical indenter probe and a flat electrode was measured. The flat electrode was then used to apply calculable electrostatic forces directly to the spherical indenter, which was mounted on a commercially available force sensor. In essence, the experiment demonstrates the direct application of a force standard to a target instrument without recourse to a transfer artifact, in much the same way calibration has been achieved by hanging known masses on springs in the past. Electrostatic forces ranging from 10 μN to 200 μN were realized with uncertainties estimated less than 1 % (k = 1). Forces were compared in an indirect fashion to deadweights through evaluation of the sensor stiffness, with the relative difference between measured stiffness recorded via the two approaches less than 1 %. In follow up papers [5,6], we have demonstrated calculable electrostatic forces nominally ranging from 320 pN to 100 nanonewtons with uncertainties of a few percent in experiments to perform a direct force calibration of colloidal probe AFM, as illustrated in Figure 4.

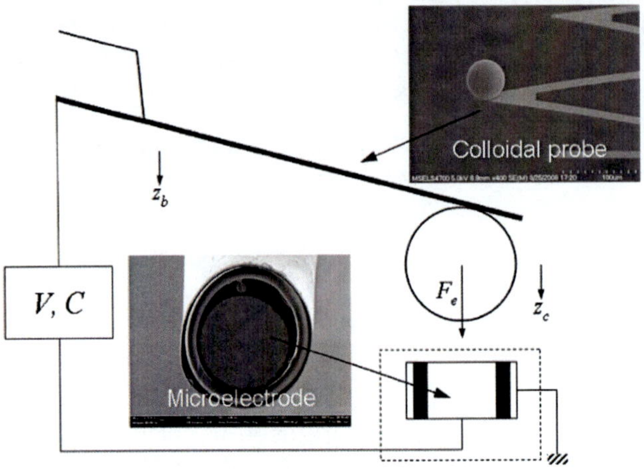

Figure 4. Schematic of an electrostatic force probe experimental setup as in References [5,6]. Insets show scanning electron microscope micrographs of devices as constructed.

CONCLUDING REMARKS

A brief overview of NIST's pioneering work in the field of intrinsic force standards has been presented. The concept of an intrinsic force standard has been introduced, and two candidate physical phenomena have been discussed to describe the current status of this measurement science at NIST. The physical measurements necessary to support the foundational single-molecule and single-atom contact experiments, namely pico-scale force and length metrologies, were reviewed. At present, NIST's Feedback Stabilized Break Junction platform has demonstrated the stability necessary to perform single-atom contact mechanics experiments with traceable force and length metrology, largely thanks to NIST innovations in the application of fiber interferometery, electrostatic force calibration, and micro-cantilever stiffness calibration. The force curve presented in Figure 2 provides a glimpse of a tantalizing future where accurately determined force transitions that occur in the tensile loading of DNA can serve as readily accessible calibration references for a variety of single-molecule science platforms, including optical and magnetic tweezers. DNA can be manufactured with atomic precision in large quantities, and could provide a very cost-effective reference material for single-molecule nanomechanical experimentation.

ACKNOWLEDGEMENTS

Thanks to our many collaborators including K-H Chung, L. P. Howard, L. Kumanchik, N. A. Burnham, L. E. Levine, A. M. Chaka, R. S. Gates, M. G. Reitsma, J. A. Kramar, and D. C. Hurley who contributed greatly to various aspects of the research described here. This work has been funded in part by the NIST Innovations in Measurement Science program.

REFERENCES

1. Smith, D. T., Pratt, J. R., Tavazza, F., Levine, L. E., and Chaka, A. M., *J. Appl. Phys.,* in press 2010.
2. Smith, D. T., Pratt, J. R., and Howard, L. P., *Rev. Sci. Inst.*, **80**:035105, 2009.
3. Pratt, J. R., Shaw, G. A., Kumanchik, L., and Burnham, N. A., *J. Appl. Phys.*, **107**:044305, 2010.
4. Chung, K-H, Scholz, S., Shaw, G. A., Kramar, J. A., and Pratt, J. R., *Rev. Sci. Inst.*, **79**:095105, 2008.
5. Chung, K-H, Shaw, G. A., and Pratt, J. R., *Rev. Sci. Inst.*, **80**:065107, 2009.
6. Chung, K-H, Shaw, G. A., and Pratt, J. R., *Proceedings of the IMEKO XIX World Congress*, Lisbon, 2009.
7. Chung, K-H, Pratt, J. R., and Reitsma, M. G., *Langmuir*, **26**(2):1386, 2009.
8. Landauer, R., IBM J. Res. Dev. **1**, 223 (1957).
9. Büttiker, M., Imry, Y., Landauer, R., and Pinhas, S., Phys. Rev. B, **31**:6207, 1985.
10. M. Brandbyge, J. Schiøtz, M. R. Sørensen, P. Stoltze, K. W. Jacobsen and J. K. Nørskov, Phys. Rev. B **52**:8499, 1995.
11. V. Rodrigues, T. Fuhrer and D. Ugarte, Phys. Rev. Lett. **85** :4124, 2000.
12. J. L. Costa-Krämer, N. García, P. García-Mochales, P. A. Serena, M. I. Marqués and A. Correia, Phys. Rev. B **55**:5416, 1997.
13. T. Frederiksen, M. Paulsson, M. Brandbyge and A.-P. Jauho, Phys. Rev. B **75**, 205413, 2007.

14. C. Untiedt, M. J. Caturla, M. R. Calvo, J. J. Palacios, R. C. Segers and J. M. van Ruitenbeek, Phys. Rev. Lett. **98** :206801, 2007.
15. Williams, M. C., Rouzina, I., Bloomfield, V. A., Thermodynamics of DNA interactions from single molecule stretching experiments, Acc. Chem. Res., **35**:159, 2002.
16. Clausen-Schaumann, H., Rief, M., Tolksdorf, C., Gaub, H. E., Biophys. J., **78**:1997, 2000.
17. Shaw, G. A., Pratt, J. R., Proc. SEM Annual Conference, 2009, Albuquerque, NM, June 1-6, 2009.
18. G. Rubio-Bollinger, S.R. Bahn, N. Agraït, K.W. Jacobsen, and S. Vieira, Phys. Rev. Lett., **87**:026101, 2001.
19. Pratt, J.R., Kramar, J. A., Newell, D.B., and Smith, D.T., *Meas. Sci. Technol.*, **16**:2129, 2005
20. R. S. Gates and J. R. Pratt, Measurement Science and Technology, 17, 2852 (2006)
21. Pratt J R, Smith D T, Newell D B, Kramar J A and Whitenton E., *J. Mater. Res.* **19:**366, 2004.
22. E.D. Langlois, G.A. Shaw, J.A. Kramar, J.R. Pratt and D.C. Hurley, Rev. Sci. Inst., **78** (9):093705, 2007.

Proceedings of the SEM Annual Conference
June 7-10, 2010 Indianapolis, Indiana USA

Magneto-Mechanical MEMS Sensors for Bio-Detection

M. Ramasamy, C. Liang and B.C. Prorok
Auburn University, Department of Mechanical Engineering
275 Wilmore Labs, Auburn University, AL 36849-5341
Email: prorok@auburn.edu

ABSTRACT: Ferromagnetic materials have shown to possess some unique and useful properties, one of which is that they are magnetomechanical transducers, that is, they exhibit a change in dimension when they are subjected to an external magnetic field and vice versa. This magneto-mechanical coupling enables magnetoelastic sensors to be driven to resonance via a modulated magnetic field to detect biological species via frequency shift by mass addition [1,2]. This work details the development of an algorithm to predict the number of captured *E. coli* cells based only upon the resonance frequency shift. This is an important issue as attaching cells influence resonance based upon their location on the sensor. It is therefore necessary to develop a statistical protocol to predict the concentration of the target agent present. A protocol was developed based upon data from microscale resonators using polystyrene beads as a simulant. The protocol was verified with numerical studies and experiments using *E. coli* cells.

INTRODUCTION: Food safety has become an important issue in the past decade after numerous intentional and unintentional contaminations. With the innovation in science and technology, there has been a remarkable increase in the development of biological weapons throughout the world. Biological agents are considered to be psychologically threatening and therefore provide more appeal to the terrorist. The human pathogenic limit of many engineered biological agents have been reduced in order of magnitude from 10^{15} to few cells [3]. This has led to more research on developing biological agents' detection techniques that are very rapid, sensitive and cost effective [4]. Magnetostrictive materials developed into mass-based acoustic wave Sensor operated on longitudinal vibration mode is one such technique. Magnetostrictive materials are soft amorphous ferromagnetic materials that change in magnetic properties when stress is applied or vice versa. These magnetostrictive materials are fashioned into acoustic wave sensors in the form of simple rectangular strips that are actuated in their longitudinal vibration mode when exposed to magnetic field. Due to the applied field the strips resonate at a specific frequency which is dependent on their mass and physical dimensions [1-4]. These devices operate similar to magnetostrictive strips used in stores as a surveillance device to prevent the theft of goods.

PRINCIPLE OF OPERATION

In this research, the principle of detection of these sensors involved measuring a resonant frequency shift as the target biological species attaches to the sensor; effectively this addition of mass dampens the resonant behavior of the sensor platform [10]. In the case of monitoring harmful biological agents, it is highly desired to detect the presence of a handful of spores/cells since many harmful agents have a very low pathogenic limit in humans, meaning it only takes a few spores/cells to infect them. When biological agents attach to these magnetostrictive sensors in small amounts they represent discrete mass additions, whereas a large number may be more analogous to mass evenly distributed on the strip. It is therefore highly desirable to better understand how spore/cell attachment location influences the resonant frequency of these magnetostrictive strips.

The basic sensor structure investigated was a freestanding beam with no fixed ends. The resonance frequency of the first harmonic mode of such a structure can be described as

$$f = \frac{1}{2L}\sqrt{\frac{E}{\rho(1-\nu)}}. \tag{1}$$

Where, L, E, and ρ are the length, Young's modulus, and density of the sensor strip, respectively. As these freestanding magnetostrictive strips are driven to their first harmonic mode, the entire strip deforms in response to the field. The resulting deformation waves propagate through the strip and reflect back from the free ends and cancel at the strip's center, which is the zero nodal posit ion for a freestanding beam resonating in its first

T. Proulx (ed.), *MEMS and Nanotechnology, Volume 2*, Conference Proceedings of the Society for Experimental Mechanics Series 2,
DOI 10.1007/978-1-4419-8825-6_2, © The Society for Experimental Mechanics, Inc. 2011

harmonic mode [17]. It should be noted that positions further from the center node of the strip move further from the node during deformation due to the accumulated deformations of all positions between it and the node. Thus, the free ends move the furthest which is clearly seen in figure1a portraying the simulation result of a mass sensor developed using Ansys® as the simulation tool. Variation in the shade from black to light shade of gray depicts the level of deformation experienced by the sensor platform due to the longitudinal vibration following the boundary conditions.

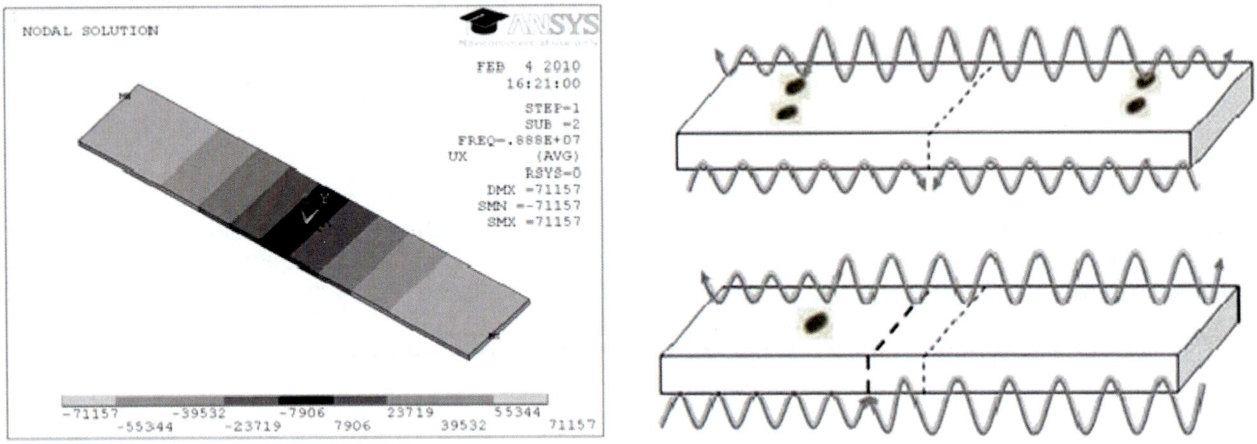

Figure 1: Deformation of Longitudinally vibrated Sensor and Mass attachment affection the wave speed of the platform

When mass becomes attached to the sensor it effectively dampens the speed of the deformation waves propagating in the strip, which reduces its resonant frequency. Figure1b represents the action of deformation waves due to the mass attachment. In the first case *E. coli* cells are attached to the sensor platform at equal distance from the nodal point [12-16]. The deformation waves subjected to this mass travel with reduced wave speed from that point and reach the central nodal position leading to the resonant frequency shift. Whereas in the second case mass is attached on one side of the sensor leading to shift in the nodal point to balance for the reduced wave speed on one side alone and hence we observe a resonant frequency shift. This paper elaborately discusses detection of biological agent considering the above cases, on the acoustic response of the sensor [11]. The frequency shift as a result of mass attachment for acoustic-based sensors such as these is described the following equation,

$$\Delta f = -f_o \left\{ \frac{\eta}{2} \right\}. \tag{2}$$

Where $\eta = \Delta m / m_{sen}$ and Δm is the mass bonded on a sensor, Δf is the resonant frequency change before and after mass attachment, m_0 and f_0 is the sensor's initial mass and resonant frequency, respectively [2-4]. Here, the bonded mass (Δm) on the sensor's surface is considered as an evenly distributed mass and is considerably small with respect to the sensor's mass. Accordingly, the change in frequency can be related to the amount of mass bonded on the sensor [8]. However, in the case when a mass is not evenly distributed or is a discrete mass bonded in a particular location this equation is no longer valid. In order to study these effects discrete masses were attached at various positions on the sensor and the response was measured and analyzed. For experimental analysis a schematic setup as shown in figure 2 involving a coil, external magnetic bias and a network analyzer was used to measure the acoustic response of the sensor as discussed in

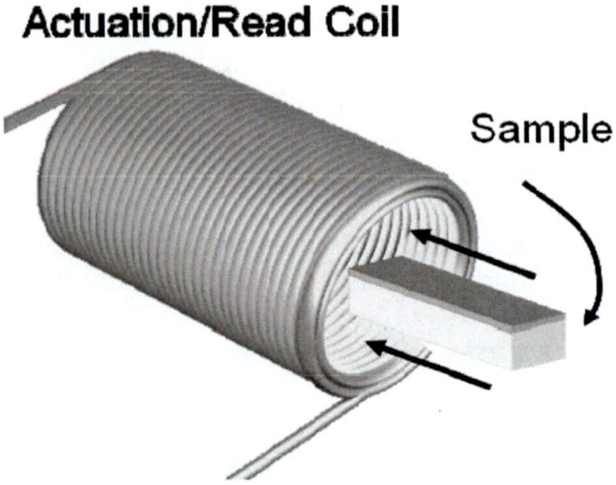

Figure 2: Schematic representation of the measurement rig.

previous work by the authors [1-3]

DESIGN AND EXPERIMENTAL METHODS: Numerical simulations were carried out using commercially available software Ansys®. Specifically, the simulations involved modal analysis on an undamped, freestanding beam with oscillations in the longitudinal mode both with and without attached mass. The structural physical (engineering) discipline is preferred for the Modal analysis of Magnetostrictive sensors. The selected element type was SOLID186. The sensor size employed here was 250 x 50 x 4 µm and the attached mass were representative of an actual *E. Coli O157:H7* cell, size of 1.43 x 0.73 x 0.73 microns and weight of 1 pictogram. The boundary conditions were set to obtain longitudinal vibration mode of the sensor platform. Prediction model was developed involving the factors influencing the resonance behavior such as (a) Mass distribution, (b) Position of the mass distributed and (c) Physical dimension of the sensor platform in compliance with the theoretical equations and simulation results. Mass of E. *coli* cells were distributed as a single layer for uniform distribution. The later was glued to the sensor platform and subjected to numerical simulation with the application of boundary condition over the whole setup. The density of the layer was modified each time for different amount of mass in uniform distribution case and the corresponding variation in frequency shift was observed. In case of the non uniform distribution the layer was split and concentrated along the free ends gradually moving towards the central nodal line and vice versa as shown in Figure 3. The darker layer in the figure is considered as the mass of E. coli cell distributed over the sensor platform.

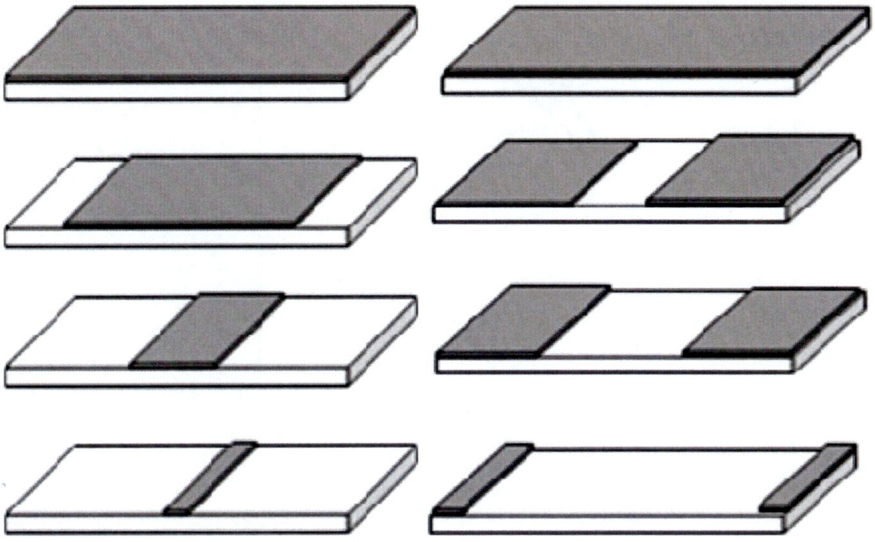

Figure 1: Non-Uniform distribution of E. *coli* cells as layer over the sensor

Glass beads attachment : Sensors with dimensions of 5 mm length and 1 mm width were cut from a 28 µm thick commercially available Metglas™ 2826MB strip. These specimens were prepared, by cleaning and drying, using the identical procedures described by the authors elsewhere. Glass beads with a diameter about 425 µm were employed to simulate the concentrated mass and were carefully loaded on to the sensor surface at prescribed locations and secured with adhesive. The average mass of a sensor and glass bead were 1066 µg and 181.5 µg, respectively. It should be noted that these experiments are aimed as assessing the position of the mass concentrations and not focused on demonstrating minimum sensitivity. Thus, significantly sized beads were employed. The amount of glue employed to affix each bead as well as its position were well controlled to minimize any errors. After a glass bead was loaded on the sensor surface, it was immobilized by drying at

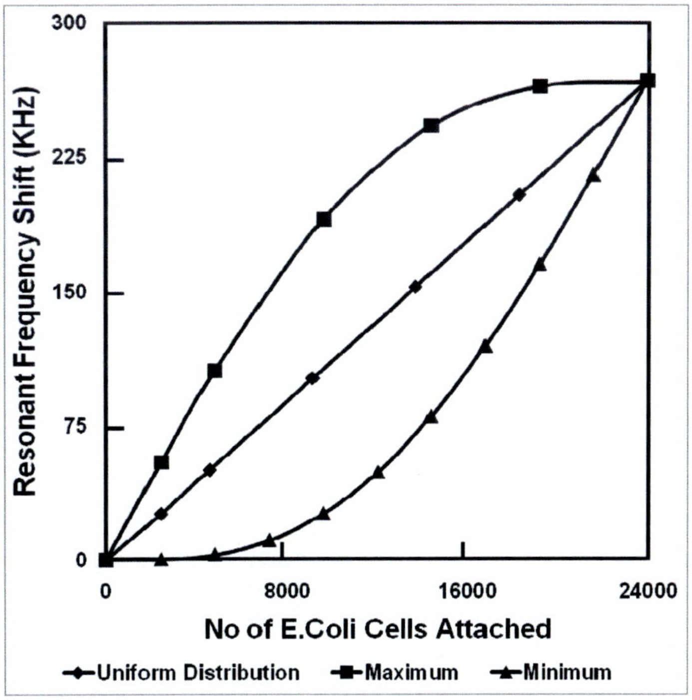

Figure 2: Simulation result of resonant frequency shift due to E. coli cells distribution

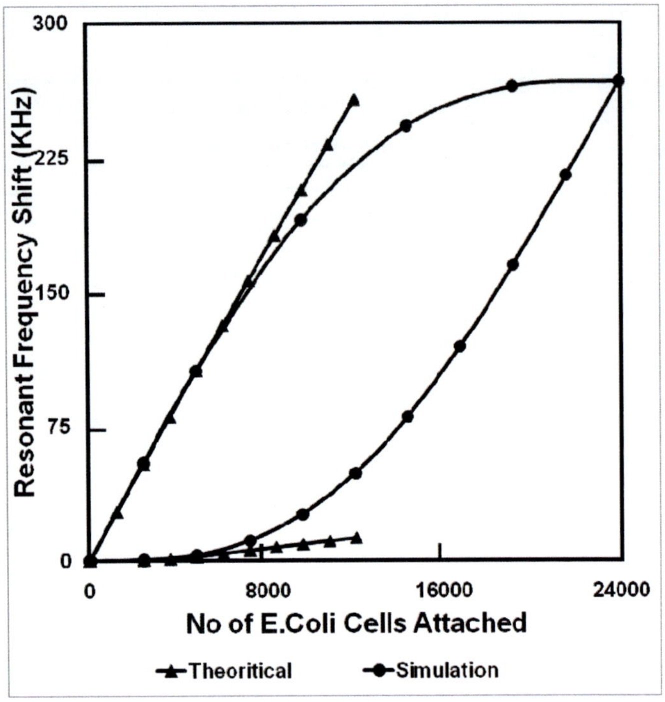

Figure 5: comparative plot of resonant Frequency shift in theoretical and simulated study.

room temperature for at least two hours. The resonant frequency of the sensor was measured before and after attachment of the glass bead in a manner identical to which is discussed in previous work by the authors.

RESULT AND DISCUSSION

(a) Resonant Frequency shift due to Mass Distribution: Frequency shift is not the same if the mass is concentrated at a particular location on the sensor platform compared to the uniform distribution of the mass. This is mainly due to reduction in wave speed of the sensor obstructed by the concentrated mass at discrete locations on the platform. Simulation results are shown in the figure 4. From the figure we can notice that uniform distribution shows a linear plot with the increase in frequency shift as the amount of mass increases gradually. The upper and lower plot in the figure explains the distribution of mass as layer as shown in figure 3 from free end to the central nodal point and vice versa. In these 2 cases the plots meet the uniform distribution linear plot at the end since that's the maximum possible mass that could be attached over the sensor platform. This mass number is based on the assumption that no two E. coli cells accumulate one over the other as demonstrated by Guntupalli, et al. and Wan, et al. in their experimental analysis. In such cases the Maximum resonant frequency shift will be observed at the free ends of the sensor platform which can theoretically determined using equation.

$$\Delta f = -f_o\left\{\left(\frac{KL}{2\pi}\right) - 1\right\}. \qquad (3)$$

Where $K = \beta\sqrt{\frac{\rho}{E}}$, and β is the angular frequency and using η we can calculate KL value for a free standing beam using the equation,

$$\eta(KL) = \tan(KL)$$

Mathematically speaking the frequency shift for certain amount of mass will have a tolerance limit based on their location and distribution. From the above equation we know that the maximum frequency shift for certain amount of mass is observed at the free ends of the sensor platform. This eventually explains the fact that the minimum frequency shift for the same amount of mass would be observed when concentrated at the centre. When mass is distributed close to the nodal point since the deformation experienced is relatively lower compared to the free ends , we observe minimum frequency shift close the centerline of the sensor platform.

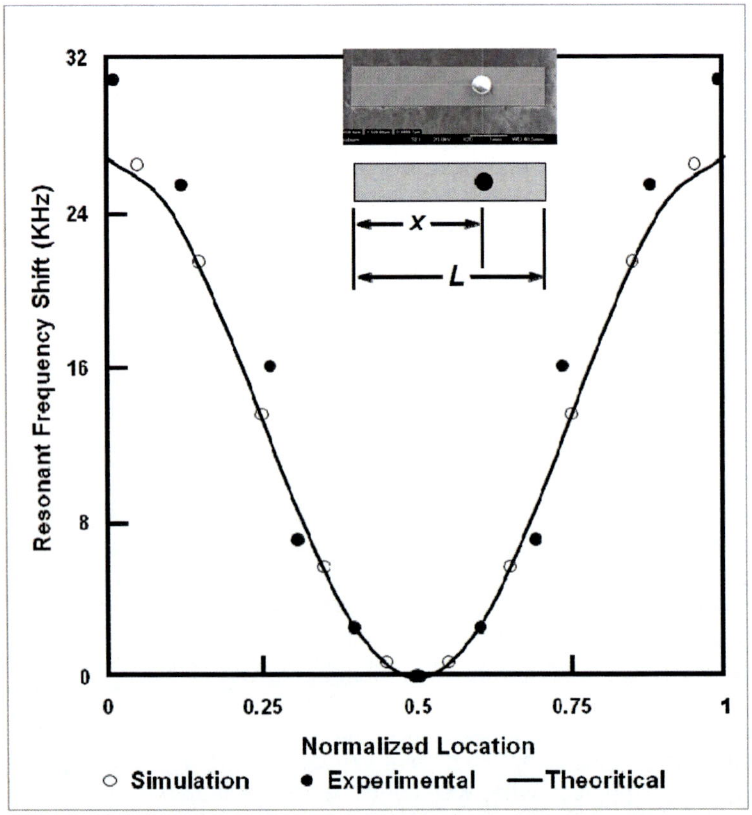

Figure 6: Position sensitive Resonant Frequency shift due to Mass attachment

$$\Delta f = -f_o\left\{\eta + \left(\frac{KL}{2\pi}\right) - 1\right\}. \tag{4}$$

We were able to formulate an equation for minimum frequency shift due to concentration of mass along the centerline of the sensor platform. These equations were verified with the help of the simulation results obtained by distributing the mass uniformly and at various discrete positions on the sensor platform as shown in Figure 5. From the figure we can also notice that after a certain amount of the mass the movement of the theoretical plot deviate in a linear manner compared to the simulation plot which is mainly due to the assumption made in accordance to the distribution of load along the free ends. but in reality as the mass increases it moves closer to the nodal point and eventually meet the uniform distribution line at a particular amount of mass which is the same in case of minimum frequency shift of the sensor due to mass distribution.

(b) Influence of position: Tolerance limit of frequency shift depends on the position of the mass attached irrespective of the amount of mass involved. In this case, the wave speed on the side of the attached mass is less than the opposite side where no mass is attached. During resonance, the wave frequency must match on either side of the sensor. In order to account for the imbalance of wave speeds, the zero nodal point must shift such that the distance the waves travel on each side enables the waves to arrive at the nodal point at the same time to ensure they are in phase. In this case it must shift towards the side with the attached mass, resulting in larger frequency shifts. The simulation and experimental results using glass beads that were attached to the sensor proves this conception. Using these results an equation was developed, to represent the influence of position on the frequency shift of the sensor.

$$\Delta f = \frac{-f_o}{2}\left(\frac{\Delta m}{m_{sen}}\right)\left\{1 - \cos\left(\frac{\pi x}{L}\right)\right\} \tag{5}$$

In this case, where the attached mass is a discrete mass bonded in a particular location, the wave speed on the side of the attached mass is less than the opposite side where no mass is attached. During resonance, the wave frequency must match on either side of the sensor. In order to account for the imbalance of wave speeds, the zero nodal point must shift such that the distance the waves travel on each side enables the waves to arrive at the nodal point at the same time to ensure they are in phase. In this case it must shift towards the side with the

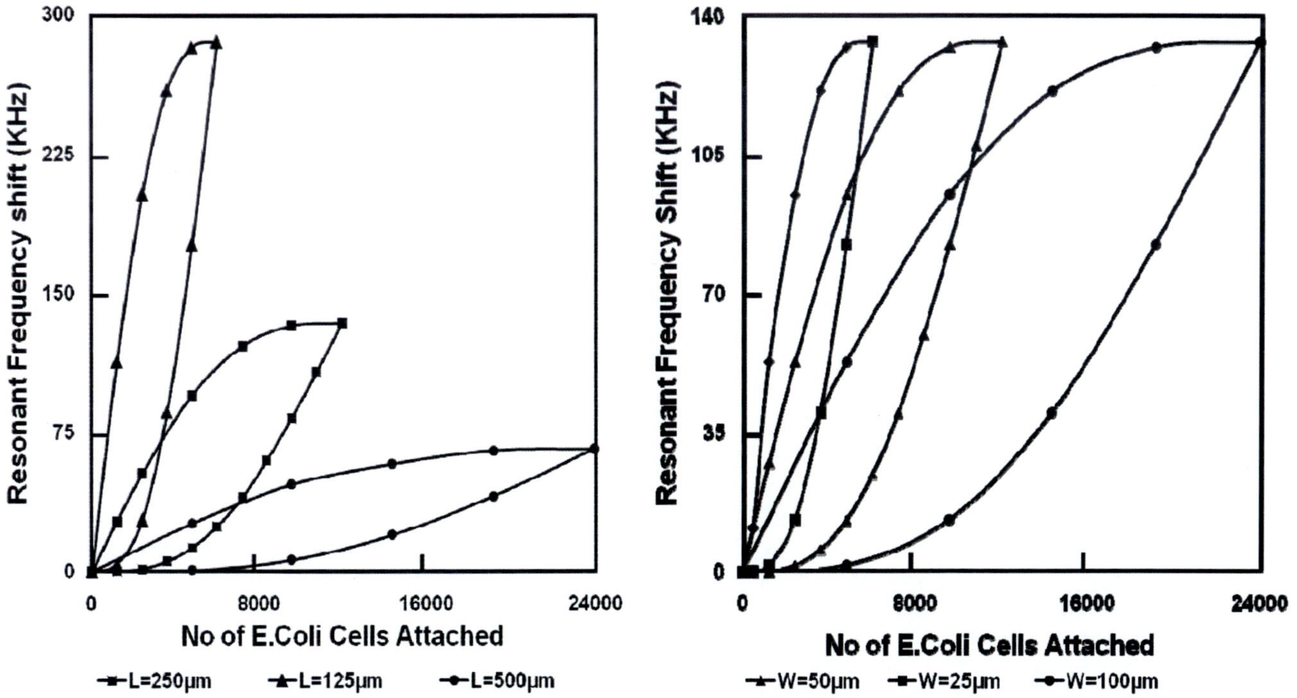

Figure 7: Plot illustrating the influence of physical dimension on the resonant frequency shift of the sensor platform

attached mass. Thus, as the mass is moved further away from the original nodal point, the wave speeds are increasingly imbalanced and the node must shift further to account for it, resulting in larger frequency shifts.

Beads were also positioned away from the longitudinal center axis of the sensor to assess the influence of lateral position for the same longitudinal position *x*. The frequency shift was relatively unaffected indicating that lateral positioning for a discrete mass addition was negligible in comparison to longitudinal positioning.

(c) Influence of dimensions on resonant frequency shift: Simulations were carried out to study the effect of variation in resonance behavior due to change in dimensions of the sensor. Figure 5(a) shows that variation in length of the sensor has a dramatic effect on the frequency shift. When the sensor is subjected to Longitudinal vibration, due to the sensor materials magnetic property the deformation occurs in the longitudinal direction along the direction of the applied field. The larger the length of the sensor greater is the sensitivity and hence greater resonant frequency shift for the same amount of mass. In figure 5(b) variation in width of the sensor platform does not have any influence on the acoustic response of the sensor. This is because lateral axis of the sensor does not experience any deformation due to their restricted boundary conditions and hence distribution of mass is considered to uniform along the lateral distance at any point on the sensor platform. This is experimentally proved with the help of polystyrene glass beads that had no influence on the resonant frequency shift when attached along their lateral position of the sensor platform.

CONCLUSION

A Model developed based on the acoustic response of Magnetostrictive Sensor actuated longitudinally with the modulated magnetic field. Numerical simulations and experimental verifications were carried out with the predicted model considering the factors influencing the resonance behavior of the sensor. The tolerance limit of frequency shift corresponding to the distribution of mass, discrete position of the mass attached and the physical dimensions of the sensor platform were determined in several ways. A Good agreement was found between these results offering a good paradigm for detecting biological agents.

References

[1] C. Liang and B. C. Prorok, "Measuring the Thin Film Elastic Modulus with a Magnetostrictive Sensor," J. Micromech. Microeng., vol. 17, pp. 709-716, 2007.

[2] C. Liang, S. Morshed, and B. C. Prorok, "Correction for Longitudinal Mode Vibration in Thin Slender Beams," Appl. Phys. Lett., vol. 90, pp. 221912, 2007.

[3] M. L. Johnson, J. Wan and B. A. Chin et al., "A Wireless Biosensor using Microfabricated Phage-Interfaced Magnetoelastic Properties",Sens. Act. A 144, 38-47, 2008.

[4] W.Shen, R.S.Lakshmanan and B.A.Chin et al., "Phage Coated Magnetoelastic Micro-Biosensors for real-time Detection of B.anthracis spores", Sens. Act.B 137,pp.501-506,2009.

[5] D. S. Ballantine, R. M. White, S. J. Martin, A. J. Rico, G. C. Frye, E. T. Zellers, and H. Wohltjen, Acoustic Wave Sensors: Theory, Design, and Physical-chemical applications. New York: Academic Press, 1997.

[6] R. Raiteri, M. Grattarola, H. J. Butt, and P. Skladal, "Micromechanical cantilever-based biosensors," Sens. Act. B, vol. 79, pp. 115, 2001.

[7] P. G. Stoyanov, S. A. Doherty, C. A. Grimes, and W. R. Seitz, "A remotely interrogatable sensor for chemical monitoring," IEEE Trans. Mag., vol. 34, pp. 1315, 1998.

[8] C. Mungle, C. A. Grimes, and W. R. Dreschel, "Magnetic field tuning of the frequency-temperature response of a magnetoelastic sensor," Sens. Act. A, vol. 101, pp. 143, 2002.

[9] C. Ruan, K. Zeng, O. K. Varghese, and C. A. Grimes, "A staphylococcal enterotoxin B magnetoelastic immunosensor," Biosens. Bioelec., vol. 20, pp. 585, 2004.

[10] N. Bouropoulos, D. Kouzoudis, and C. A. Grimes, "The real-time, in situ monitoring of calcium oxalate and brushite precipitation using magnetoelastic sensors," Sens. Act. B, vol. 109, pp. 227, 2005.

[11] S. Schmidt and C. A. Grimes, "Characterization of nano-dimensional thin-film elastic moduli using magnetoelastic sensors," Sens. Act. A, vol. 94, pp. 189, 2001.

[12] L. S.Q., L. Orona, Z. M. Li, and Z.-Y. Cheng, "Biosensor Based on Magnetostrictive Microcantilevers as a Sensor Platform," Appl. Phys. Lett., vol. 88, pp. 073507, 2006.

[13] J. Wan, H. Shu, S. Huang, B. Fiebor, I.-H. Chen, V. A. Petrenko, and B. A. Chin, " Phage-Based Magnetoelastic Wireless Biosensors for Detecting Bacillus Anthracis Spores," IEEE Sens. J., vol. 7, pp. 470-477, 2007.
[14] T. Thundat, E. A. Wachter, S. L. Sharp, and R. J. Warmack, "Detection of mercury vapor using resonating microcantilevers," Appl. Phys. Lett., vol. 66, pp. 1695, 1995.
[15] G. Y. Chen, R. J. Warmack, T. Thundat, D. P. Allison, and A. Huang, "Resonance response of scanning force microscopy cantilevers," Rev. Sci. Instr., vol. 65, pp. 2532, 1994.
[16] B. Ilic, D. Czaplewski, M. Zalalutdinov, C. H.G., P. Neuzil, C. Campagnolo, and C. Batt, "Single cell detection with micromechanical oscillators," J. Vac. Sci. Tech. B, vol. 19, pp. 2825, 2001.
[17] R. D. Blevins, Formulas for natural frequency and mode shape. Malabar, FL. USA: Krieger Publishing Company, 2001.M. D. Ward and D. A. Buttry, "Insitu Interfacial Mass Detection with Piezoelectric Transducers," Science, vol. 249, pp. 1000, 1990.

Proceedings of the SEM Annual Conference
June 7-10, 2010 Indianapolis, Indiana USA

Recent Progress of Piezoelectric MEMS for Energy Harvesting Devices

Dong-Joo Kim and Jung-Hyun Park, Materials Research and Education Center, Department of Mechanical Engineering, 275 Wilmore Labs, Auburn University, Auburn, AL 36849
*dkim@eng.auburn.edu

ABSTRACT: With higher integration, smaller size, and automated processes, sensors and wireless devices have seen dramatic enhancements to their quality, robustness, and reliability. Recent efforts have been made toward developing autonomous, self-powered remote sensor systems that can offer enhanced applicability and performance with cost savings. With the decrease in power requirements for wireless sensors, the application of piezoelectricity to energy harvesting has become viable. The technological challenge of realizing such a system lies in the construction and fabrication of a miniaturized vibration energy harvester. The current design of MEMS-scale devices comprises a seismic mass made of silicon connected to the substrate by a thin PZT cantilever beam. Factors relating to power improvement and reliability of the device are discussed by addressing the shape of the cantilever beam, piezoelectric mode, MEMS process, and environmental temperature.

INTRODUCTION: During recent years, energy harvesters that generate electricity from vibration and motion sources have attracted much interest as micro power sources. The main applications are wireless communication and sensor nodes. Mechanical vibration and motion have been considered as promising sources due to their wide permeation and efficient converting mechanism when light energy is not available [1]. The vibration energy can be converted into electricity by electrostatic [2], electromagnetic [3], or piezoelectric transducers [4]. Among them, the piezoelectric transducer has become favorable to many researchers since it has high electromechanical coupling, simple configuration without external voltage source, and easy integration. Recently, studies of piezoelectric energy harvesters have focused on maintaining their conversion efficiency during scaling down via the micromachining process [5-9]. In our previous work, we demonstrated a successful microelectromechanical systems (MEMS) energy harvester using sol-gel deposited PZT, which generated 160 mV and 3272 μWcm^{-3} at 19.6 ms^{-2} acceleration. However, such a device may not be implemented due to insufficient power output for current sensor devices. It is considered that the scaling down of the generated output power in the piezoelectric MEMS device results from single-layered unimorph structures and clamping of piezoelectric layers by massive silicon-based substrate. In addition, the micromachining process can cause degradation of the piezoelectric properties of the thin films.

As a way to improve the output power in vibration energy harvesters, several approaches can be considered. A cantilever design has been explored to change the strain distribution on piezoelectric films. The rectangular cantilever beam is commonly chosen due to simple design for microfabrication, but most of its stress is driven to near the anchor of the cantilever beam, and then converted into power; therefore, it is dominated by the anchor area. In addition, such non-uniform distribution can weaken the strength of the cantilever structure during higher vibration conditions. Baker et al. have demonstrated that the trapezoidal shaped-cantilever increased the power output by as much as 30% compared with a rectangular cantilever in bulk scale [10]. Although much research has focused on investigating the strain distribution with the change of the cantilever shape, less attention has been paid to analysis and experimental verification of the devices in MEMS scale. Another important factor in thin film piezoelectric harvesters is the choice of the electrode type, for example 31 or 33 mode. A 31 mode PZT harvester with parallel plate capacitors exhibits a large capacitance resulting in low voltage outputs, while a 33 mode harvester constructed by using top-side interdigitated electrodes shows a lower capacitance value due to longer distances between electrodes. Since vibration harvesting devices have an AC current output that needs to be rectified for energy storage in a battery, it may need at least 500 mV for efficient rectification when considering current rectifying semiconductor devices [11]. Therefore, 33 mode harvesters have advantages in building a circuit. However, higher voltage for poling and difficulty in controlling PZT orientation in 33 mode will be issues, which may result in a non-uniform poling direction and lower output power [12]. In 31 mode, lower output voltage can be multiplied by building multiple arrays. Besides the above two main parameters, the issues relating to the

DOI 10.1007/978-1-4419-8825-6_3, © The Society for Experimental Mechanics, Inc. 2011

reliability of the PZT MEMS energy harvesters such as process-induced degradation and environment-dependent properties should be carefully considered.

In this study, PZT MEMS devices were fabricated and the effects of the cantilever geometry, electrode design, and other factors with the reliability of the devices on the device performance such as resonance frequency, output power and voltage, and optimal resistance load were investigated.

DEVICE FABRICATION: The MEMS fabrication process mainly consists of thin film deposition and etching with patterns. An inductive coupled plasma reactive ion etching (ICP RIE) system for anisotropic etching was used to etch out all layers and to release the cantilever beam with the Si proof mass. A Si supporting layer under the cantilever beam is necessary to withstand the internal stress and to improve the mechanical strength of the beam. The thickness of this Si supporting layer can be an important factor to determine the resonant frequency of a device when other dimensions are fixed because the resonant frequency is highly sensitive to the thickness of a cantilever beam. Since it is hard to precisely control the thickness of the Si supporting layer by the RIE process, the whole fabrication process was based on a silicon-on-insulator (SOI) wafer, and the buried SiO_2 layer was used as an etching stop layer. Four masks were used in the fabrication of the MEMS PZT cantilever with a proof mass. The fabrication process began with a 100 mm SOI wafer. The thicknesses of the first silicon layer, the thermal silicon dioxide layer, and the second silicon layer are, respectively, 20 μm, 500 nm, and 450 μm. A layer about 500 nm thick of SiO_2 was grown on both sides of the SOI wafer by a wet O_2 method. The silicon dioxide on the front surface will be used to balance the inertial stress of the PZT film, and the one on the back-side surface could be used to mask areas of the surface to prevent damage during the back-side etching. For 31 mode electrodes, interlayer Ti (10 nm) and bottom electrode Pt (120 nm) were deposited sequentially by magnetron sputtering, and then about 1 micrometer thick of PZT were coated by sol-gel method. In case of 33 mode, ZrO_2 layers was coated instead of bottom Pt electrode, and PZT layer was then deposited. Both 31 and 33 mode devices, the top electrode was obtained by a liftoff process after Pt deposition on a layer of photo resist (PR) patterned. After top electrode patterning, RIE process was performed to form bottom electrode windows for wire bonding, to define a cantilever beam, and to etch backside proof mass.

DEVICE MEASUREMENT: The output behavior of the device was evaluated using an experimental setup described previously [4]. The device was mounted on an electromagnetic shaker. The vibration frequency and acceleration amplitude were controlled by a function generator and an amplifier, and the simultaneous acceleration amplitude was monitored by an accelerometer mounted on the shaker. A resistive load was connected with the device, and the voltage measured across the resistive load was recorded by an oscilloscope. A sine wave signal was generated by the function generator used as a vibration source. The output voltages of the device at different vibration accelerations, different vibration frequencies, and different resistive loads were then systematically measured. The amplitudes of the output AC voltages were recorded and the corresponding output average powers were calculated.

RESULTS AND DISCUSSION:

SHAPE OF CANTILEVER BEAM

Figure 1 shows the dimensions of cantilevers to investigate the effects of the cantilever shape on the output powers. These cantilever beams have 4x2 (L x W) for the beam and 3x2x0.5 (L x W x H) for the mass in the unit of mm. The trapezoidal cantilever had a reduced width of 1 mm at the contact point with the proof mass. The output power was measured at different electrodes for both cantilevers.

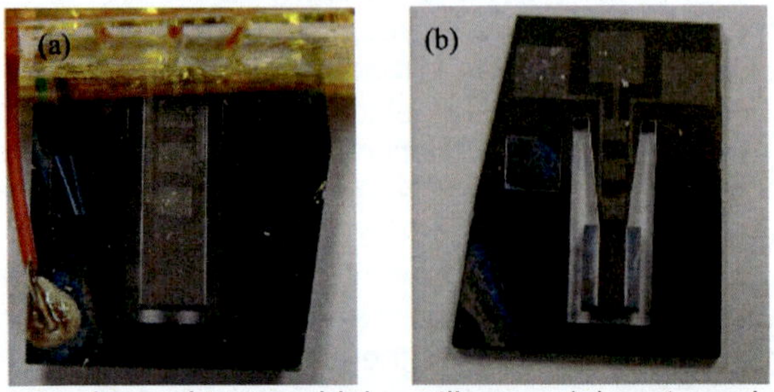

Figure 1: The photograph of rectangular and trapezoidal cantilevers: (a) rectangular; (b) trapezoidal.

Figure 2(a) and 2(b) show the normalized powers at a given resistive load for each electrode under 2.45 ms^{-2} of acceleration. The total power was normalized based on the real size of the actual electrode area of 1 mm^2 after excluding the output power from a patterned wire electrode on a PZT layer. There were two limitations on this normalization. The first is that the generated charge on a piezoelectric material is not linearly proportional to the amount of strain due to the nonlinearity of PZT. The nonlinear piezoelectric coefficient, however, was neglected in this study. The other limitation is that the maximum power was shown at the optimal resistive loads of 20±1 kΩ of the device. Their small differences of the optimal resistive loads between electrodes resulted from the contribution of the patterned wire on a PZT layer to the output power. The maximum power obtained from each electrode was used to compare them with average strain values of each electrode for both cantilevers, as shown in Figure 2(c) and 2(d). The electrodes near the anchor area showed the highest power (denote normalized power without notice from now on) in both cantilevers, while the output power at electrodes near the mass area had the smallest value. It is common that a large amount of strain occurs near the anchor and the strain is reduced near the mass. The output power from three electrodes showed a nonlinear relationship, indicating nonlinear strain distribution. The power at the middle electrode of the rectangular cantilever showed a smaller value compared to the average value calculated from the anchor and mass electrodes, while the trapezoidal cantilever showed a bit larger value. If it is assumed that the materials in a cantilever structure exhibit linear, elastic, and isotropic with an average stress applied along the 1-1 direction, the output power can be expressed as

$$P = \frac{V^2}{R_L} = \frac{1}{R_L}\left\{ g_{31} E \varepsilon \left(\frac{x}{L} \right) L \right\}^2 \tag{1}$$

where x is the location from the clamped end, L is the length of the beam, g is the piezoelectric voltage constant, E is the Young's modulus, and ε is the strain. In Figure 2(c), the relationship between strain and power has good agreement with equation (1). Conversely, the trapezoidal cantilever has a negative parabolic relationship. The reason is because the strain distribution has spread out into the region of the middle electrode in the trapezoidal cantilever structure. It is also considered that the nonlinear piezoelectric coefficient (g_{31}) might increase the output power at that region of PZT thin film. More accurate quantitative analysis of these results is under investigation. All electrodes were connected together, and their output power was re-measured for both cantilevers to compare the total power without normalization, as shown in Figure 3. The maximum power was obtained at the resistive load of 7 kΩ for both cantilevers. The output power from the trapezoidal cantilever shows 39% higher than that of the rectangular cantilever.

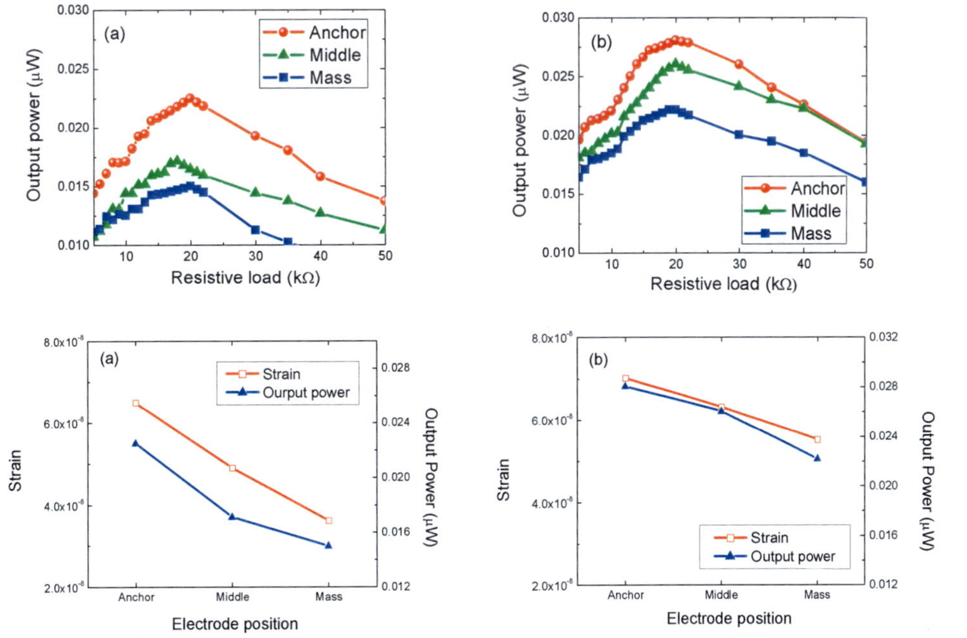

Figure 2: The output powers of different electrodes for (a) the rectangular and (b) the trapezoidal cantilever across the resistive load. Their maximum power was compared with the FEM modeling strains for (c) the rectangular and (d) the trapezoidal cantilever.

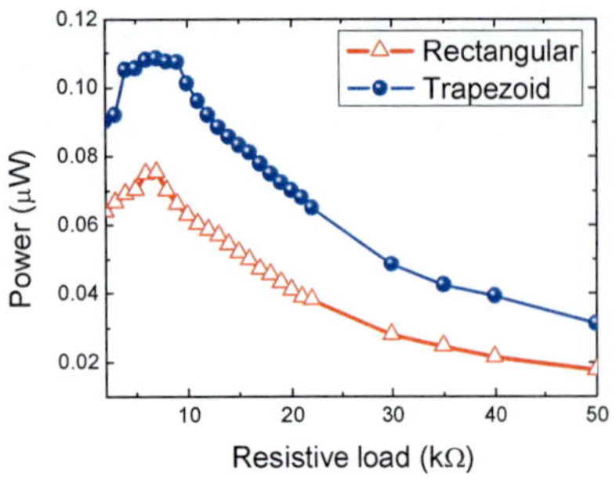

Figure 3: The generated total powers from rectangular- and trapezoid-shaped cantilevers.

ELECTRODE DESIGN

To compare the two modes of electrode design, cantilevers with the same dimensions were fabricated where the cantilever beam has 7 x 2 mm^2 size and the proof mass has 3 x 2 x 0.5 mm^2 size as shown in Figure 4. The 33 mode has the interdigitated electrodes with 20 μm width and 40 μm gaps in Figure 4(b). Both devices were designed to have about 125 Hz of resonant frequency, but the measured resonant frequency was 130.8 Hz for the 31 mode and 118.1 Hz for the 33 mode. Although the dimensions of the cantilever beam were same, a slight difference in resonant frequencies occurs due to the ZrO_2 buffer layer in 33 mode device.

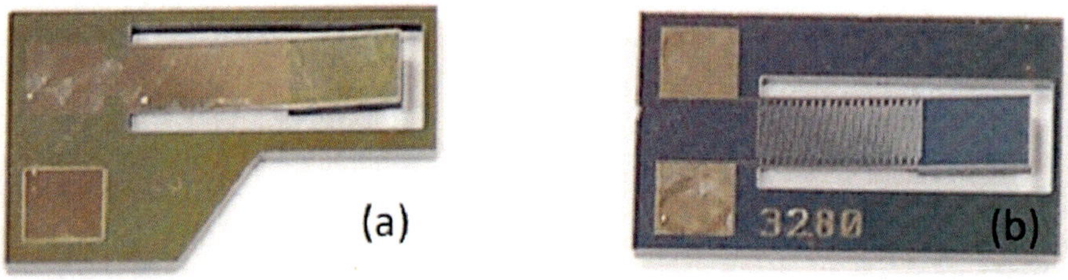

Figure 4: Photographs of the fabricated energy harvesting devices that have (a) 31 mode and (b) 33 mode.

The 31 mode device shows the output voltage of 1.35 V_{p-p} at the resonant frequency with open circuit conditions under a 0.5 g acceleration force. Figure 5(a) shows that the maximum power of 6.72 μW at 11 kΩ resistive load where the output voltage was 0.769 V_{p-p} and the current was 70 μA. The 33 mode device shows 7.725 V_{p-p} under the open circuit condition. In Figure 5(b), 4.675 V_{p-p} was obtained at the optimal resistive load of 4.5 MΩ. The maximum power output was 0.47 μW and the current was 1 μA. A 31 mode PZT harvester exhibits low voltage outputs, while a 33 mode harvester shows a higher voltage. Such a high voltage may be beneficial to rectify the generated AC current output for energy storage in a battery. However, non-textured orientation may cause a non-uniform poling direction and potentially lower output power. A possible approach for voltage amplification of 31 mode devices will be constructing arrays. By serial connection of each cantilever, the output voltage can be amplified. Figure 6 shows an example of the increase in output voltage with serial connection of two cantilevers.

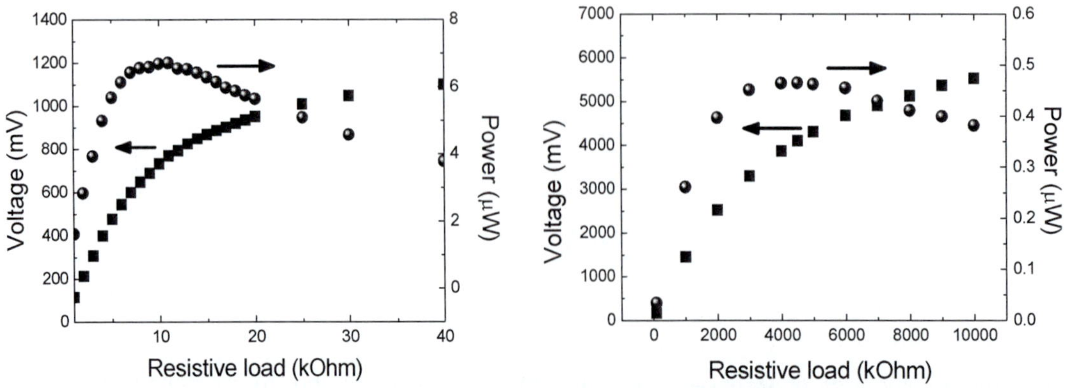

Figure 5: Output voltages and powers generated from two modes of piezoelectric energy harvesting devices: (a) 31 mode and (b) 33 mode. The measurement was performed at 0.5-g acceleration.

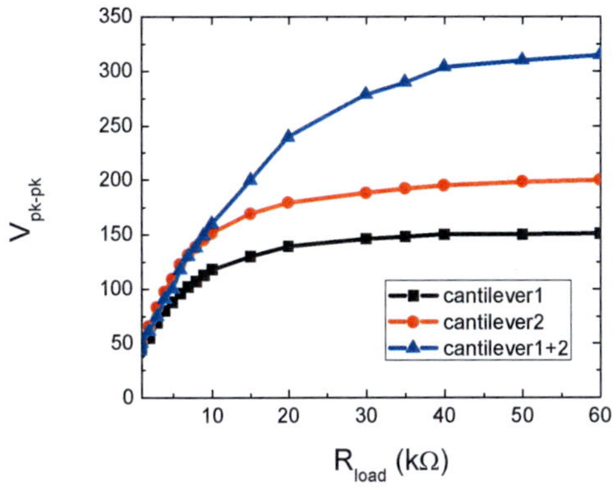

Figure 6: The output voltage from the cantilever beams by serial connection (The evaluation was performed at 0.25-g acceleration).

PERFORMANCE RELIABILITY

Fracture due to cyclic loading: The PZT MEMS cantilever can exhibit brittle fracture at higher gravimetric force due to higher strain at the beam. In general, the beam experiences maximum strain near the clamping area where higher probability of fracture. The microstructural investigation of fractured cantilever beam was performed as shown in Figure 7. PZT cantilever was broken due to cyclic loading at higher accelerations. The failure occurred near the anchor, i.e. clamping area, and the wide range of fractured surface was examined. It is reported that Si has about 1.94 MPa√m of toughness, and PZT thin film has around 0.5 MPa√m [13,14]. The fatigue propagation may be nucleated and grown from PZT surface if they are under same amount of cyclic load, but the crack was initiated from the Si surface and generated failure in fast time. When the fractured surface was examined, it appears that the origin of fracture is originated from the base side across SiO_2 layer. From this point, the water marks were found at entire of cantilever. The etching surface by RIE is not perfectly smooth, which may work as a crack initiation source. The optimization of the etching process is required to obtain a flat etched surface, or using a coating polymer on the surface might be another possible option to smooth the surface.

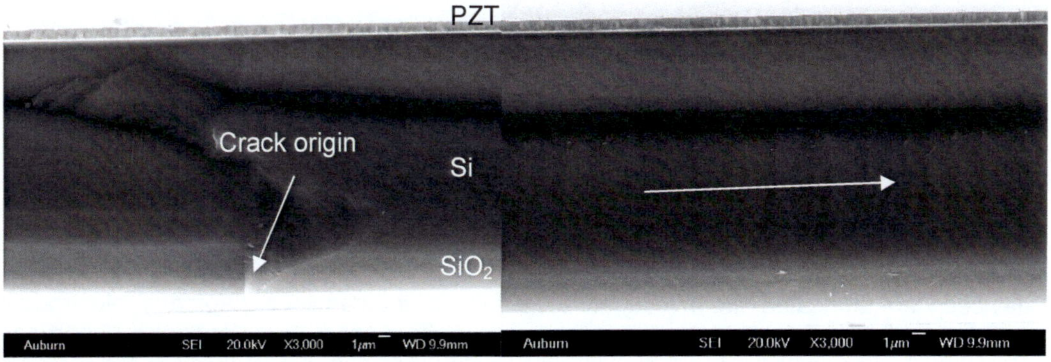

Figure 7: SEM observation of the fractured PZT cantilver beam.

Depoling due to cyclic loading: PZT film degrades its piezoelectric properties due to electric and/or mechanical cyclic loading. The cyclic loading by vibration can also cause depoling of PZT domains. In order to examine the effect of cyclic loading on the performance, a PZT MEMS device was resonated for three days at 1-g condtion, and the output voltage was measured in open circuit conditions. Figure 8 shows the output voltage of the device with cycling time. The temperature of the PZT with cycling was measured, but no heat was built up during the entire measurement period. After $5x10^6$ cycles, the output voltage started to decrease from 280 mV. Since no change of the resonant frequency was observed, the decrease of the output voltage was considered due to depoling of the domains in PZT films resulting in degradation of piezoelectric properties. In order to confirm the effect of poling on the output voltage, the PZT film was poled again at $2.3x10^7$ cycles. After poling, the output voltage was recovered.

Depoling due to environment temperature: The output voltage as temperature was also investigated, as shown in Figure 9. The output voltage was measure with 1 MΩ of resistive load during temperature goes up and down. When the temperature increased, the output voltage decreased. When the device was measured with cooling down, the output voltage could not be recovered to its initial value. It seems that the piezoelectric properties of PZT experienced degradation, mostly depoling at higher temperature. After the re-poling process, the output voltage was recovered, which confirms the role of poling on the output voltage of the PZT device.

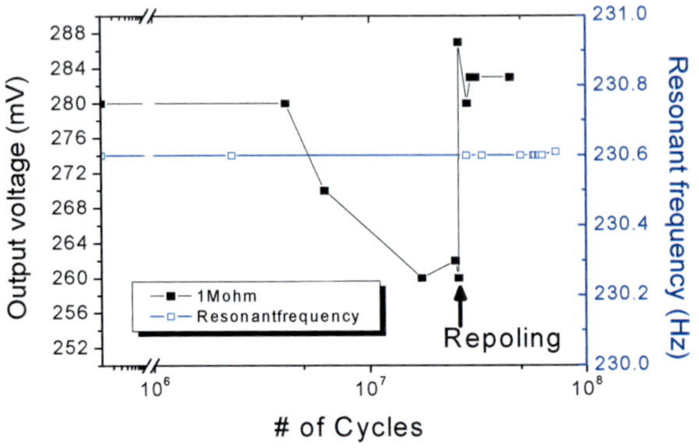

Figure 8: Output voltage and resonant frequency of a PZT cantilever device as a function of cyclic loading.

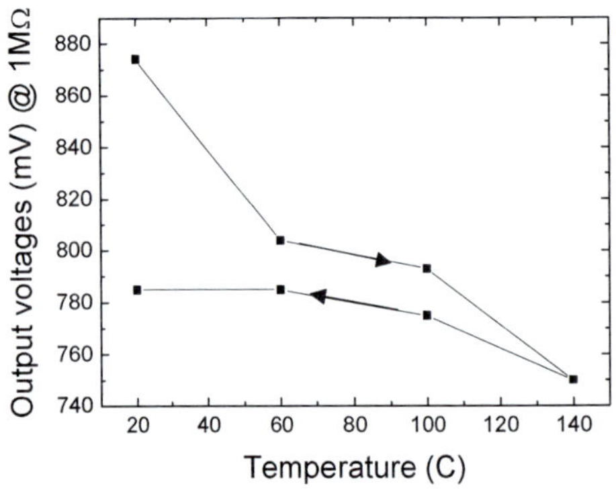

Figure 9: Output voltage change as a function of environment temperature. The voltage was measured under open circuit conditions.

CONCLUSION:

PZT MEMS devices were fabricated and investigate the effects of the cantilever geometry, electrode design, and other factors with the reliability of the devices on the device performance. In a trapezoid-shaped cantilever, the strain distribution with higher values along the cantilever beam was verified compared to the rectangular shape. The examined power from individual electrodes located near the anchor, middle, and mass has good agreement with the strain distribution. Two types of PZT devices, 31 and 33 mode were constructed and investigated. The 33 mode device exhibit higher voltage compared to 31 mode, which may provide benefit to build convenient rectifying circuit. However, the advantage of 31 mode in terms of power density is not clear. Besides higher output power from devices, the reliability of the devices is also of concern. The results show that cyclic loading leads to mechanical failure of the cantilever structure. The depoling due to the cyclic loading and environment temperature should also be considered to design vibration energy harvesting devices based on piezoelectric PZT film.

ACKNOWLEDGMENTS: This research was partially supported by the Auburn University Detection and Food Safety Center funded by USDA-CSREES.

REFERENCES:

1. Roundy S, Wright PK, Rabaey, J (2003) A study of low level vibrations as a power source for wireless sensor nodes. Comput Commun 26:1131-1144.
2. Mitcheson PD, Green TC, Teatman EM, Holmes AS (2004) Architectures for vibration-driven micropower generators. J Micro Sys:13:429-440.
3. Torah RN, Beeby SP, Tudor MJ, O'Donnel T, Roy S (2006) Kinetic energy harvesting using micro scale electromagnetic generators. In Proceedings of the 17th MicroMechanics Europe, UK.
4. Shen D, Choe SY, Kim DJ (2007) Analysis of piezoelectric materials for energy harvesting devices under high-g vibrations. Jap J Appl Phys 46:6755-6760.
5. Shen D, Park JH, Ajitsaria J, Choe SY, Wikle HC, Kim DJ (2008) Design, fabrication and evaluation of a mems pzt cantilever with an integrated si proof mass for vibration energy harvesting. *J Micromech Microeng* 18:055017.
6. Shen D, Park JH, Noh JH, Choe SY, Kim SH, Wikle HC, Kim DJ (2009) Micromachined PZT cantilever based on SOI structure for low frequency vibration energy harvesting. Sens Actuators A 154:103-108.
7. Marzencki M, Ammar Y, Basrour S (2007) Integrated power harvesting system including a mems generator and a power management circuit. In Proceedings of International Conference on Solid-State Sensors, Actuators and Microsystems:887-890.
8. Fang HB, Liu JQ, Xu ZY, Dong L, Wang L, Chen D, Cai BC, Liu Y (2006) Fabrication an dperformance of mems-based piezoelectric power generator for vibration energy harvesting. Microelect J 37:1280-1284.
9. Jeon YB, Sood R, Jeong JH, and Kim SG (2005) MEMS power generator with transverse mode thin film pzt. Sen and Actuators A 122:16-22.
10. Baker J, Roundy S, Wright P (2005) Alternative geometries for increasing power density in vibration energy scavenging for wireless sensor networks. In Proceedings of the 3rd International Energy Conversion Engineering Conference. San Freansico, USA.
11. Torah R, Glynne-Jones, P, Tudor M, O'Donnell T, Roy S, Beeby S (2008) Self-powered autonomous wireless sensor node using vibration energy harvesting. Meas Sci Technol 19:125202.
12. Lee BS, Lin SC, Wu WJ, Wang XY, Chang PZ, Lee CK (2009) Piezoelectric MEMS generators fabricated with an aerosol deposition PZT thin film. J Micromech Microeng 19: 065014.
13. Ando T, Li X, Nakao S, Kasai T, Tanaka H, Shikida M, Sato K (2005) Fracture toughness measurement of thin-film silicon. Fatigue Fract Eng Mater Struct 28: 687-694.
14. Zheng X, Zhou Y, Yan Z (2003) Dependence of crystalline, ferroelectric and fracture toughness on annealing in $Pb(Zr_{0.52}Ti_{0.48})O_3$ thin films deposited by metal organic decomposition. J Mat Res 6: 551-556.

Proceedings of the SEM Annual Conference
June 7-10, 2010 Indianapolis, Indiana USA

Performance of Piezoelectric Power Generator in Environmental Conditions

Seon-Bae Kim, Jung-Hyun Park, Hosang Ahn, and Dong-Joo Kim

Materials Research and Education Center, Auburn University, Auburn, AL, 36849, USA

E-mail: szk0022@auburn.edu

ABSTRACT

Piezoelectric power generators can be effective power suppliers for small devices by collecting energy from ambient vibration and converting it into electrical energy. Since such small devices are usually mobile, their operation environments can vary widely. For example, the conditions such as vibration frequency and temperature energy highly affect harvesting efficiency of piezoelectric power generators. If the ambient vibration frequency is away from the resonance frequency, there will be a decrease of output power. The ambient temperature can be a possible factor to change the resonance frequency of the power generating device since temperature influences material constants of the constituent components in a power generator such as dielectric constant, piezoelectric strain coefficient, and the stiffness Therefore, the effect of temperature on output power is studied. Since the common piezoelectric material is PZT and it is known that hard- or soft-type PZT exhibits different dependence of materials properties as a function of ambient temperature, this paper investigates the output power of the soft and hard PZT-based power generator depending on ambient temperature.

INTRODUCTION

Piezoelectric devices are widely studied as a potential power source for small devices. Most research on the characterization of piezoelectric power generators are performed at room temperature, but there are many applications where they have to operate at temperatures above the ambient. Therefore, it is important to know the behavior of their main parameters at different temperatures in order to make the proper design for any specific application. Piezoelectric devices should be maintained as a stable power source to ensure reliable operation. Although power change can be compensated by a circuit or program when environment temperature varies, it is required to investigate how its output power changes with the temperature and which parameters are mostly influenced by the temperature.

For piezoelectric power generators, the most popular material is lead zirconate titanate (PZT) because of its high performance originated from a high dielectric constant and piezoelectric strain coefficient. PZT can be divided into soft and hard PZT depending on the valence state of dopants. Hard PZT is made by doping of lower valance state ions and soft PZT by higher valent dopants[1]. The dielectric constant and

T. Proulx (ed.), *MEMS and Nanotechnology, Volume 2*, Conference Proceedings of the Society for Experimental Mechanics Series 2,
DOI 10.1007/978-1-4419-8825-6_4, © The Society for Experimental Mechanics, Inc. 2011

piezoelectric strain coefficient of soft PZT are known to exhibit stronger temperature dependence compared to those of hard PZT due to lower phase transition temperature. Therefore, the power generator made by hard PZT may have a wider range of operation temperature. However, lower piezoelectric and coupling coefficients may result in lower power output. In literature, their temperature dependence on materials properties are studied[2][3][4]. However, no systematic report on the ambient temperature dependence on the piezoelectric power generator is published. Therefore, this work investigates the behavior of the output power from two different types of PZT materials as a power generator depending on the temperature ranging from room temperature to 150 ^{o}C.

EXPERIMENTAL METHODS

PZT materials for constructing power generator were purchased from piezo.com. The cantilever structure consisting of two layers of PZT with brass between PZTs was used as a design for a power generator device. The dimension of cantilever is 0.0247 x 0.0032 x 0.00038(m) (length x width x thickness). The proof mass is attached at the free end of cantilever and its resonance frequency is around 120 Hz where the range of ambient vibration source is between 50 and 200 Hz[5].Samples were wired and fixed between metal holders . They were vibrated in the chamber designed to change temperature ranging from room temperature to 150 °C. Samples were kept for 10 minutes for thermal equilibrium at each temperature point and vibrated to obtain output voltage at various resistances. As temperature increased, elastic modulus of the cantilever decreased, which results in the decrease of the resonance frequency of the cantilever. The vibration frequency of a shaker is tuned to resonate with the resonance frequency of the cantilever at each temperature to measure the highest output power

THEORY

Piezoelectric power generation can be described as equivalent circuit models where piezoelectric energy conversion from mechanical to electrical energy is considered as a change of voltage and current through coil in a transformer[6]. A schematic illustration of a cantilever and this model are shown in fig. 1 and fig. 2. The left part of fig. 2 indicates a mechanical system, and the right part shows the electrical part including capacitance of cantilever and resistive load. Mass, mechanical damping during vibration and stiffness of cantilever are modeled as inductor, resistor, and capacitor, respectively. Because voltage in an electric circuit loop is summation voltage in each element, vibration input in the mechanical part is divided into vibration of mass, damping, and stiffness. The moment of inertia, and moment of a cantilever are calculated as in equation (1) and (2), respectively. Based on equation (2), stress is calculated as in equation (3), where w, t_c, b, η_s, t_{sh}, m, $\ddot{y}$, $\ddot{z}$, l_b, lm, and x are width of cantilever, thickness of piezoelectric layer, distance from center of shim to center of piezoelectric layer, ratio of Young's modulus of piezoelectric layer to that of shim, thickness of shim, mass of proof mass, acceleration value of vibration stage, length from fixed part of cantilever to proof mass, length of mass, and distance from fixed part of cantilever. To calculate strain in the piezoelectric layer from the deflection of proof mass, Euler beam theory is used. Also to calculate across variable, voltage in the electrical aspect and stress in mechanical aspect, Kirchoff's Voltage Law (KVL) is used. Voltage due to a resulting strain is calculated as in equation (4), where δ, k_{sp}, b_m, b**, d, b*, and ζ are strain, spring constant of cantilever, relation between stress and strain in damping element, $\frac{2I}{b(2l_b+l_m-l_e)}$, piezoelectric strain coefficient, $\frac{3b}{l_b^2}\frac{(2l_b+l_m-l_e)}{(2l_b+\frac{3}{2}l_m)}$, and damping ratio. For comparison between soft and hard PZT samples, same damping ratio is used because it depends on dimension, stiffness and proof mass, and these are same to prepare soft and hard PZT cantilevers. The damping ratio is experimentally calculated by measuring the quality factor because ζ is approximately1/2Q if ζ<0.05[7]. The damping ratio is C/C_c, where C is damping factor and C_c is critical damping factor. The damping factor is proportional to angular frequency, ω, which is identical to the resonance frequency, and can be calculated as in equation (5)[8]. The critical damping ratio is $2\sqrt{km}$, where k and m are spring constant and mass, respectively, when a cantilever is modeled as spring-mass system, and k is $\frac{3YI_0}{l^3}$. By combining terms for the damping ratio, it can be known that theoretically the damping ratio does not change as a function of temperature except for the density change of air. Therefore, the damping ratio change is calculated based on the data based on piezo.com. Equation (4) is expressed in time domain, and to calculate power in frequency domain, Laplace transformation is used. By assuming the cantilever is operated at resonance frequency, maximum power is calculated as in equation (6).

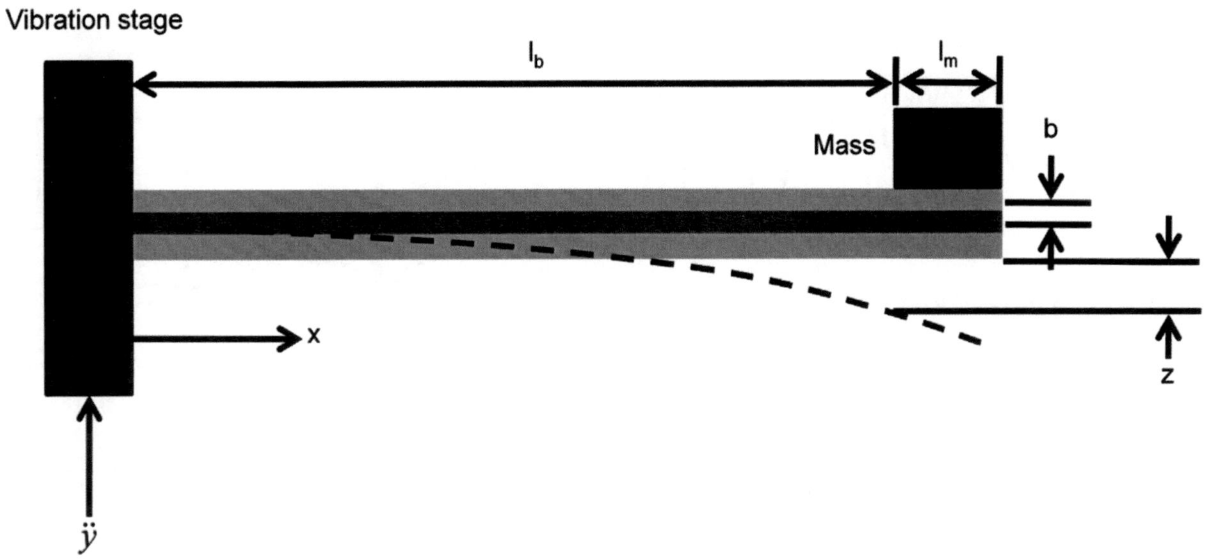

Fig. 1 Schematic diagram of cantilever.

$$I = 2\left[\frac{wt_c^3}{12} + wt_c b^2\right] + \frac{\eta_s wt_{sh}^3}{12} \quad -(1)$$

$$M(x) = m(\ddot{y} + \ddot{z})\left(l_b + \frac{1}{2}l_m - x\right) \quad -(2)$$

$$\sigma = \frac{1}{l_e}\int_0^{l_e} \frac{M(x)b}{I}dx = m(\ddot{y} + \ddot{z})\frac{b(2l_b + l_m - l_e)}{2I} \quad -(3)$$

$$\begin{bmatrix}\dot{\delta}\\ \ddot{\delta}\\ \dot{V}\end{bmatrix} = \begin{bmatrix} 0 & 1 & 0 \\ -\frac{k_{sp}}{m} & -\frac{b_m b^{**}}{m} & \frac{k_{sp} d}{2mt_c} \\ 0 & -\frac{2dY_c t_c}{\epsilon} & -\frac{1}{RC_p}\end{bmatrix}\begin{bmatrix}\delta\\ \dot{\delta}\\ V\end{bmatrix} + \begin{bmatrix}0\\ b^*\\ 0\end{bmatrix}\ddot{y} \quad -(4)$$

$$\omega_n = \sqrt{\frac{YI_0 k_n^4}{m}} \quad -(5)$$

$$P = \frac{1}{\omega^2}\frac{RC_p^2\left(\frac{2Y_c dt_c b^*}{a\epsilon}\right)^2}{(4\zeta^2 + k^4)(RC_p\omega)^2 + 4\zeta k^2(RC_p\omega) + 2\zeta^2} \quad -(6)$$

$$R = \frac{1}{\omega C_p} \frac{2\zeta}{\sqrt{4\zeta^2 + k^4}} \quad -(7)$$

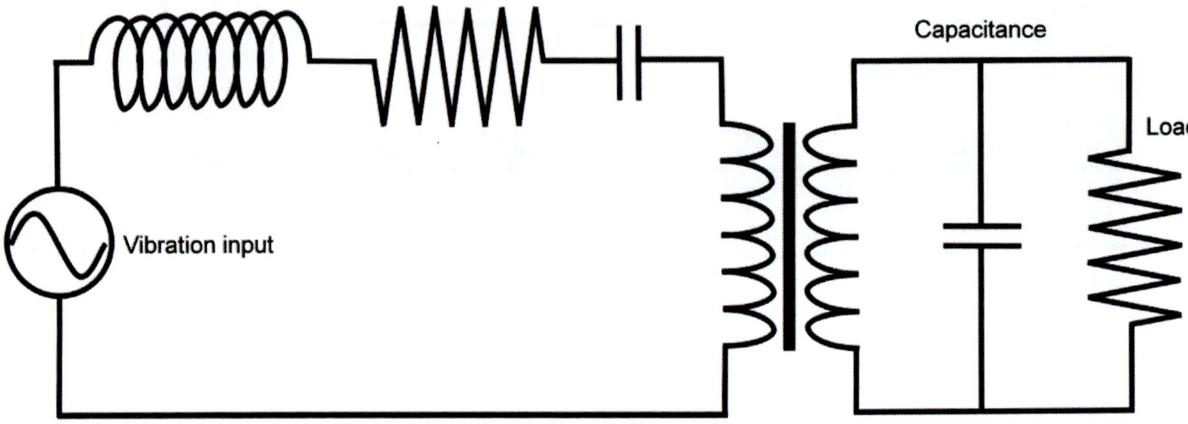

Fig. 2 Electrically modeled piezoelectric cantilever.

RESULTS AND DISCUSSION

Experimentally measured and calculated powers as a function of temperature are shown in fig. 3 and 4, respectively. The measured value is about 20% of calculated value. The difference is attributed to mismatch in damping and nonlinear effect that is ignored for the simplification [9]. The adhesion layer between PZT and shim is ignored, which was reported as a cause of discrepancy [10]. Commonly, they exhibit the same order of power and decreasing tendency in power. Thus, it is reasonable to use equation (6) for estimating output power from a cantilever. By substituting equation (7) into (6), and separating the terms that are different from materials, equation (8) yields, where C_0 is $\frac{AY^2t(b^*)^2}{\omega^2\zeta}$, and A is area of electrode in cantilever. Therefore the output power of a cantilever can be proportional to $1/\omega^2$, Y^2, $1/d^2$, $1/\varepsilon$, $1/\zeta^2$. In both hard and soft PZT-based cantilevers, the term with dielectric constant is more significant to others since its known strong dependence on the temperature [reference]. The dielectric property is contributed by 180° and non-180° polarization domains. The materials constant of soft PZT changes more than that of hard PZT because of faster domain motion[2]. A change of those values according to temperature is normalized by its value at room temperature to compare temperature dependence of the values as shown fig. 5 and 6. In both types of PZT cantilevers, the predominant factor is considered as the damping ratio because it has the highest slope change. Therefore as temperature increases the damping ratio becomes the most significant in determining the output power of the piezoelectric cantilever. Among the materials constants of PZT, the dielectric constant is the biggest factor. Therefore, a smaller value with less change of dielectric constant is recommended for higher power and stable power generation at various environment temperatures.

$$P = C_0 \frac{d^2\epsilon}{3\sqrt{4\epsilon^4\zeta^2 + d^8Y^4 + 4d^4Y^2}} \quad -(8)$$

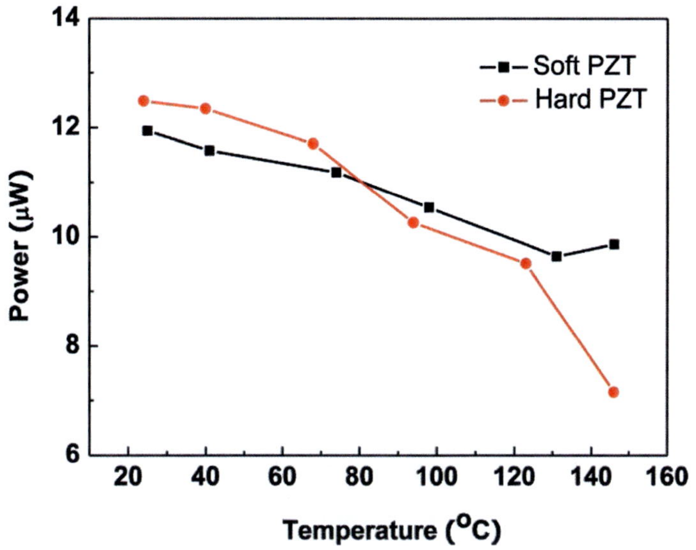

Fig. 3 Experimentally measured output power of cantilever.

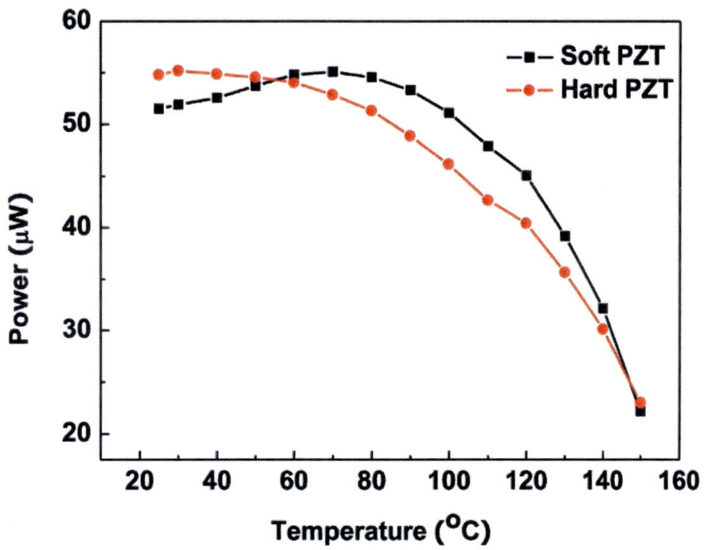

Fig. 4 Calculated power of cantilever.

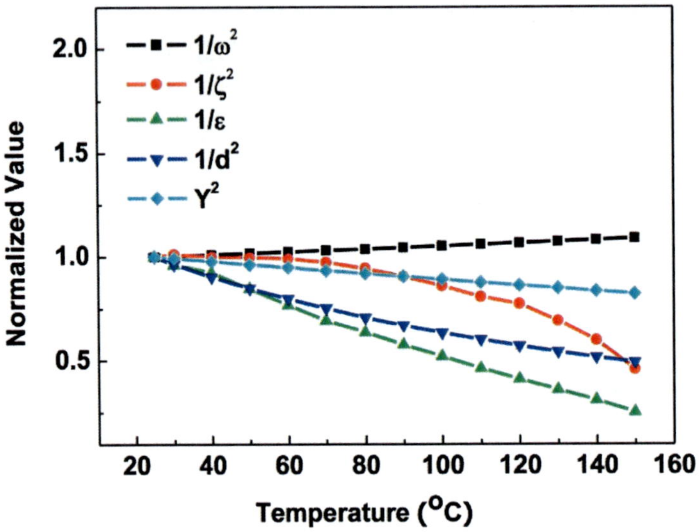

Fig. 5 Change of factors influencing power according to temperature in soft PZT.

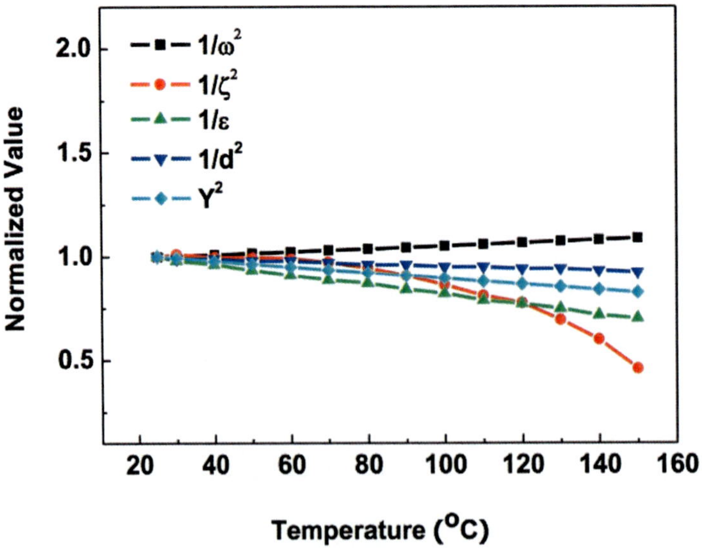

Fig. 6 Change of factors influencing power according to temperature in hard PZT.

CONCLUSIONS

The factors critical to piezoelectric power generation devices are studied by examining the output powers from two types of soft and hard PZT-based cantilevers. When considering only material constants, the increase of the dielectric constant is the dominant factor for power decrease at an increased temperature. Therefore to minimize the loss of output power with the temperature, it will be necessary to suppress the increase of the dielectric constant by dopants resulting in preventing domain motion. In terms of design factor, the effect of damping on power decrease becomes significant. Controlling the damping ratio is an important design factor for piezoelectric cantilevers to operate stably at higher temperature.

ACKNOWLEDGMENTS

This research was partially supported by Auburn University Detection and Food Safety Center funded by USDA-CSREES.

REFERENCES

[1] Tan, Qi, Li, Jie-Fang, and Viehland, Dwight, Ferroelectric behaviours dominated by mobile and randomly quenched impurities in modified lead zirconate titanate ceramics, Philosophical Magazine B, Vol. 76 No. 1 pp. 59-74, 1997.
[2] Sabat, Ribal Georges et al, Temperature dependence of the complete materials coefficients matrix of soft and hard doped piezoelectric lead zirconate titanate ceramics, J. Appl. Phys., Vol. 101 064111, 2007.
[3] Wolf, R. A., Trolier-McKinstry, S., Temperature dependence of the piezoelectric response in lead zirconate titanate films, J Appl Phys Vol. 95 pp. 1397-1406, 2004.
[4] Julian, Christelle and Viehland, Dwight, Comparison of polarization switching in "hard," "soft," and relaxor ferroelectrics, J. Appl. Phys., Vol. 95 No. 8 pp. 4316-4318, 2004.
[5] Roundy, Shad, Wright, Paul K., and Rabaey, Jan M., A study of low vibrations as a power source for wireless sensor nodes, Computer Communications, Vol. 26 pp. 1131-1144, 2003.
[6] Roundy, Shad, Wright, Paul K., and Rabaey, Jan M., Energy Scavenging for Wireless Sensor Networks, Kluwer Academic Publishers, pp. 51-85, 2004.
[7] Rao, Singiresu S., Mechanical Vibrations 4th ed., Pearson Prentice Hall, p. 234, 2004.
[8] Chen, Shih-Nung, Wang, Gou-Jen and Chien, Ming-Chun, Analytical modeling of piezoelectric vibration-induced micro power generator, Mechatronics, Vol. 16 pp. 379-387, 2006.
[9] Shen, Dongna, Choe, Song-Yul and Kim, Dong-Joo, The design, fabrication and evaluation of a MEMS PZT cantilever with an integrated Si proof mass for vibration energy harvesting, Jpn. J. Appl. Phys., Vol. 46 pp. 6755-6760, 2007.
[10] Gubinyi, Z. et al., Electrical properties of PZT piezoelectric ceramic at high temperatures, J. Electroceram., Vol. 20 pp. 95-105, 2008.

Proceedings of the SEM Annual Conference
June 7-10, 2010 Indianapolis, Indiana USA

Comparison of Transduction Efficiency for Energy Harvester between Piezoelectric Modes

Jung-Hyun Park[a], Hosang Ahn[a], Seon-Bae Kim[a], Seung-Hyun Kim[b], and Dong-Joo Kim[a,*]

[a]Materials Research and Education Center, Auburn University, AL, 36849, USA,
[b]Inosteck, Inc., Ansan, Gyeongii 425-791, Republic of Korea
*E-mail: dkim@eng.auburn.edu

ABSTRACT

The efficiency of micro-electro-mechanical systems (MEMS) energy harvesters using transverse and longitudinal piezoelectric modes was studied. Since the emergence of piezoelectric MEMS energy harvesters, studies have mainly focused on improving the conversion efficiency at a given strain on a transduction layer. The transverse piezoelectric mode has been applied to use a higher d_{33} piezoelectric coefficient of a lead zirconate titanate oxide (PZT) material, but few studies have verified its distinguished efficiency compared to that of transverse mode using d_{31}. PZT thin films were deposited on $Pt/Ti/SiO_2$ and ZrO_2/SiO_2 substrates as the mode by a chemical solution deposition method. The rectangular cantilever with a Si proof mass was fabricated on an SOI wafer, and it has 20 μm of Si device layer. The cantilevered energy harvesters in the same size were respectively fabricated for longitudinal {3-1} and transverse {3-3} mode using parallel electrode plates and interdigitated electrodes (IDE). The generated voltage and power were analyzed considering piezoelectric constants of the PZT films in transverse and longitudinal modes. Since the efficiency of the transverse energy harvester seems to be strongly affected by IDE configuration, a study of this is accompanied. This study might give the direction of choosing the piezoelectric mode for the higher efficiency of MEMS transducers based on the material study.

INTRODUCTION

Recently, there has been a growing interest in micro-power energy harvesting from the environment using energy harvesting devices, which have a small volume, light weight, and long lifetime for miniaturized electronics such as wireless sensors requiring 10-100's of μW in low power VLSI design [1, 2]. While traditional batteries have a short life time using limited chemical energy sources, the energy harvesting concept has been developed by collecting usable electricity from permanent energy sources such as solar power, thermal gradients, radio frequency, and mechanical vibration [1-4]. Most of all, mechanical vibration is one of the potential energy sources that has wide permeation and high conversion efficiency. The successful demonstration of piezoelectric energy harvester (PEH) in bulk scale was introduced, which generated few volts and hundreds of μW [5]. Energy harvesters using piezoelectric films have promising alternatives with their high conversion efficiency compared with other conversion mechanisms such as electrostatic and electromagnetic [3, 6, 7]. The miniaturizing of PEH for practical uses with small volume and light weight is the current issue and challenge.

The simple configuration of PEH enabled to access to the microelectromechnical systems (MEMS) and some prototypes of MEMS PEH have been introduced [8-13]. They have a similar structure to the piezoelectric transducer layer on the silicon cantilever frame with a buffer layer between them. A proof mass at the end of the tip can accelerate the gravity under seismic motion and adjust the resonant frequency into a low level, as shown in figure 1. $Pb(Zr_x,Ti_{1-x})O_3$ (PZT) thin film is usually used due to its excellent piezoelectric constant compared with others [14]. A different electrode configuration is available in {3-1} and {3-3} piezoelectric mode. When the piezoelectric film is laminated by platinum electrodes, the {3-1} mode in figure 1(a) has perpendicular coordination

T. Proulx (ed.), *MEMS and Nanotechnology, Volume 2*, Conference Proceedings of the Society for Experimental Mechanics Series 2,
DOI 10.1007/978-1-4419-8825-6_5, © The Society for Experimental Mechanics, Inc. 2011

between stress and electric field. The reported PEH generated output powers and peak-to-peak voltages were below 2.2 μW and 600 mV under optimized resistive load and {3-1} mode, which is still insufficient for their application.

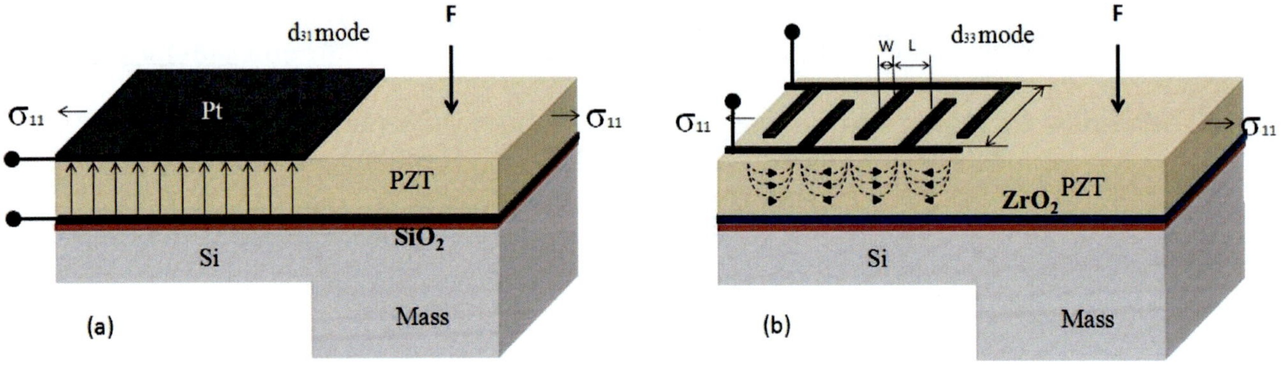

Figure 1 PEH configurations: (a) {3-1} mode, (b) {3-3} mode

An interdigitated electrode (IDE) has been used to increase the output voltage. The IDE configuration enables the {3-3} mode, which has parallel coordination between stress and electric field, as shown in figure 1(b). Since the piezoelectric constants d_{33} or g_{33} are two times higher than d_{31} and g_{31}, a higher efficiency might be expected. Moreover, a higher output voltage can be designed by using a large distance between finger electrodes by equation (1). The open circuit voltage can be explained by [13]:

$$V_{oc} = \sigma_{11} g_{33} L \tag{1}$$

$$V_{oc} = \sigma_{11} g_{31} t \tag{1}$$

where σ_{11} is the stress on piezoelectric film, g_{33} and g_{31} are the piezoelectric constant, L is the distance between electrodes, and t is the thickness of PZT film. Since the AC output signal from the PEH should be rectified to DC signal for its applications, a voltage level over threshold voltage of the rectifiers is required to be generated from PEH. The reported {3-3} mode generated a higher output peak-to-peak voltage up to 2 V and a lower power up to 1.2 μW than that of the {3-1} mode. While the output voltage from the {3-1} mode is restricted by the thickness of 1~2 μm by equation (2), the {3-3} mode has the advantage of high voltage generation. However, a relatively small area and longer distance between electrodes in IDE configuration induces very high optimal resistive load and small current. The optimal resistive load of IDE can be evaluated by equation (3~5) [15, 16]:

$$R_{opt} = \frac{1}{2\pi \cdot f \cdot C} \tag{3}$$

$$C = \varepsilon_r \varepsilon_0 \frac{A}{l} \tag{4}$$

$$A = Wt(N-1) \tag{5}$$

where f is resonant frequency, C is capacitance, ε_r is dielectric constant, ε_0 is permittivity of vacuum, A is electrode area, l is distance between electrodes, W is width of finger electrode, t is film thickness, and N is the number of finger electrodes. Moreover, a dead zone is formed due to non-uniform poling direction, which is perpendicular to the electrode coordination under electrodes area. The entire PZT film cannot be contributed to the piezoelectric conversion so that its volumetric efficiency decreases.

In this paper, the PEHs in both {3-1} and {3-3} modes will be fabricated using a thin PZT transducer layer of the same dimensions. When they have been poled and saturated, their output performance as energy harvesters will be compared in voltage, current, power, and optimal resistive load.

DESIGN AND FABRICATION

The PEH structure is composed of a buffer layer/thick silicon layer to support the PZT film and a proof mass was integrated at the end tip of cantilever. A 20 μm silicon layer could compensate internal stress as an experimental result and give mechanical strength to the device. The Si proof mass enabled low frequency design and larger displacement at a given force with easy integration by pattering and strong atomic bonding to the beam. PEHs should be designed to have targeted resonant frequency of its environment. The resonant frequency was evaluated by equation (6), which is based on our previous work [8].

$$f_n = \frac{v_n^2}{2\pi}\sqrt{\frac{0.236D_p w}{(l-\frac{l_m}{2})^3[m_e+\Delta m]}} \quad (6)$$

where f_n is the nth mode resonant frequency, v_n is the nth mode eigenvalue, D_p is bending modulus per unit width and function of the Young's moduli and thicknesses of the PZT and Si, w is the width of the cantilever beam, l is length of the cantilever, l_m is the length of the proof mass, m_e is effective mass, and Δm is mass of the proof mass.

The microfabrication was performed patterning with 3 masks and micromachining with a Deep Reactive Ion Etcher (DRIE). Figure 2 shows the fabrication process and the layer information for a {3-1} mode PEH. The {3-3} mode PEH has a mostly similar structure, but it has the ZrO_2 buffer layer instead of the bottom Pt layer and the top Pt layer has IDE. After patterning the Pt top electrode by lift-off process in figure 2(b), the cantilever structure was formed by etching with a thick photoresist mask in figure 2(c). A backside photoresist mask was built up, including a proof mass, as shown in figure 2(d). Since the buried SiO_2 effectively worked as an etching stop with a 100-times slower etching rate than the silicon, the etching profile beneath the cantilever was very flat when backside etching was finished. Its flat surface made the resonance frequencies of the devices close to the estimated values.

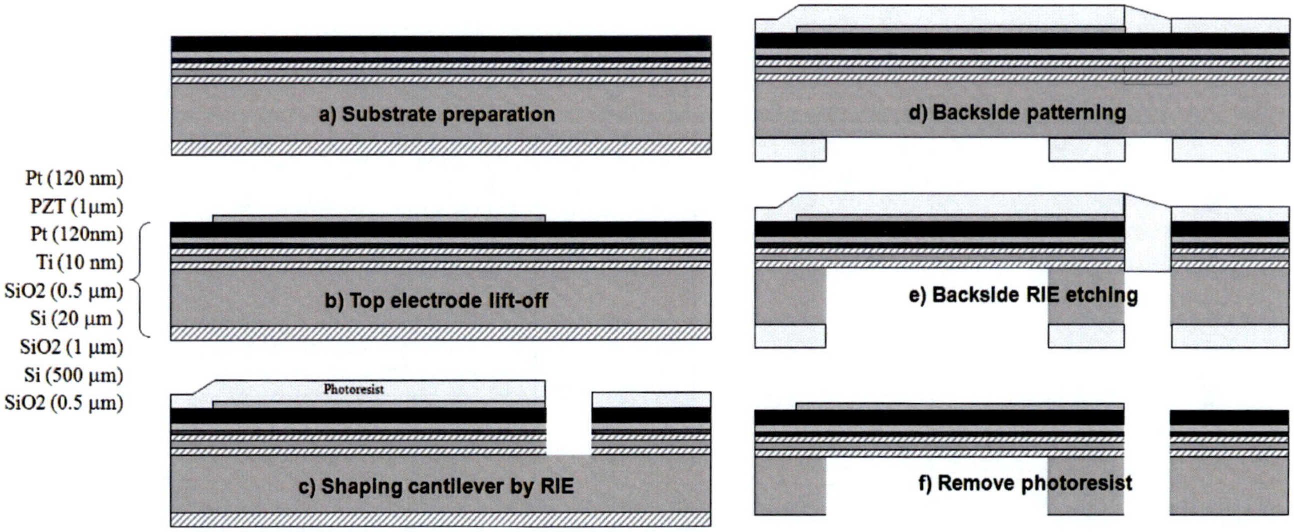

Figure 2 Microfabrication process of PEH. (a) PZT film was prepared on SOI wafer, (b) Pt top electrode patterned by lithography, (c) cantilever was shaped by RIE using photoresist masking, (d) after protecting front side with photoresist, backside was patterned, including patterning proof mass, (e) backside was etched out by RIE, (f) completed fabrication after cleaning.

The PZT films, with thicknesses of 1 μm, were prepared by Inosteck, Inc. on SOI wafer that has 20 μm Si device layer. $Pb(Zr_{0.52}Ti_{0.48})O_3$ was deposited on $Pt/SiO_2/SOI$ for the {3-1} mode and on $PbTiO_3$ seed layer/$ZrO_2/SiO_2/SOI$ for the {3-3} mode. When the polarization properties of the PZT thin film for the {3-1} mode were analyzed by a TF Analyzer 2000 (axiACT), the P-E hysteresis loop had a remnant polarization (P_r) and coercive field (V_c) of 20 μC/cm2 and 43 kV/cm with symmetry. The electric properties of the PZT film on ZrO_2 for the {3-3} mode were unfortunately unavailable at this time due to the restriction of the instrument, which should provide few hundreds voltages for wide gap electrodes in IDE configuration.

Both PZT films had a dense and columnar structure as shown in figure 3 (a) and (b). Their grain sizes are on average 150 nm and 2 μm for PZT/Pt and PZT/ZrO_2 respectively, which were confirmed by SEM photographic analysis. The crystallographic orientation of PZT films were investigated by X-ray Diffraction (XRD) in figure 4 (a) and (b). Both films were well crystallized into a perovskite phase without a pyrochlore phase. Both films were randomly oriented. The PZT film on Pt substrate exhibited the highest intensity on (100) and (111) orientation, while the PZT film on ZrO2 has high (100) peak intensity. Since the XRD analysis was performed after completed fabrication, it is suggested that no serious damage occurred during the microfabrication processes.

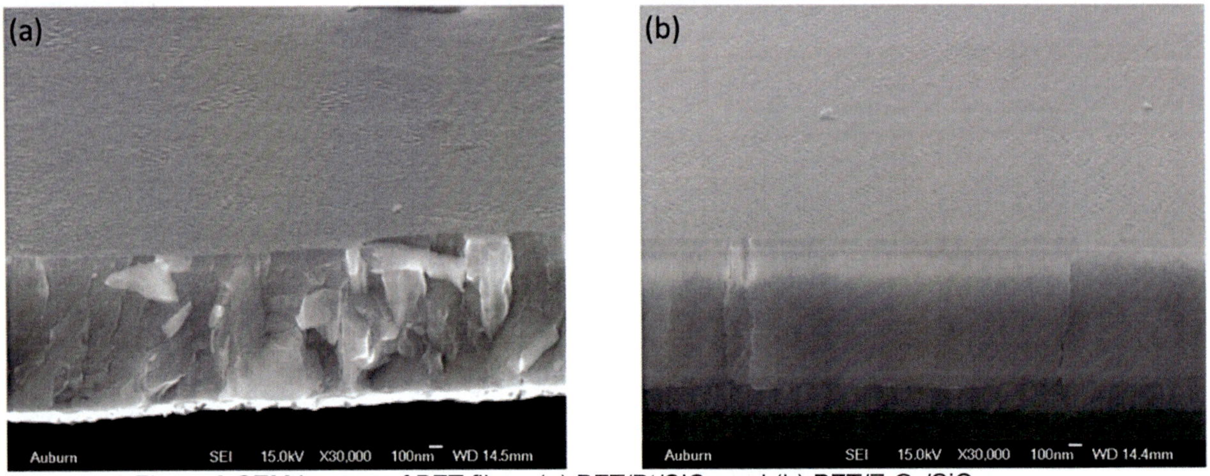

Figure 3 SEM images of PZT films: (a) $PZT/Pt/SiO_2$ and (b) $PZT/ZrO_2/SiO_2$

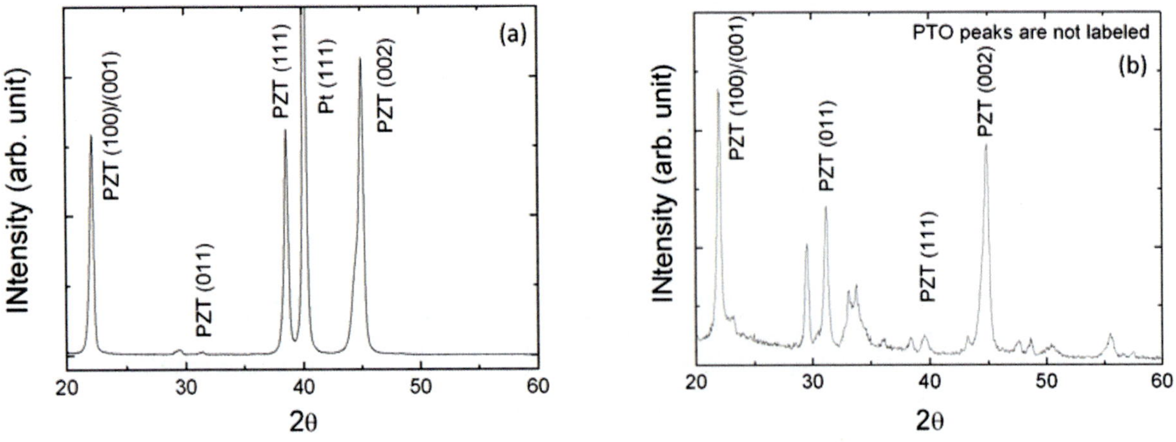

Figure 4 XRD analysis of PZT films: (a) $PZT/Pt/SiO_2$ and (b) $PZT/ZrO_2/SiO_2$

EXPERIMENTS

Output voltages, connected to various resistive loads, were obtained when a device was resonating under 0.5 g acceleration conditions. The peak-to-peak output voltages (V_{p-p}) were read from an oscilloscope (Tektronix TDS3014B), and the optimal resistive load was found where the calculated power was maximized. When the output power values were calculated by Ohm's law $P=V^2/R_{load}$, Root Mean Square (RMS) voltages were used. An electromagnetic shaker (Labworks ET-132-203) was used to vibrate the PEH device under its resonance frequency and certain acceleration force using a function generator (Agilent 33220A) and an amplifier. The acceleration of the generated vibration was monitored by an accelerometer (PCB Piezoelectronics 352C65) that was mounted on the shaker and connected to the oscilloscope. The setup is illustrated in figure 5.

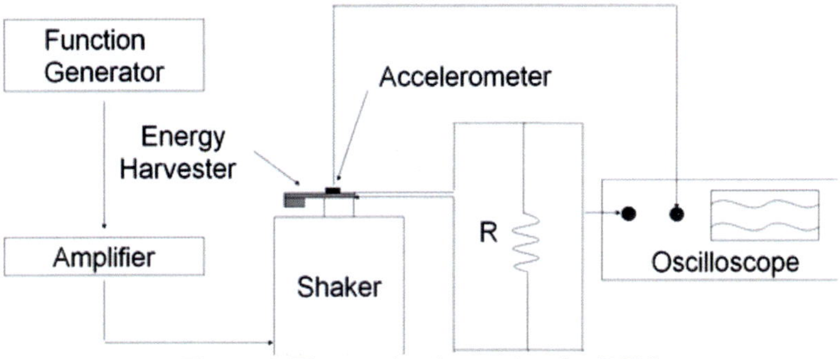

Figure 5 The evaluation setup for PEH

RESULTS AND DISCUSSIONS

The two modes of PEH were fabricated in the same dimensions of 7 x 2 mm^2 size of cantilever beam and 3 x 2 x 0.5 mm^2 size of proof mass, as shown in figure 6. The {3-3} mode has the interdigitated electrodes with 20 µm width and 40 µm gaps in figure 6(b). The resonant frequency was decided where the output voltage has maximum value. The measured resonant frequency was 130.8 Hz for the {3-1} mode and 118.1 Hz for the {3-3} mode so that they were in range of ambient vibration sources. The difference in resonant frequencies caused from the different electrode structure and added ZrO_2 buffer layer. The estimated resonant frequency from equation (6) was 125.6 Hz because the equation was simplified and never considered a thin Pt electrode layer. More accurate modeling, including electrode and buffer layers, will be continued in a future.

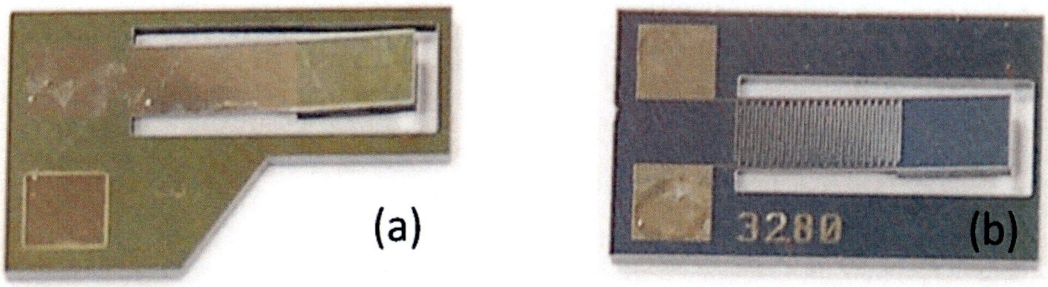

Figure 6 Pictures of fabricated PEHs: (a) {3-1} mode and (b) {3-3} mode

When the PZT thin film of the {3-1} mode PEH was poled at 120 °C under a 200 kV/cm^2 for 30 minutes, it generated 1.35 V_{p-p} at the resonant frequency with open circuit conditions under a 0.5 g acceleration force. Figure

7(a) shows that the maximum power of 6.72 μW was obtained with 11 kΩ resistive load. The output voltage was 0.769 V_{p-p} and the current was 70 μA calculated by Ohm's law. In terms of power density in area considering g force, MEMS PEH has 96 $\mu W/cm^2 \cdot g$, which is still 3~4 time less than that of bulk PEH but close to the requirement of wireless applications [5].

The PZT thin film of the {3-3} mode PEH was poled at 120 °C under a 60 kV/cm^2 for 30 minutes. When investigating the poling condition, which induced saturating polarization of PZT thin film within 40 μm gap of IDE, an electric field higher than 60 kV/cm^2 could not increase the output voltage and further increase up to 100 kV/cm^2 induced breakdown. The output peak-to-peak voltage under the open circuit condition was 7.725 V_{p-p}. In figure 7(b), 4.675 V_{p-p} was generated when the optimal resistive load was 4.5 MΩ. The maximum power output was 0.47 μW and the current was 1 μA.

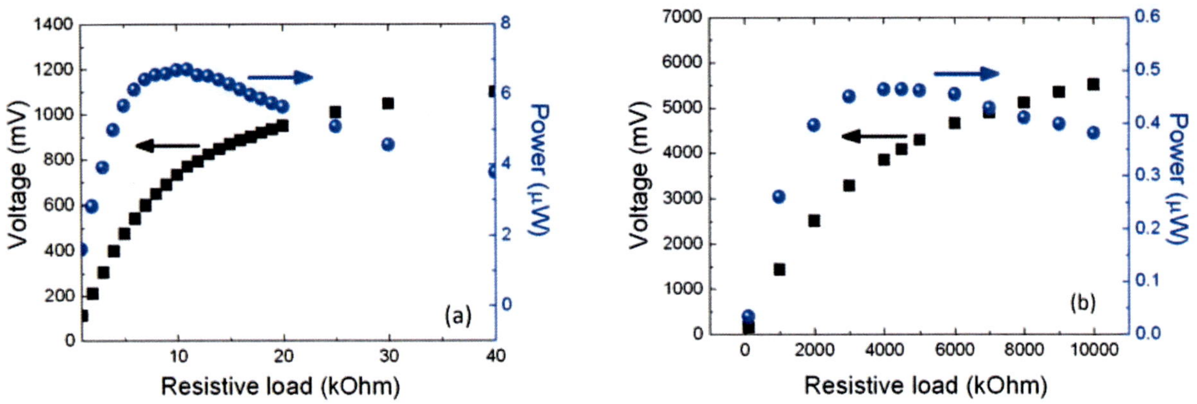

Figure 7 Output peak-to-peak voltages and powers from PEHs: (a) {3-1} mode and (b) {3-3} mode

Table 1 Comparison of performance for {3-1} and {3-3} mode PEHs

Mode	Resonant frequency (Hz)	V_{p-p} (V) (open circuit)	Max. power (μW)	V_{p-p} (V) (optimal R_{load})	Optimal R_{load} (kΩ)	Current (μA)
{3-1}	118.1	1.350	6.72	0.769	11	70
{3-3}	130.8	7.725	0.47	4.675	4500	1

The output voltage of {3-3} mode under an open circuit condition was 7.725 V_{p-p}, which is 6 times higher than that of the {3-1} mode. However, it is much less than the estimation by equation (1). Based on equations (1) and (2), the {3-3} mode should generate 80-times higher voltage than that of {3-1} mode under the same stress on the films because it has 2 times higher piezoelectric constant (g_{33}) and 40 times larger distance between electrodes in IDE configuration. From the experimental results, both g_{31} and g_{33} piezoelectric response should be explained under successful poling process.

The experimental results were summarized for both piezoelectric mode PEHs in table 1. The large output voltage is important for the energy harvester to rectify and store the current onto a battery. In this point of view, the {3-3} mode has the advantage of easy access to adjusting the output voltage by the gap of electrodes. However, its low capacitance largely decreased the current and power output. It might cause a very slow charging time as shown in its very small output power. On the contrary, the {3-1} mode has been developed in increasing its output voltage by successful integration of piezoelectric film into MEMS devices. With the development of VLSI wireless sensors, 272 V_{rms} of output voltage at optimal resistive load is higher than the required threshold voltage.

CONCLUSIONS

In this paper, piezoelectric micro power generators with two different piezoelectric modes were designed and fabricated for their comparison. Well-crystallized PZT thin films were confirmed by XRD and microstructural analysis. When they were fabricated in the same dimensions and resonated under matching impedance conditions at 0.5 g acceleration force, their output voltage, current, and power were compared. The {3-1} mode generated 0.769 $V_{p\text{-}p}$ where a maximum output power of 6.72 μW and 70 μA was obtained under 11 kΩ of optimal resistive load. Within comparison, the {3-3} had higher output voltage as 4.675 $V_{p\text{-}p}$ but its power and current were too low for its applications. Although the {3-3} mode using interdigitated electrodes still requires further study, such as optimizing IDE configuration and poling conditions, the {3-1} mode satisfies the requirements for now.

ACKNOWLEDGMENT

The authors would like to acknowledge Charles Ellis for his valuable advice in microfabrication. This work was supported by AUDFS.

REFERENCES

[1] D. Steingart, S. Roundy, P.K. Wright, and J.W. Evans, "Micropower materials development for wireless sensor networks", MRS Bulletin, 33, 408-409 (2008)
[2] S.J. Roundy, “Energy scavenging for wireless sensor nodes with a focus on vibration to electricity conversion”, Dissertation in Mechanical Engineering, University of Californica, Berkeley (2003)
[3] P.D. Mitcheson, T.C. Green, E.M. Yeatman, and A.S. Holmes, "Architectures for vibration-driven micropower generators", Journal of Microelectromechanical Systems 13, 429-440 (2004)
[4] S. Roundy, P.K. Wright, and J. Rabaey, "A study of low level vibrations as a power source for wireless sensor nodes", Computer Communications 26, 1131-1144 (2003)
[5] D. Shen, S.-Y. Choe, and D.-J. Kim, "Analysis of piezoelectric materials for energy harvesting devices under high-g vebrations", J. J. A. P. 46(10A),6755-6760 (2007)
[6] S. Roundy and P.K. Wright, "A piezoelectric vibration based generator for wireless electronics", Smart Mater. Struct. 13, 1131–1142 (2004)
[7] S.P. Beeby, R.N. Torah, M.J. Tudor, P. G. Jones,T. O'Donnell, C.R. Saha and S. Roy, "A micro electromagnetic generator for vibration energy harvesting", J. Micromech. Microeng. 17 1257–1265 (2007)
[8] D. Shen, J.-H. Park, J. Ajitsaria, S.-Y. Choe, H.C. Wikle, and D.-J. Kim, "The design, fabrication and evaluation of a mems pzt cantilever with an integrated Si proof mass for vibration energy harvesting", J. Micromech. Microeng 18 055017(2008)
[9] M. Marzencki, B. Charlot. S. Basrour, M. Colin, and L. Valbin, "Design and fabrication of piezoelectric micro power generators for autonomous microsystems", DTIP symposium on design testing integration and packaging of MEMS/MOEMS 229-302 (2005)
[10] H.B Fang, J.Z. Liu, Z.Y. Xu, L. Dong, L. Wang, D. Chen, B.C. Cai, and Y. Liu, "Fabrication and performance of mems-based piezoelectric power generator for vibration energy harvesting" J. Microelectronics 37 1280-1284 (2006)
[11] Y.B. Jeon, R. Sood, J.-h. Jeong, S.-G. Kim, "Mems power generator with transverse mode thin film pzt" Sensors and Actuators A 12216-22.(2005)
[12] M. Renaud, T. Sterken, A. Schmitz, P. Fiorini, C. Hoof, and R. Puers , "Piezoelectric harvesters and mems technology: Fabrication , modeling and measurements ", Proc. Int. conf. on Solid-State Sensors, Actuators and Microsystems 891-894 (2007)
[13] B.S Lee, S.C. Lin, W.J. Wu, X.Y. Wang, P.Z. Chang and C.K. Lee, "Piezoelectric mems generators fabricated with an aerosol deposition PZT thin film", J. Micromech. Microeng. 19 065014 (2009)
[14] P. Muralt, "PZT thin films for microsensors and actuators: Where do we stand?", IEEE trans. on ultra. ferro. freq. cont. 47(4),903-915 (2000)
[15] H.-C. Song, H.-C. Kim, C.-Y. Kang, H.-J. Kim, S.-J. Yoon, and D.-Y. Jeong, "Multilayer piezoelectric energy scavenger for large current generation", J Electroceram, DOI 10.1007/s10832-008-9439-9 (2008)
[16] Q.Q. Zhang, S.J. Gross, S. Tadigadapa, T.N. Jackson, F.T. Djuth, and S. Trolier-McKinstry, "Lead zirconate titanate films for d_{33} mode cantilever actuators", Sensors and Actuators A 105 91-97 (2003)

Proceedings of the SEM Annual Conference
June 7-10, 2010 Indianapolis, Indiana USA

Size Effects Associated with Microcompression Experiments on Single-Crystal Magnesium

Cynthia M. Byer*, K.T. Ramesh*
*Johns Hopkins University, 3400 N. Charles Street, Baltimore, MD 21218

ABSTRACT

Microcompression is becoming an increasingly popular technique to investigate the orientation dependence and size effects associated with single crystals under uniaxial compression. Literature shows that for certain materials, as the diameter of these micro-scale pillars decreases, yield stresses and strain hardening rates may increase; however, this phenomenon has not yet fully been investigated for hexagonal close packed (hcp) materials. In this study, microcompression experiments are conducted on micropillars that are fabricated using focused ion beam (FIB) milling. These single crystal magnesium specimens are loaded in compression along the [0001] c-axis, and the stress-strain curves reveal that there are no significant size effects.

INTRODUCTION

Hexagonal close-packed (hcp) Mg and its alloys are very attractive for automotive and aerospace applications because of their low density (1740 kg m^{-3} for pure Mg). However, due to the low symmetry, the deformation behavior of these materials is much more complicated than that of high-symmetry face-centered cubic (fcc) metals, such as aluminum and copper. Due to this complexity, prior work on the fundamental deformation mechanisms has often been controversial or inconclusive.

A related issue is that of size effects in hcp metals, and in Mg in particular. Many microcompression studies on fcc metals have shown that as pillar diameters decrease, the flow stress increases. Decreasing diameters have sometimes been shown also to increase the amount of strain hardening micropillars exhibit. Similar studies have been conducted on bcc metals, as well as on other materials such as alloys, intermetallics, nanocrystalline metals, nanoporous foams, and metallic glasses. The explanation for these phenomena is still under debate. We have performed microcompression experiments on Mg to investigate the possibility of a size effect in hcp metals, given that their deformation processes are very different from those of fcc crystals.

EXPERIMENTAL TECHNIQUE

Micropillars, ranging in size from 0.6 to 10 micrometers, are fabricated on a 99.999% pure single crystal of magnesium (1 cm in diameter and 2 mm thick), oriented along the [0001] c-axis, using a dual-beam FEI Nova 600 focused ion beam (FIB). Rough pillars are milled using a series of concentric circular cuts with decreasing diameters and currents. However, this fabrication process alone yields

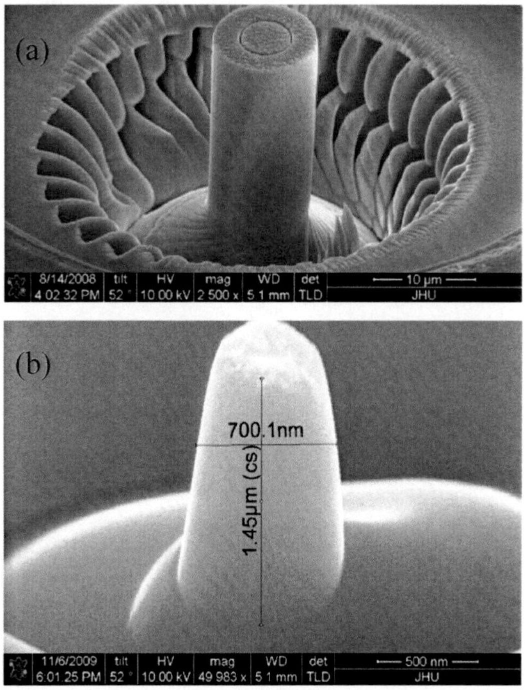

Figure 1: (a) Pillar with no taper, made by lathe-type milling. (b) Pillar made by top-down annular milling, resulting in taper. Note the magnifications are not the same for the two images.

DOI 10.1007/978-1-4419-8825-6_6, © The Society for Experimental Mechanics, Inc. 2011

tapered pillars, which could result in inaccurate measurements of elastic modulus, hardening, and yield stress. In order to eliminate taper in the final specimens, lathe-type secondary milling is performed using a program developed by Uchic and Dimiduk. This procedure is not precise enough for our pillars smaller than 2.5 μm, so care is taken to fabricate these specimens using top-down concentric annular cuts, resulting in micropillars with tapered sides of a few degrees. Final micropillars for these experiments have a length that is twice that of the diameter, resulting in an aspect ratio of 2. Using an aspect ratio above 2 could potentially result in buckling, while using anything lower than 2 may result in non-uniform stress measures.

In order to prevent oxidation, the specimen is stored in a vacuum chamber and is only exposed to air for the duration of the tests. The experiments are performed with an MTS nanoindenter, using a 35 μm by 35 μm square flat punch diamond tip, at a nominal strain rate of 10^{-4}. The samples are examined post mortem using a scanning electron microscope.

RESULTS AND DISCUSSION

The raw data from the nanoindenter is the load history and crosshead displacement history. This data is corrected for both machine compliance and base compliance, and from this corrected data, we compute stress and strain. Data from experiments

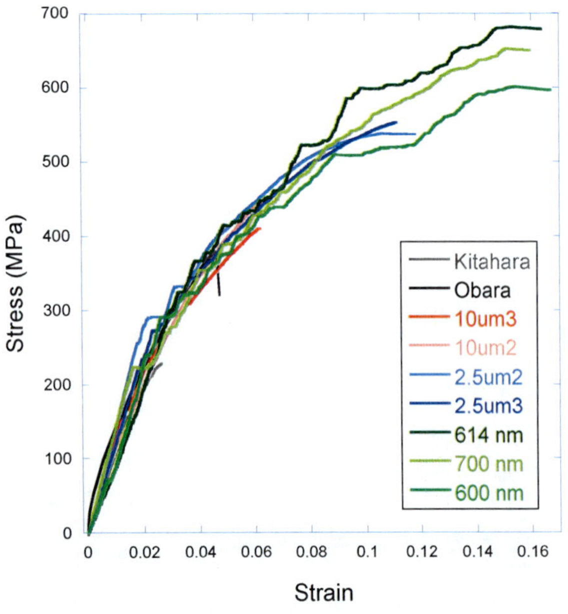

Figure 2: Engineering stress-strain curves for compression experiments on single-crystal Mg along the c-axis, for samples ranging over the orders of nanometers to millimeters in length scale.

on micropillars 0.6 to 10 μm in diameter are plotted in Figure 2. Data from Kitahara et al. and Obara et al. are also plotted in Figure 2, so that we can compare our microcompression data to that of bulk compression tests. Kitahara et al. conducted quasistatic compression experiments on 3 x 3 x 6 mm single crystals along the [0001] orientation. Obara et al. also performed quasistatic compression texts along the c-axis on single crystals that were 5 mm in diameter and 5 mm in length.

The obvious similarities in the data sets for the macroscale and microscale tests imply that there are no significant size effects on the strength of Mg for these length scales, and all curves have similar levels of strain hardening. Note that we are able to reach larger strains in these small specimens. The millimeter-scale specimens of Kitahara et al. and Obara et al. fractured at the end of their stress-strain curves presented in Figure 2, whereas no fracture occurred in our experiments. The stress-strain curves end when the pillar is unloaded. This larger apparently strain to failure is attributed to the likelihood of fewer defects in our specimens due to their smaller volumes.

REFERENCES

1. Bei, H. et al., "Compressive Strengths of Molybdenum Alloy Micro-pillars Prepared Using a New Technique," Scripta Materialia 57 (2007) 397.
2. Byer, Cynthia et al., "Microcompression of Single-Crystal Magnesium," Scripta Materialia 62 (2010) 536.
3. Greer, Julia et al., "Size Dependence of Mechanical Properties of Gold at the Micron Scale in the Absence of Strain Gradients," Acta Materialia 53 (2005) 1821.
4. Kitahara, Takaaki et al., "Deformation Behavior of Magnesium Single Crystals in C-Axis Compression," Key Engineering Materials 345 (2007) 129.
5. Lilleodden, Erica, "Microcompression Study of Mg (0001) Single Crystal," Scripta Materialia 62 (2010) 532.
6. Obara, T. et al., "{11-2-2}<-1-123> Slip Systems in Magnesium," Acta Metallurgica 21 (1973) 845.
7. Uchic, Michael et al., "Sample Dimensions Influence Strength and Crystal Plasticity," Science 305 (2004) 986.
8. Uchic, Michael et al., "A methodology to investigate size scale effects in crystalline plasticity using uniaxial compression testing," Materials Science & Engineering A 400 (2005) 268.
9. Zhang, H. et al., "The Design of Accurate Microcompression Experiments," Scripta Materialia 54 (2006) 181.

Proceedings of the SEM Annual Conference
June 7-10, 2010 Indianapolis, Indiana USA

Mechanically probing time-dependent mechanics in metallic MEMS

J.P.M. Hoefnagels[1*], L.I.J.C. Bergers[1,2], N.K.R. Delhey[1], M.G.D. Geers[1]

[1] Eindhoven University of Technology, Department of Mechanical Engineering,
P.O.Box 513, 5600MB, Eindhoven, The Netherlands

[2] Foundation for Fundamental Research on Matter (FOM),
P.O.Box 3021, 3502 GA, Utrecht, The Netherlands

[*] E-mail: j.p.m.hoefnagels@tue.nl, tel:+31-40-2475894, Fax:+31-40-2447355

ABSTRACT

The reliability of metallic micro-electromechanical systems (MEMS) depends on time-dependent deformation such as creep. To this end, a purely mechanical experimental methodology for studying the time-dependent deformation of free-standing microbeams has been developed. It is found most suitable for the investigation of creep due to the simplicity of sample handling and preparation and setup design, whilst maximizing long term stability and displacement resolution. The methodology entails the application of a constant deflection to a μm-sized free-standing aluminum cantilever beam for a prolonged period of time. After this load is removed, the deformation evolution is immediately recorded by acquiring surface height profiles through confocal optical profilometry. Image correlation and an algorithm based on elastic beam theory are applied to the full-field beam profiles to yield the tip deflection as a function of time. The methodology yields the tip deflection as function of time with ~3 nm precision.

1. RELIABILITY AND TIME-DEPENDENT MECHANICS

The application of metals as structural components in MEMS is common for 'radio-frequency MEMS' (RF-MEMS). Figure 1 shows an example. The reliability of these devices has been shown to critically depend on time-dependent mechanics, such as fatigue and creep [1]. Fatigue affects the device life time through its limitation on the number of device operation cycles, e.g. the number of open/closed cycles of an RF-MEMS switch. Creep can directly affect the operational characteristic, e.g. through a shift in pull-in voltage of an RF-MEMS switch which results in a reduced power handling [2]. Whereas fatigue effects may pose less of a problem than expected at small geometrical length scales [3], the detrimental influence of creep seems to increase upon miniaturization [3].

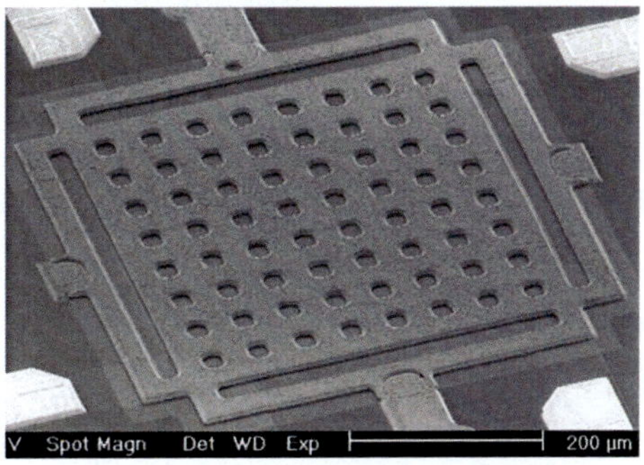

Figure 1: Scanning electron micrograph of an RF-MEMS switch (courtesy of EPCOS Netherlands B.V.).

The difference between micro- and macroscale creep is generally attributed to the size-effect: the interaction between microstructural length scales and dimensional length scales [4,5]. The physical micro-mechanisms of creep in these free-standing microbeams are, however, poorly understood, let alone implemented in models. In the literature some reports can be found discussing creep and relaxation effects in thin aluminum films [6-11].

However, specifically for free-standing thin films not much research has focused on determining size-effects in time-dependent material behavior [12]. Therefore, there is a clear need for detailed studies into the physical micro-mechanisms underlying the size-effects in creep in metallic MEMS.

As a first step towards such studies, the goal of the current work is to construct and validate an experimental mechanical methodology to quantify creep, of µm-sized free-standing cantilever beams. To this end, first the factors are discussed that determine the choices for the design of the experiment, followed by the explanation of the methodology. Next, proof-of-principle measurements are conducted and the results are presented. Finally, conclusions are drawn with respect to the goal of this work and the further application of this work to the study of size-effects in creep in metallic-MEMS.

2. DESIGN OF EXPERIMENT

Performing mechanical tests on specimens that are free-standing and have dimensions in the order of a few micrometers or less is not trivial. Aspects of sample preparation, handling, loading, load and deformation measurement and control have to be carefully addressed [13]. Considering the sample preparation and handling, it is highly preferred to test free-standing, micron-scale on-wafer samples that have been fabricated in the same micro-fabrication process as that of the actual MEMS device. This simplifies the handling of µm sized samples to handling 20x20 mm^2 dies/wafer pieces. It also guarantees that the results obtained in an experiment are fully applicable to the corresponding MEMS device. A small and simple displacement controlling mechanism is most suited to yield long term stability of the load and displacement control and measurement: this is especially critical for prolonged creep measurements. Therefore, a so-called 'micro-clamp' is designed, which is a simple horizontal knife-edge attached to an elastic hinge, see Figure 2. A differential screw is used to control the height of the edge with respect to the substrate to ~50 nm, whilst simple adjustment screws are used for in-plane adjustment with ~5 µm accuracy. The compact and near-monolithic fully mechanical design minimizes drift due to temperature and electrical effects. With optical confocal profilometry the out-of-plane deformation is measured, yielding full-field deformation information, with 10-nm RMS repeatability (on an ultra-flat surface) for the high magnification super long working distance objective used here. Interferometric profilometry is not suitable as a large working distance (>5 mm) is imposed by the micro-clamp and large local surface angles are associate with the relatively high surface roughness of the investigated metallic MEMS surface. The measurement precision will benefit from the full-field deformation imaging combined with simple image processing and analysis with elastic beam-bending theory. Another advantage of this methodology is that multiple samples can be deflected in parallel simply by aligning all samples on the chip and using a broad knife-edge. Furthermore the setup does not involve any highly specialized parts or instruments which allows for easy and inexpensive parallelization. This is particularly useful for prolonged creep experiments. The only drawback is the lack of a direct measurement of force. Given the benefits of this methodology, i.e. simplicity, long term stability, deflection resolution, it is found most suitable.

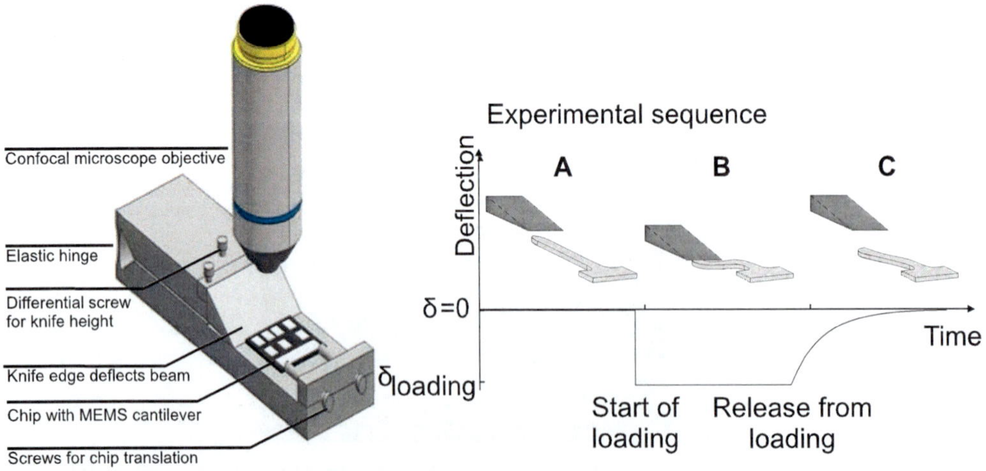

Figure 2: (Left) Schematic of the MEMS cantilever deflection experiment with a micro-clamp under a confocal optical profilometer, which monitors the deflection as function of time. (Right) Schematic representation of the measurement sequence of the cantilever deflection recovery experiment: (A) the knife-edge of the micro-clamp approaches the cantilever; (B) the knife-edge deflects the cantilever to a depth of $\delta_{loading}$ and holds it there for a certain period of time; and (C) the edge is raised, releasing the cantilever, after which the deflection recovery is measured over time using a surface profilometer.

Based on these considerations, the choice has been made to design an experiment measuring the time-dependent deflection recovery of a cantilever after constant deflection. Figure 2 shows the deflection sequence. The initial deflection is controlled by the 'micro-clamp'. Directly after release of the deflection, a time series of full-field deformation maps is acquired with a commercial confocal optical profilometer. An algorithm based on elastic beam theory is then applied to the deformation data to extract the evolution of the deflection as function of time. This methodology thus allows simple sample handling and MEMS-device fabrication, and precise control and measurement of the deformation over a longer period of time.

3. EXPERIMENTAL PROCEDURE

The deflection of the microbeam test structure is controlled by the micro-clamp. This is done under the optical profilometer, a Sensofar Plu2300. The surface profilometer is operated in the confocal microscopy mode: 470 nm LED light, 100x long working distance objective with N.A. of 0.7, height of scan of 20 μm, data acquisition time ~85 s. In this configuration, the RMS repeatability is <10 nm, although data acquisition and processing will improve the overall precision. Finally, the (optically flat) substrate is leveled to within 0.01° by using the interferometric mode of the optical profilometer and manually adjusting a 2-axis tilt stage. The setup is stabilized on an active vibration isolation table. The measurements are conducted in an environmentally controlled room, with $T_{ambient}$ = (21.0±1.0) °C, RH = (14±1)%.

Figure 3 shows an example of the cantilever beam test structures investigated here. The microbeams are produced from Al-Cu (1 wt%) on a test-wafer having undergone a production process scheme as used for actual RF-MEMS switches. The Young's modulus is 66.8 GPa, determined by eigenfrequency measurements and finite element analysis of electro-statically actuated cantilevers [14,15]. The film thickness is (4.8±0.2) μm and has a surface roughness R_A = 45 nm. The film consists of columnar, through-thickness grains with an average grain diameter of 20 μm (assuming cylindrical grain shapes) and a strong {111} surface orientation, as determined with electron backscatter diffraction. Several beams having a width of 25 μm and lengths between 45 and 500 μm are located on the test-wafer. Cross-sections are trapezoidal due to the thin film patterning. For the current work, beams of 65 μm length are utilized. This choice is a trade-off between practical imaging, deflection and induced stress levels. The experiment is conducted as follows. The wafer is placed on the micro clamp and the beam is aligned to the knife–edge (with ~5 μm lateral precision) under the microscope. Leveling (to within 0.01°) is conducted with interferometry. A surface topography of the unaltered beam (reference height position) is acquired in the confocal profilometer mode. The knife edge is then lowered to deflect the beam to a depth of approximately 800 nm (much less that the microbeam-substrate gap of 3 μm). This level is chosen as it corresponds to $\sigma_{bend,max}$ = 50 MPa for this length. This stress level corresponds to ~30 % of the yield stress of the material and is a lower level encountered in RF-MEMS device operation. The beam is then kept deflected at this position for 48 hours, after which the unloading procedure is started. The edge knife is raised and immediately thereafter, at every 85 s, surface topographies are obtained for a duration of 10000 s. Finally images are taken every 900 s for another 10000 s. In this manner a set of surface profiles is obtained of the recovering deformation in the beam.

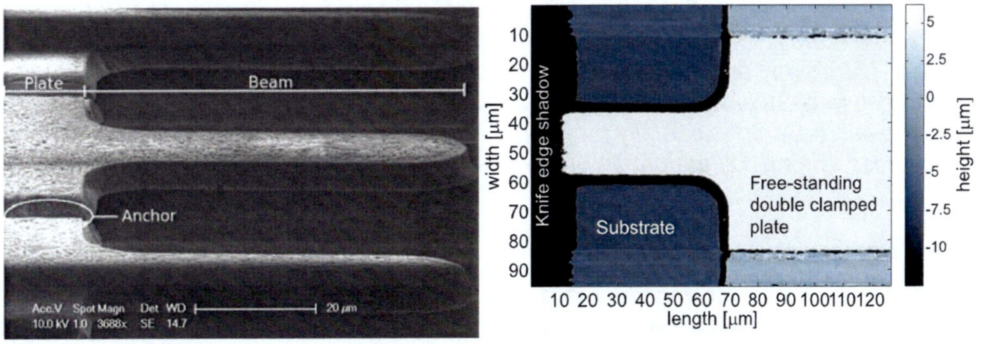

Figure 3: (Left) A SEM-image of a test cantilever beam that is attached to a free-standing plate clamped on three sides by an anchor. (Right) A (mirrored) contour map of a profile of a beam placed under the micro-clamp obtained with confocal profilometry.

The surface profiles do not immediately yield the tip deflection as function of time. Data processing is required to deal with image translations and tilt due to thermal and various other drifts in the optical profilometry set-up. Therefore, an analysis procedure is implemented in Matlab to correlate x-y translations with pixel resolution and level with a linear fit. The tip deflection is determined by fitting standard beam bending theory to the full-field deformation data of the beam. This uses the intrinsic assumptions that the bending profile is elastic and that the double clamped plate effectively fixes the beam. Area bins along the beam's length are created, in which the average height is determined to minimize height measurement error. These are the plot to form a profile, to which the equation for a single clamped beam is fit. Figure 4 shows this result.

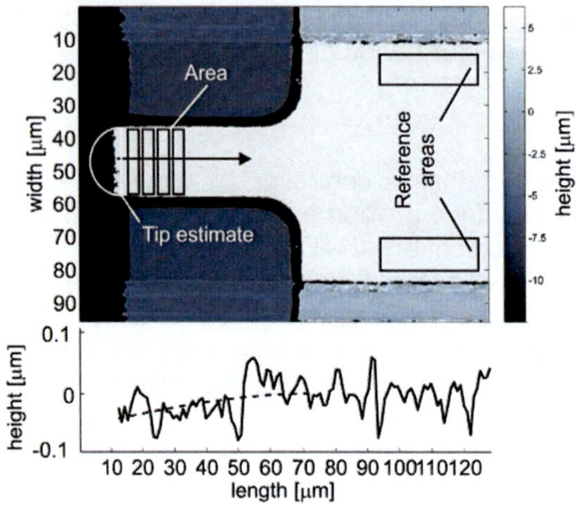

Figure 4: Surface profile schematically illustrating the area bins and selected reference areas, all of which are schematically drawn here for clarification purpose. The mean height of these boxed areas is evaluated to form a profile. Linear plane fitting is applied based on the reference areas to correct for tilt around x and y axis. These areas are then used as reference height and set to zero height. The profiles are referenced to this height. Standard beam bending equations are fit to the profile as indicated by the dashed line in the lower graph. For clarification, a semi-circle is drawn at the location of the tip.

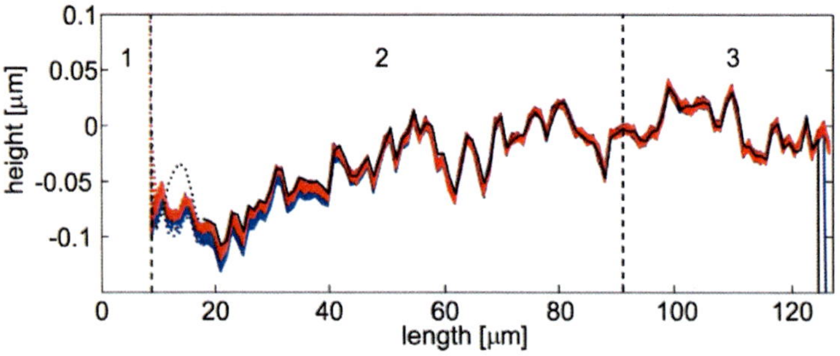

Figure 5: Results of the obtained profiles after deflection at ~800 nm for a duration of ~48 hours. Note that the surface roughness of the metallic MEMS is much higher than the change in tip deflection, which is of interest. The initial profile (acquired before any applied deflection) is plotted in black. The first profile acquired after unloading is plotted in blue and subsequent profiles are plotted with a gradient from blue to red. The dotted lines represent unreliable data, as the micro-clamp is partially blocking the reflection of light. Three sections are distinguished: (1) the micro-clamp, (2) the profile of the beam, and (3) the reference height region.

4. EXPERIMENTAL RESULTS

The result of the measured deflection recovery is a sequence of profiles, shown in Figure 5, from which the tip deflection as a function of time is determined, as shown in the left graph of Figure 6. After an instantaneous initial spring back of more than 95% of the loading depth, the beam shows a relatively small, though clear monotonous increase in height of ~20 nm over a time period of about 3 hours. This is evidence of time-dependent recovery of the cantilever. Next to this, a permanent deflection of ~10 nm is observed. To asses the precision of the measurement a fit is made to the tail of the graph where the deflection has saturated. A histogram based on the difference between average and measured deflection then shows that the tip deflection measurement has a precision (1 σ) of ~3 nm, as shown in the right graph of Figure 6. This precision is quite substantial considering the ~45-nm RMS surface roughness and 10-nm RMS repeatability of the confocal profilometry (on a perfectly flat surface). The measurement has been repeated for several similar beams, all revealing similar results.

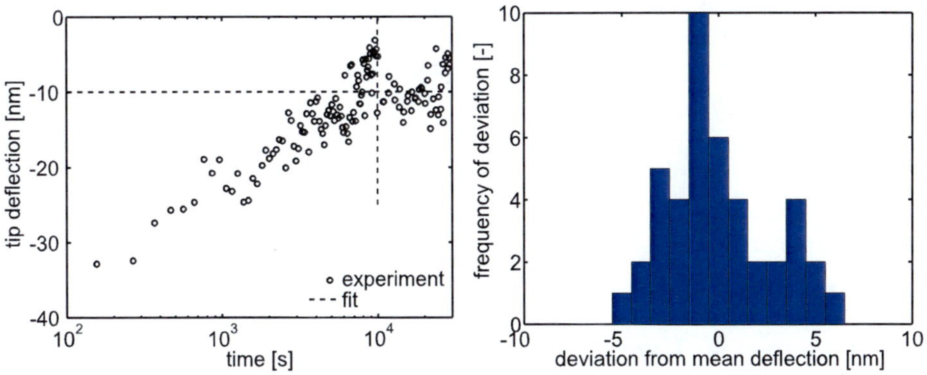

Figure 6: (Left) The displacement of the tip versus the logarithm of time. A fit (dashed line) is made for the time segment from 10.000-25.000 seconds. (Right) Histogram based on the data of the last 10000-25000 seconds to evaluate the precision of the measurement. From this a standard deviation in the tip deflection measurement of ~3 nm is estimated.

5. CONCLUSIONS

An experimental methodology is presented together with measurement results to measure time-dependent deformation in µm-sized free-standing aluminum-copper alloy cantilever beams. As a first result the deformation recovery of a microbeam is measured over a period of ~6 hours, showing a recovery of 20 nm during this period after initial spring back. A permanent deflection of ~10 nm is observed. The precision (1 σ) is ~3 nm, which is quite substantial considering the relatively high surface roughness and unavoidable experimental limitations.

The measurement is possible due to the design of the experiment: measuring time-dependent recovery with confocal profilometry in a simple mechanical setup with minimal sample handling. A micro-clamp deflects on-chip micro-beams to a depth that can be controlled with ~50 nm precision (but can be measured again with ~3 nm precision). At the same time, the methodology is insensitive to thermal fluctuations.

The next step in this research is to apply time-dependent material models incorporated into FEA to extract physically meaningful parameters, the procedure of which will be published in the future. With this full methodology in place, grain size and grain orientation effects can be probed by coupling measured differences in grain size and grain orientation distributions between different microbeams of the same geometry (on the same wafer) to changes in the observed creep behavior. Furthermore, experiments can be conducted at various temperatures to accelerate or probe different creep mechanisms. Ultimately, device structural dimensions will be varied to probe more size-effects in creep of these structures.

ACKNOWLEDGMENTS

Dr.Ir. Marcel van Gils and Ir. Jeroen Bielen at EPCOS Netherlands B.V. are greatly acknowledged for their cooperation, support and fruitful discussions in this work. This research is carried out under project number M62.2.08SDMP12 in the framework of the Research Program of the Materials Innovation Institute M2I and the Foundation for Fundamental Research on Matter (FOM), which is financially supported by the Netherlands Organization for Scientific Research (NWO).

REFERENCES

1. Van Spengen W. M., "MEMS reliability from a failure mechanisms perspective," Microelectron. Reliab. **43**, 7, pp. 1049-1060, 2003.
2. Van Gils M., J. Bielen, and G. McDonald, "Evaluation of creep in RF MEMS devices," proceedings of the EuroSimE 2007 conference, London, 2007.
3. Douglas M. R., "Lifetime estimates and unique failure mechanisms for a Digital Micromirror Device", proceedings of the 36th Annual International Reliability Physics Symposium, Reno, U.S.A., pp. 9-16, 1998.
4. Dehm G., C. Motz, C. Scheu, H. Clemens, P. H. Mayrhofer, and C. Mitterer, "Mechanical size-effects in miniaturized and bulk materials," Adv. Eng. Mater. **8**, 11, pp. 1033-1045, 2006.

5. Arzt E., "Size effects in materials due to microstructural and dimensional constraints: A comparative review," Acta Mater. **46**, 16, pp. 5611-5626, 1998.
6. Lee H. J., P. Zhang, and J. C. Bravman, "Stress relaxation in free-standing aluminum beams," Thin Solid Films **476**, 1, pp. 118-124, 2005.
7. Kalkman A. J., A. H. Verbruggen, and G. C. A. M. Janssen, "Young's modulus measurements and grain boundary sliding in free-standing thin metal films," Appl. Phys. Lett. **78**, 18, pp. 2673-2675, 2001.
8. Kalkman A. J., A. H. Verbruggen, G. C. A. M. Janssen, and S. Radelaar, "Transient creep in free-standing thin polycrystalline aluminum films," J. Appl. Phys. **92**, 9, pp. 4968-4975, 2002.
9. Hyun S., W. L. Brown, and R. P. Vinci, "Thickness and temperature dependence of stress relaxation in nanoscale aluminum films," Appl. Phys. Lett. **83**, 21, pp. 4411-4413, 2003.
10. Modlinski R., A. Witvrouw, P. Ratchev, R. Puers, J. M. J. Den Toonder, and I. De Wolf, "Creep characterization of al alloy thin films for use in mems applications," Microelectron. Engg. **76**, 1-4, pp. 272-278, 2004.
11. Modlinski R., P. Ratchev, A. Witvrouw, R. Puers, and I. D. Wolf, "Creep-resistant aluminum alloys for use in MEMS," J. Micromech. Microengg. **15**, 7, p. S165-S170, 2005.
12. Connolley T., P. E. Mchugh, and M. Bruzzi, "A review of deformation and fatigue of metals at small size scales," Fatigue Fract. Eng Mater. Struct. **28**, 12, pp. 1119-1152, 2005.
13. Hemker K. J. and W. N. Sharpe Jr, "Microscale characterization of mechanical properties," Ann. Rev. Mater. Res. **37**, pp. 92-126, 2007.
14. N. K. R. Delhey, "An experimental methodology to characterize time-dependent deformation in free-standing aluminum thin-films." Eindhoven University of Technology, 2009.
15. Bielen J., J. Stulemeijer, D. Ganjoo, D. Ostergaard, and S. Noijen, "Fluid-electrostatic-mechanical modeling of the dynamic response of RF-MEMS capacitive switches," proceedings of the EuroSimE 2008 conference, Freiburg im Breisgau, 2008.

Proceedings of the SEM Annual Conference
June 7-10, 2010 Indianapolis, Indiana USA

Small mass measurements for tuning fork-based atomic force microscope cantilever spring constant calibration

Gordon A. Shaw, Jon R. Pratt, Zeina J. Jabbour
Manufacturing Engineering Laboratory
U.S. National Institute of Standards and Technology

Abstract

Cutting edge mass sensors are capable of discriminating mass changes as small as several dozens of atoms, however the smallest mass commercially available from NIST with a calibration traceable to the International System of Units (SI) is 0.5 mg. To bridge this gap, new metrological techniques are being developed. A mass change from the electrochemical dissolution of tungsten wire has been measured using a commercial microbalance, and applied to a dynamic calibration of the spring constant of a tuning fork oscillator designed for use in frequency modulated atomic force microscopy (FMAFM). The spring constant measured using the dynamic method agreed within experimental uncertainty with that determined using an instrumented indenter, however an improved model for the indenter's contact mechanics will be necessary to validate the assumptions used in the dynamic method to less than 10 %.

Introduction

Currently, the smallest mass artifact available that is traceable to the international system of units (SI) has a mass value of 0.5 mg. A wide variety of manufactured products require the use of materials with mass values less than 500 μg, however. For example, a 10 nm titanium adhesion layer deposited on a 100 mm wafer in the semiconductor industry has a mass of less than half a microgram. All traceable mass measurements are based on the International Prototype Kilogram (IPK) stored at the Bureau International des Poids et Mesures (BIPM), although the definition of the kilogram may soon change to one based on fundamental quantum invariants [1]. The IPK is compared against a set of replica kilograms stored at various National Measurement Institutes, including the U.S. National Institute of Standards and Technology (NIST). To obtain smaller traceable mass values, the kilogram must be subdivided [2], and this process introduces additional uncertainty in the mass value. The total relative expanded uncertainty of mass for a 1 mg artifact is approximately 3×10^{-4}, and by linear extrapolation as shown in Figure 1, for a 1 μg artifact this uncertainty could increase to 1×10^{-2}. In practice, the minimum mass of commercially available ultramicrobalances is approximately half a microgram, leading to possible uncertainties approaching 50 % for 1 μg mass measurements.

T. Proulx (ed.), *MEMS and Nanotechnology, Volume 2*, Conference Proceedings of the Society for Experimental Mechanics Series 2,
DOI 10.1007/978-1-4419-8825-6_8, © The Society for Experimental Mechanics, Inc. 2011

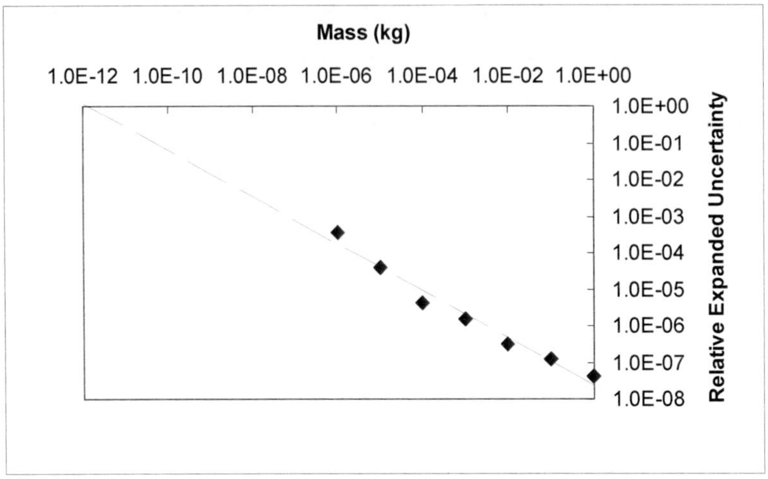

Figure 1. Projected uncertainties in small mass metrology. Points indicate current uncertainty at various mass levels, and the dashed line is a linear fit to these data points.

Two approaches can be followed to extend the range of traceable mass measurement. The first is to develop new traceability paths for mass. This type of work is already underway through the electronic kilogram [1] and small force [3,4] metrology programs at NIST. Another approach is to examine how currently available mass metrology techniques scale to smaller values of mass. This approach has been followed in previous studies [5] and shown the feasibility of mass metrology down to 100 micrograms with a combined standard uncertainty of 0.38 micrograms limited primarily by the statistical uncertainties associated with the balance used for measurement. This is still impractical for mass metrology in the microgram range. The development of a mass sensing method better suited to the small mass range could improve resolution, repeatability, and reproducibility in the mass measurement process, and decrease the uncertainty associated with small mass values.

Dynamic methods offer an attractive option for mass measurement, provided that mass can be correlated with a frequency measurement, since frequency measurements can be performed with very small uncertainty. For this study, a small quartz tuning fork resonator was chosen for examination as a mass sensor. These oscillators are inexpensive, readily available, and designed as frequency reference devices. They have also been used extensively in FMAFM, producing sub-atomic imaging resolution [6], and atomic-scale force measurements [7]. The quantitative measurement of forces and potentials at this scale requires the evaluation of the tuning fork spring constant, k. Although methods have been described for the measurement of k for the tuning fork systems [8-10], no method yet exists that provides SI-traceability. The following study provides a possible method for the calibration of this spring constant based on SI-traceable mass metrology similar to the Cleveland method [11]. This method is then compared against the result from an instrumented indentation method previously described, and the values from the two methods are shown to agree within experimental uncertainty.

Experimental Methods

Tuning fork sensors were obtained from Omicron [12] pre-mounted on a ceramic chip attached to a metal tripod base such that one tine of the tuning fork was firmly affixed to the ceramic. This creates a rectangular cantilever beam out of the other tine. A tungsten wire tip had been mounted at the end of the free tine with conductive epoxy, creating a conductive pathway from one leg of the tripod base to the tungsten wire tip(i.e., a “qPlus” sensor [13].)

The tuning fork was then placed on a piezo stack under an inspection microscope equipped with a Polytech laser Doppler vibrometer (LDV) as shown in Figure 2. The sensor was shaken by the piezo, which was attached to the oscillator output of an Agilent 35670A dynamic signal analyzer. The oscillator frequency was swept through the sensor's resonance at constant excitation voltage amplitude of 20 mV, and the output of the LDV simultaneously monitored. From this data, a frequency response function was generated by dividing the LDV RMS output amplitude by the oscillator amplitude and plotting against oscillator frequency. A natural frequency, f_n, was extracted from this data using the following algorithm

$$f_n = f_{\max} - 2\xi \tag{1}$$

where f_{max} is the frequency corresponding to the maximum amplitude in the frequency response function, and ξ is the damping calculated using the resonance's full width at half maximum. This process was repeated five times, and the tuning fork was removed from and replaced onto the piezo stack between each trial to randomize systematic boundary effects at the contact between the piezo shaker and the tuning fork's tripod mount.

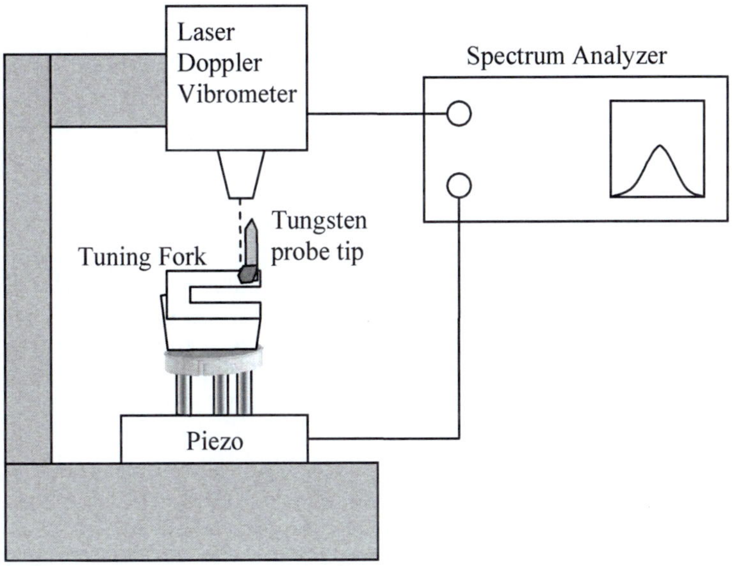

Figure 2. Experimental apparatus for determining sensor resonant frequency. The qPlus sensor is shaken by the piezo, which is excited using the sine wave oscillator from the dynamic signal analyzer. The motion of the tip is measured using the laser Doppler vibrometer, and used to construct a frequency response function.

Subsequently, the mass of the tuning fork, and its attached mount were determined using a Mettler XP2U microbalance. Prior to all measurements, an internal self-calibration procedure was used to calibrate the balance. The balance's mass measurements after this internal calibration were then checked against two NIST-calibrated wire masses bracketing the mass of the tuning fork and mount. A mass reading was taken before placing any objects on the balance pan, then the object to be tested was placed approximately at the center of the balance pan and another mass reading was taken. This process was repeated five times, and then a final mass reading was taken with no object on the balance pan. The mass of the object was then calculated calculating the difference between the balance reading when the object was on the pan and the average of the two baseline measurements before and after that reading. The average of these five mass measurements agreed with the mass value provided by NIST for the wire masses tested within the standard deviation of the five measurements described above, although the standard deviation of the five measurements was greater than the uncertainty of the NIST values.

After determining the mass and $f_{n,}$ of the tuning fork sensor, the tungsten wire tip was partially etched away using 1 mol/L potassium hydroxide solution by immersing the tip in the solution and applying a 0.5 V potential between the tip and a platinum mesh counter electrode for a period of 8 minutes. The mass and f_n of the tuning fork sensor were then remeasured, and the etch and weigh cycle repeated once more.

Instrumented indentation testing of the tuning fork stiffness was carried out in a Hysitron Triboscope indenter. A Berkovich indenter tip was positioned near the center of the tuning fork next to the area occupied by the conductive epoxy holding the tip onto the free end of the cantilevered tuning fork tine. The indenter tip was preloaded to 250 μN, then a series of 5 linear load and unload cycles from 265 μN to 350 μN were carried out while applied force and displacement of the indenter tip were monitored. These force and displacements can be used to provide a stiffness measurement, and the use of this apparatus to provide traceable spring constant measurements has been discussed elsewhere [14].

Results and Discussion

Previous work has shown that a microfabricated atomic force microscope (AFM) cantilever's spring constant can be determined by placing micromasses near the cantilever tip [12]. Resonant frequency of a mass-loaded cantilever in the case of a single degree of freedom harmonic oscillator scales with added mass according to

$$M = k\omega^{-2} - m^{*} \qquad (2)$$

where M is the value of the mass change from the micromass positioned at the cantilever's free end, $\omega = 2\pi f_{n,}$, and m^{*} is the effective mass of the cantilever. Figure 3 shows a graph of the mass changes resulting from the tip etch process vs. the resonant circular frequency of the tuning fork. According to Equation 3, the slope of this line provides a value for the static spring constant of the cantilevered tuning fork tine. However, in contrast to the previous implementations of the technique, this approach eliminates the uncertainty associated with the position of the mass relative to the tip, since it is the mass of the tip itself that is changing. The method also contrasts with the destructive process of gluing spheres onto cantilever tips, in that the tip etch process enhances the tip's usefulness for scanning probe applications. Equation 2 is based upon a simplified representation of the cantilevered tuning fork, however.

In order to test the validity of the assumptions of the dynamic model mentioned above, the spring constant determined using Equation 3 can be compared to one determined using the quasi-static indentation method described previously. A portion of the instrumented indentation data is shown in Figure 4. The slope of the force displacement curve is the static spring constant of the cantilever at the point of contact with the indenter. The dynamic stiffness measurement described above defines stiffness at the point where the mass is varied. As a result, the stiffness from the instrumented indenter must be corrected to collocate it with the mass of the tip according to

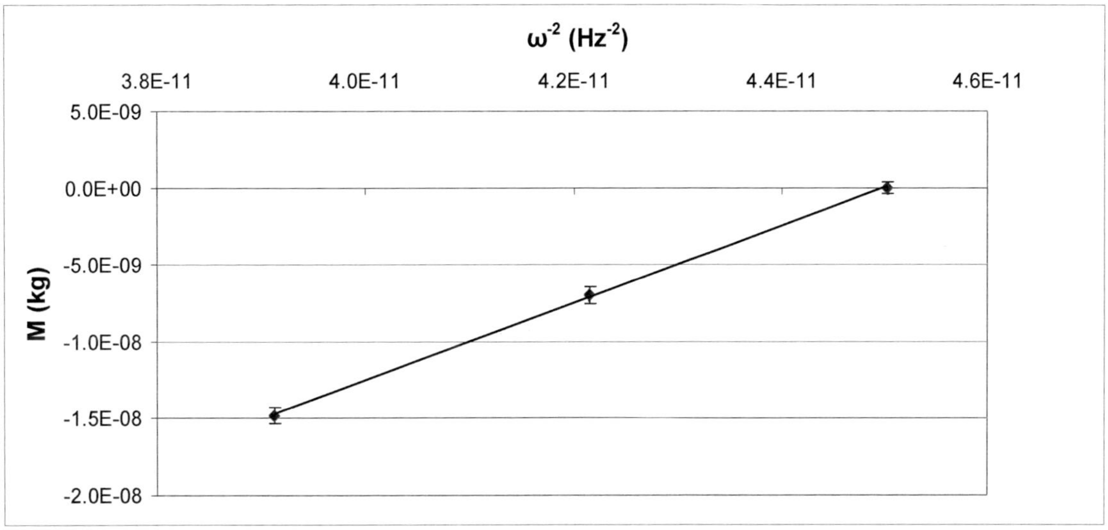

Figure 3. Sensor resonant frequency as a function of mass change of the tungsten probe tip. Error bars indicate the root-sum-of-squares from the standard deviation of five measurements of mass and frequency before and after etching part of the tungsten tip off. Note that frequency error bars are very small on this scale.

$$k_T = k_{ind}\left(\frac{L-\Delta L}{L}\right)^3, \tag{3}$$

where k_T is total system compliance, as discussed further below, k_{ind} is the spring constant determined using indentation, L is the distance from the tuning fork cantilever base to the tungsten wire tip, and ΔL is the distance between the tip and the point of contact between the cantilever and indenter tip. An additional correction must be made for the effects of contact compliance, c_c, machine compliance, c_m, and the angle of repose of the tuning fork sensor, θ, such that

$$k = \cos^2\theta\left(k_T^{-1} - c_c - c_m\right)^{-1}. \tag{4}$$

Machine compliance was measured by indenting a fused quartz reference sample [15], and contact compliance was estimated by assuming the contact area could be calculated from the hardness of the gold surface, using

$$c_c = \frac{\sqrt{\pi}}{\beta E_r\sqrt{F_m / H}}, \tag{5}$$

where β is a geometric factor equal to 1.034 for a Berkovich indenter, F_m is the maximum force applied to the indenter, H is the indentation hardness of gold, E_r is reduced modulus calculated using

$$\frac{1}{Er} = \frac{1-\nu^2}{E} + \frac{1-\nu_i^2}{E_i}, \tag{6}$$

where E and E_i are the Young's modulus of the gold film coating the end of the tuning fork sensor at the contact point, and diamond indenter tip, respectively. Likewise v and v_i are the Poisson's ratios of the film and indenter materials [16].

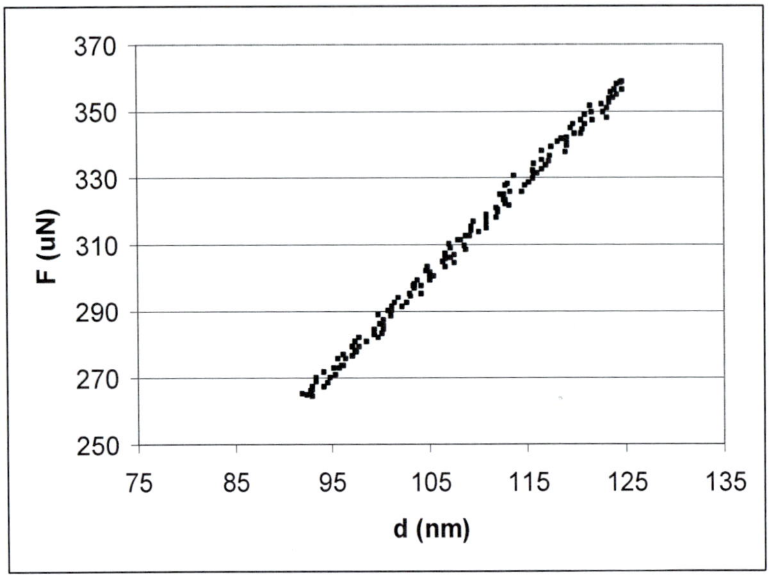

Figure 4. Indentation measurement of sensor spring constant. This data represent the combined qPlus sensor, contact, and machine compliance.

The values and uncertainties of the measured sensor spring constants are shown in Table 1 for the dynamic and indentation methods. Combined standard uncertainty is reported as the root sum of squares of all uncertainty components [17]. For the dynamic method, the curve linearity was evaluated using the methods of Taylor [18] from an analysis of the linear least squares fit to the data shown in figure 4, and is a type A uncertainty. The type A uncertainty in frequency is calculated by first determining the standard deviation of f_n from the sets of five measurements performed as described above, and then calculating the average of these standard deviations for all the tests performed. The type A uncertainty in mass was likewise calculated from the standard deviation of the weighing results. The mass correction for buoyancy of the tip mass in air is also a small contributing uncertainty [2]. For the Indentation method, uncertainties due to indenter force, displacement, cantilever length and indenter linearity have been previously described [14]. Repeatability is the standard deviation of the slopes of the force displacement curves such as that shown in Figure 4. The type B uncertainty in machine compliance was evaluated by conservatively assuming a 100 % uncertainty in the determined value of 2 nm/mN. Type B uncertainties from Hardness and Young's modulus values were determined by propagation through equations Equations 4-6. It was assumed that the contact sampled only the mechanical properties of the gold film on the surface of the tuning fork sensor. There is a substantial range of literature values available for Hardness and Young's modulus. The bulk value of Young's modulus for gold is 80 GPa, however low-force indentation experiments have reported values as high as 119 GPa for E [19]. To calculate k, the average of these two numbers was used. The uncertainty was estimated from the difference between the mean and either of these values. There is similar spread in the value of Hardness determined for gold thin films [20], and therefore the same approach is followed in determining uncertainty.

In order for the dynamic sensor described here to be used for mass metrology, the assumption that the tuning fork oscillator behaves as a single degree of freedom harmonic oscillator must be tested in a more precise fashion than described above. With a better procedure for determining contact stiffness, the uncertainty in the indentation measurement of *k* can be decreased substantially. The dynamic method is primarily limited by the uncertainty of the mass measured using the microbalance, since the values of the masses measured are close to the balance's resolution. The uncertainty in this measurement can be decreased by examining larger mass changes, by performing more experimental repetitions, or by using an alternate SI-traceable method for determining the mass [4]. The continued development of small mass metrology will help ensure SI traceable standards are available as the ultimate physical limits of measurement are probed [21,22].

Table 1. Measurement results and uncertainty components for qPlus sensor spring constant.

Dynamic		**Indenter**	
k (N/m)	$2.5x10^3 \pm 0.2x10^3$	k (N/m)	$2.5x10^3 \pm 0.2x10^3$
Source of uncertainty	**Magnitude of uncertainty**	**Source of uncertainty**	**Magnitude of uncertainty**
Curve linearity	$3.4x10^{-2}$	Force	$1x10^{-2}$
Frequency	$7x10^{-5}$	Displacement	$1x10^{-2}$
Mass change	$8.1x10^{-2}$	Repeatability	$7x10^{-4}$
Buoyancy	$1x10^{-4}$	Indenter linearity	$1x10^{-3}$
		Machine compliance	$2x10^{-3}$
		Young's modulus	$3.3x10^{-2}$
		Hardness	$9.0x10^{-2}$
		Cantilever length correction	$6.9x10^{-2}$
Combined Standard Uncertainty	$8.7x10^{-2}$	Combined Standard Uncertainty	$9.7x10^{-2}$

Conclusion

In order to determine the accuracy of the assumption that a cantilevered quartz tuning fork mechanical resonator (i.e. a qPlus sensor) behaves as a simple harmonic oscillator, a comparison of the values of its spring constant were determined with a dynamic and a quasi-static method. The spring constant values determined agreed within the experimental uncertainties of approximately 10 %. There is, however, substantial potential to reduce this uncertainty with further development of the measurement procedures described above.

Acknowledgements
The authors wish to thank Stuart Smith and Barton Nowakowski of The University of North Carlolina-Charlotte, and Patrick Abbott of NIST for helpful discussions.

References

1. Jabbour, Z. J., Abbott, P., Williams, E., Liu, R., Lee, V., Linking air and vacuum mass measurement by magnetic levitation, *Metrologia*, **46** 339-344, 2009
2. Jabbour, Z. J., Yaniv, S. L., The kilogram and measurements of mass and force, *J. Res. Natl. Inst. Stand. Technol.*, **106** 25-46, 2001
3. Pratt, J. R., Kramar, J. A., Newell, D. B., Smith, D. T., Review of SI-traceable force metrology for instrumented indentation and atomic force microscopy, *Meas. Sci. Technol.*, **16**, 2129-2137, 2005

4. Jones, C. W., Kramar, J. A., Davidson, S., Leach, R. K., Pratt, J. R., Comparison of NIST SI force scale to NPL SI mass scale, *Proc. ASPE Annual Meeting*, Portland, OR, 2008
5. Madec, T. Gaëlle, M., Meury, P.-A., Rabault, T., Micro-mass standards to calibrate the sensitivity of mass comparitors, *Metrologia*, **44** 266-274, 2007
6. Giessibl, F. J., Hembacher, S., Bielefeldt, H., Mannhart, J., Subatomic features on the Silicon (111)-(7x7) surface observed by atomic force microscopy, *Science*, **289** 422-425, 2000
7. Ternes, M., Lutz, C. P., Hirjibehedin, C. F., Giessibl, F. J., Heinrich, A. J., The force needed to move an atom on a surface, *Science*, **319** 1066-1069, 2008
8. Sader, J. E., Jarvis, S. P., Accurate formulas for interaction force and energy in frequency modulation force spectroscopy, *Appl. Phys. Lett.*, **84** 1801-1803, 2004
9. Qin, Y., Riefenberger, R., Calibrating a tuning fork for use as a scanning probe microscope force sensor, *Rev. Sci. Instrum.*, **78** 063704, 2007
10. Simon, G. H., Heyde, M., Rust, H.-P., Recipes for cantilever parameter determination in dynamic force spectroscopy: spring constant and amplitude, *Nanotechnology*, **18** 22503, 2007
11. Cleveland, J. P., Manne, S., Bocek, D., Hansma, P. K., A nondestructive method for determining the spring constant of cantilevers for scanning force microscopy, *Rev. Sci. Instrum.,* **64** 403-405, 1993
12. **Disclaimer:** This article is authored by employees of the U.S. federal government, and is not subject to copyright. Commercial equipment and materials are identified in order to adequately specify certain procedures. In no case does such identification imply recommendation or endorsement by the National Institute of Standards and Technology, nor does it imply that the materials or equipment identified are necessarily the best available for the purpose.
13. Giessibl, F. J., Hembacher, S., Herz, M., Schiller, C., Mannhart, J., Stability considerations and implementation of cantilevers allowing dynamic force microscopy with optimal resolution: the qPlus sensor, *Nanotechnology*, **15** S79-S86, 2004
14. Shaw, G. A., Kramar, J. A., and J. R. Pratt, SI-traceable spring constant calibration of microfabricated cantilevers for small force measurement, *Exp. Mech.*, **47** 143-151, 2007
15. Joslin, D. L. and Oliver, W.C., A new method for analyzing data from continuous depth sensing microindentation tests, *J. Mater. Res.*, **5** 123-126, 1990
16. Oliver, W. C., Pharr, G. M., An improved technique for determining hardness and elastic modulus using load and displacement sensing indentation experiments, *J. Mater. Res.*, **7** 1564-1583,1992
17. Taylor, B. N., Kuyatt, C. E., Guidelines for evaluating and expressing the uncertainty of NIST measurement results, NIST Technical Note 1297, 1994
18. Taylor, J. R., An introduction to error analysis, University Science Books, Sausalito, CA, 181-207, 1997
19. Li, X., Nardi, P., Baek, C.-W., Kim, J.-M., Kim, Y.-K., Direct nanomechanical machining of gold nanowires using a nanoindenter and an atomic force microscope, *J. Micromech. Microeng.*, **15** 551-556, 2005
20. Lilleodden, E. T., Nix, W. D., Microstructural length-scale effects in the nanoindentation behavior of thin gold films, *Acta Materialia*, **45** 1583-1593, 2006
21. Ekinci, K. L., Yang, Y. T., Roukes, M. L., Ultimate limits to inertial mass sensing based upon Nanoelectromechanical systems, J. Appl. Phys., **90** 2682-2689, 2004
22. Naik, A. K., Hanay, M. S., Hiebert, W. K., Feng, X. L., Roukes, M. L., Towards single-molecule nanomechanical mass spectrometry, Nature Nanotechnology, **4** 445-450, 2009

Proceedings of the SEM Annual Conference
June 7-10, 2010 Indianapolis, Indiana USA

Linear and Nonlinear Mass Sensing Using Piezoelectrically-Actuated Microcantilevers

Vijay Kumar[1,2,3,*], J. William Boley[1], Hendrik Ekowaluyo[1,2], Jacob K. Miller[1,2,3]
George C. Marvin[1,2], George T.-C. Chiu[1,2,3], Jeffrey F. Rhoads[1,2,3]

[1] *School of Mechanical Engineering, Purdue University, West Lafayette, IN 47907*

[2] *Birck Nanotechnology Center, Purdue University, West Lafayette, IN 47907*

[3] *Ray W. Herrick Laboratories, Purdue University, West Lafayette, IN 47907*

* *Corresponding author: kumar2@purdue.edu*

Abstract

Chemical and biological sensors based on resonant microcantilevers offer distinct utility due to their small size, low power consumption, high sensitivity, and, when bulk fabricated, comparatively-low cost. Modern resonant mass sensors typically utilize chemomechanically-induced frequency shifts in linear resonators for analyte detection. Recent results, however, have indicated that nonlinear sensors, which actively exploit dynamic transitions across sub-critical or saddle-node bifurcations in the device's frequency response, have the potential to exhibit improved performance metrics and operate effectively at smaller scales. This relatively-novel sensing approach directly exploits chemomechanically-induced amplitude shifts, instead of frequency shifts, for detection. Accordingly, it has the potential to eliminate the need for numerous power-consuming signal processing components in final sensor implementations. The present work details the ongoing development of low-cost, linear and nonlinear, bifurcation-based mass sensors founded upon selectively-functionalized, piezoelectrically-actuated microcantilevers. Specifically, the work describes the modeling, analysis, micro-inkjet functionalization, and experimental characterization of these devices.

1 Introduction

Chemical and biological sensors based on resonant micro- and nanocantilevers offer distinct utility due to their small size, low power consumption, high sensitivity, and, when bulk fabricated, comparatively-low cost (see, for example, [1–6]). Collectively, these benefits have spurred the implementation of resonant micro- and nanosensors in medical diagnostic, environmental safety, and national security settings [7–11]. Though a number of transduction mechanisms have been utilized for small-scale mass sensing, piezoelectrically-actuated microcantilevers have garnered significant interest of late, due to their self-sensing capability [12] and maturity in the related technical areas of force sensing and scanning probe microscopy [13]. The study detailed herein seeks to build upon prior research in each of the aforementioned areas with the development of nonlinear, bifurcation-based mass sensors based on piezoelectrically-actuated microcantilevers.

Traditionally, resonant mass sensors have exploited shifts in a linear resonator's natural frequency for analyte detection [see Fig. 1(a)]. In these devices, the adsorption/absorption of a target analyte onto a selectively-functionalized surface layer results in a measurable change in the system's mass (or stiffness). This, in turn, leads to a measurable change in the resonator's natural frequency, which can be directly correlated to an analyte detection event, and, in certain cases, used to approximate the analyte's relative concentration in the test environment [1]. While this linear sensing approach offers proven utility, recent work has demonstrated that improved sensor metrics may be achievable through the active exploitation of nonlinear system behaviors. Dai et al., for example, highlighted the benefits of resonant mass sensors that utilize shifts in the system's nonlinear resonant frequency for detection [14]. Likewise, Younis and Alsaleem recently investigated an amplitude shift based sensing system which leverages the pull-in dynamics of electrostatically-actuated microresonators to realize improved performance [15]. These nonlinear approaches have particular appeal as they are not only functional at the microscale, but potentially facilitate resonant sensing in nanowire and nanotube-based systems, which commonly suffer from a limited linear dynamic range [16].

The present work details the ongoing development of linear and nonlinear mass sensors based on piezoelectrically-actuated microresonators. The latter sensors, described at length in Section 2.1, are founded upon a novel bifurcation-based sensing approach, which utilizes amplitude shifts resulting from chemomechanically-induced dynamic transitions across saddle-node bifurcations to signal analyte detection [see Fig. 1(b)]. The work begins with the development of a

T. Proulx (ed.), *MEMS and Nanotechnology, Volume 2*, Conference Proceedings of the Society for Experimental Mechanics Series 2,
DOI 10.1007/978-1-4419-8825-6_9, © The Society for Experimental Mechanics, Inc. 2011

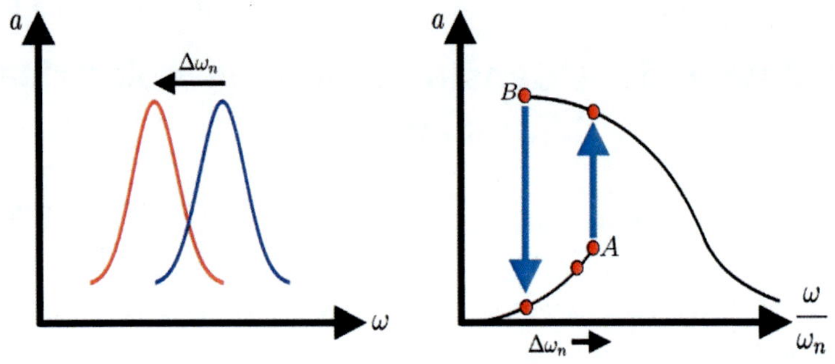

(a) (b)

Figure 1: (a) Linear frequency response associated with a representative piezoelectrically-actuated microresonator. Here a chemomechanical interaction between the sensor and a target analyte, induces a measurable frequency shift. (b) Nonlinear frequency response associated with a representative piezoelectrically-actuated microresonator plotted in terms of the system's normalized excitation frequency. Note that a small shift in the system's natural frequency, caused by a chemomechanical interaction between the sensor and a target analyte, drives the system across the saddle-node bifurcation, inducing a dynamic transition between the co-existing stable states. The dramatic increase in response amplitude attendant to this transition can be directly correlated to an analyte detection event.

comprehensive nonlinear, distributed-parameter model for a piezoelectrically-actuated microcantilever. This model is reduced to a lumped-mass analog and subsequently analyzed using numerical continuation methods. Predicted nonlinear behaviors are experimentally validated using Veeco DMASP probes. These devices are functionalized for sensing purposes and tested in an a controlled environment. The work concludes with a brief summary and an overview of ongoing research efforts.

2 Modeling and Analysis

Figure 2 highlights, schematically and pictorially, the piezoelectrically-actuated microbeam system of interest. Here, an applied potential across the system's piezoelectric layer serves to actuate the device. The equation of motion governing the piezoelectrically-actuated microbeam can be recovered through the use of classical energy methods. Though a complete derivation of this governing equation is beyond the scope of the present work, it can be briefly developed as follows.

To begin, axial and transverse deformations $u(s,t)$ and $v(s,t)$, and the derivative operators $(\dot{\bullet})$ and $(\bullet)'$, taken with respect to time and the arc length variable s, respectively, are introduced. Assuming small (but not necessarily linear) deflections, these deformations render a kinematic constraint on the microbeam's angular deflection (ψ) given by

$$\tan\psi = \frac{v'}{1+u'}. \tag{1}$$

Likewise, the axial strain associated with the microbeam, ε_{11}, is given by

$$\begin{aligned} \varepsilon_{11} &= -(y-y_n)\psi' \quad \text{for} \quad s < l_1, \\ &= -y\psi' \quad\quad\quad \text{for} \quad l_1 < s < l, \end{aligned} \tag{2}$$

where, y_n is the location of neutral axis from the origin for the region $s < l_1$ and is defined to be

$$y_n = \frac{E_p w_p t_p (t_p + t_b)}{2(E_p t_p w_p + E_b t_b w_b)}. \tag{3}$$

The relationships between the induced stress and the strain in the silicon cantilever and the piezoelectric patch are derived using classical constitutive relations. For the cantilever, the constitutive equation is given by $\sigma_{11}^b = E_b \varepsilon_{11}^b$, while, for the piezoelectric patch, the constitutive relations are given by [17, 18]

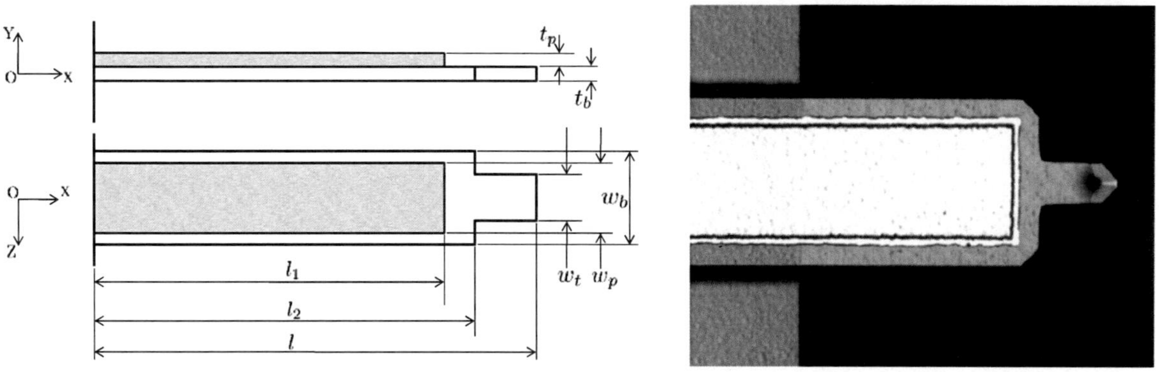

(a) (b)

Figure 2: (a) Schematic of the piezoelectrically-actuated microcantilever. (b) A representative piezoelectrically-actuated Veeco DMASP probe.

$$\begin{aligned}\sigma_{11}^{p} &= E_p\varepsilon_{11}^{p} - h_{31}Q_3 + \frac{\alpha_1}{2}(\varepsilon_{11}^{p})^2 + \frac{\alpha_2}{2}(Q_3)^2 - \alpha_3\varepsilon_{11}^{p}Q_3,\\ D_3 &= h_{31}\varepsilon_{11}^{p} + \beta_{31}Q_3 + \frac{\alpha_3}{2}(\varepsilon_{11}^{p})^2 + \frac{\alpha_4}{2}(Q_3)^2 - \alpha_2\varepsilon_{11}^{p}Q_3,\end{aligned} \tag{4}$$

where, the superscripts b and p denote the beam and the piezoelectric patch, respectively, and σ_{11}, E_b, E_p, Q_3, D_3, h_{31} and β_{31} represent the axial stress, the elastic modulus of the beam, the elastic modulus of the piezoelectric material, the applied electric field, the electrical displacement, the effective coupling coefficient that relates the electrical and mechanical displacements, and the effective permittivity coefficient, respectively. α_i $(i = 1, 2, 3, 4)$ are the nonlinear material coefficients. Note that the electric field is the gradient of the applied potential with respect to the thickness variable.

Using the aforementioned relationships, the kinetic energy T and the potential energy U associated with the system can be expressed as

$$T = \frac{1}{2}\int_0^l m(s)\left[\dot{u}^2 + \dot{v}^2\right] ds, \tag{5}$$

$$\begin{aligned}U =& \frac{1}{2}\int_0^{l_1}\iint_A \sigma_{11}^{p}\varepsilon_{11}^{p} dAds + \frac{1}{2}\int_0^{l_1}\iint_A \sigma_{11}^{b}\varepsilon_{11}^{b} dAds - \frac{1}{2}\int_0^{l_1}\iint_A Q_3D_3 dAds\\ &+ \frac{1}{2}\int_{l_1}^{l_2}\iint_A \sigma_{11}^{b}\varepsilon_{11}^{b} dAds + \frac{1}{2}\int_{l_2}^{l}\iint_A \sigma_{11}^{b}\varepsilon_{11}^{b} dAds,\end{aligned} \tag{6}$$

where $m(s)$ and A represent the mass density and cross-sectional area of the composite system. Using these as a basis for study, the Lagrangian of the system can be assembled and used in conjunction with extended Hamilton's principle to recover the equations governing the system's transverse and longitudinal vibrations. The equation governing the system's longitudinal dynamics can be use to recover the embedded Lagrange multiplier, introduced to maintain inextensibility. Substituting this expression into the equation governing transverse vibrations yields a single, distributed-parameter equation of motion for the system. Expanding this equation, retaining terms of up to third order, and non-dimensionalzing the result, renders a final distributed parameter model.

The complex distributed-parameter model referred to above can be reduced to a lumped-mass analog by decomposing the transverse displacement $\hat{v}$ into its spatial and temporal components

$$\hat{v}(\hat{s}, \hat{t}) = w(\hat{t})\Phi(\hat{s}), \tag{7}$$

and projecting the result onto the first mode shape of the cantilever. This yields a lumped-mass equation of motion for the system given by

$$\ddot{w} + 2\zeta\omega_n\dot{w} + (\omega_n^2 + \Delta\phi\lambda_1 + \Delta\phi^2\gamma_1)w + (k_2 + \Delta\phi\lambda_2)w^2 + (k_3 + \Delta\phi\lambda_3 + \Delta\phi^2\gamma_3)w^3 + \chi(w\dot{w}^2 + w^2\ddot{w}) = \eta_1\Delta\phi. \tag{8}$$

The nondimensional coefficients in Eqn. (8) are strong functions of the system's geometry, resonant modeshapes, and constituent material properties. The effective linear, quadratic, and the cubic stiffnesses (ω_n^2, k_2, k_3), and the nonlinear

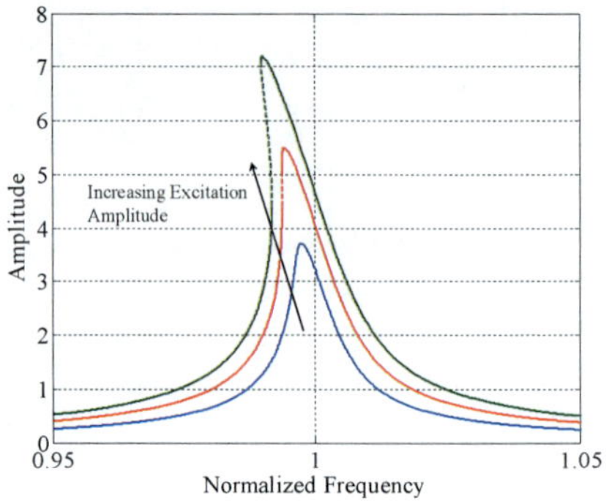

Figure 3: Simulated nonlinear frequency response of a representative piezoelectrically-actuated cantilever. The frequency response exhibits further softening as the excitation amplitude is increased.

term (χ) vary strongly with system's geometry and elastic material properties and exhibit a weak dependence on material nonlinearities. In contrast, the other terms in the expression are strong functions of the linear and nonlinear piezoelectric material coefficients. Note that though nonlinear constitutive relations were used to develop this model, if the constitutive relations are assumed to be linear, the system resembles a classic Duffing resonator. Detailed expressions for each of the terms presented here are omitted for the sake of brevity.

Equation (8) is highly nonlinear and fails to feature a tractable closed-form solution. Accordingly, the equation is analyzed using a numerical continuation program, AUTO [19]. This preliminary analysis is carried out by assuming that all the nonlinear material coefficients, with the exception of α_1, are zero. This follows the approach previously detailed in [17]. Figure 3 depicts the frequency response structure associated with the system, under various excitation amplitudes, for a representative value of α_1. As evident at comparatively-large excitation amplitudes, the system exhibits a softening-like behavior, with coexistent stable solutions and hysteresis. Accordingly, if the system is excited at a comparatively-low excitation frequency and frequency is swept to a higher value, the response amplitude will slowly increase until the excitation frequency crosses the (lower) point at which the stable and unstable solution branches meet (a saddle-node bifurcation). At this point, the response amplitude jumps to the large-amplitude response from the low amplitude state, and then decreases with increasing excitation frequency. In contrast, if the system is excited at a comparatively high excitation frequency and frequency is swept to a lower value, the response amplitude will rapidly increase until the excitation frequency crosses the (upper) point at which the stable and unstable solution branches meet (a second saddle-node bifurcation). At this point, the response amplitude jumps to the small-amplitude response from the high amplitude state, and then decreases with decreasing excitation frequency.

2.1 Bifurcation-Based Mass Sensing

As previously noted, in a traditional linear resonance based mass sensor, chemomechanically-induced shifts in the system's resonance frequency are tracked and used to signal an analyte detection event. In contrast, the proposed bifurcation-based resonant mass sensors exploit the rich nonlinear frequency response structure detailed above. Specifically, these sensors are designed to operate near the saddle-node bifurcation point designated A in Fig. 1(b). As with linear sensors, chemomechanical interactions with a target analyte are used to alter the effective mass of the resonator. However, in this instance, the added mass serves to not only shift the system's natural frequency, but also to drive the system across the saddle-node bifurcation. This yields a rapid and dramatic jump in the system's response amplitude which can be easily detected, even in the absence of significant readout electronics.

3 Experimental Investigation

To validate the feasibility and merits of the proposed nonlinear, bifurcation-based mass sensors a succinct experimental investigation was initiated. To begin, the nonlinear frequency response behavior of a representative device was characterized via laser vibrometry. Select microcantilevers were then selectively functionalized and tested within a carefully controlled environment using a custom test apparatus. Veeco DMASP probes, initially designed for use in scanning probe microscopy applications, were used as a test platform. These devices consist of a silicon cantilever and an integrated piezoelectric actuator incorporating a ZnO layer sandwiched between two Au/Ti electrodes.

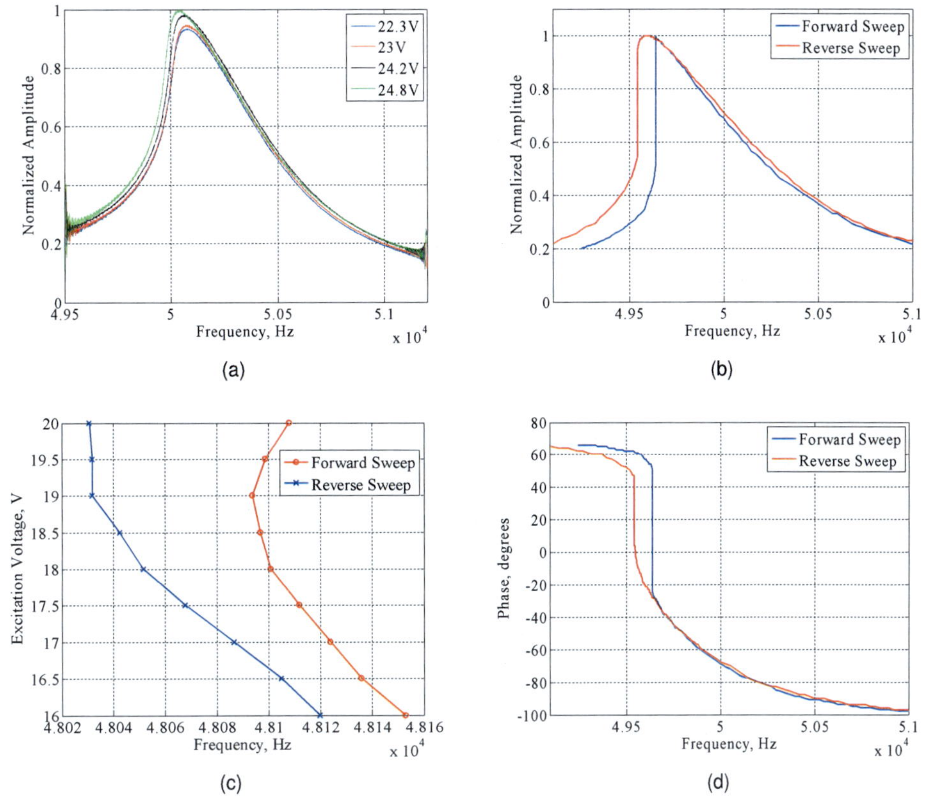

Figure 4: (a) Frequency response of the system under different excitation voltages. In this subfigure, 20.48 second sweep times were utilized. Note that the response becomes increasingly nonlinear as the excitation voltage is increased. (b) and (d) Amplitude and phase responses recovered with a 20.2 V_{pp} excitation using a total sweep time of 400 seconds. Note the clear bifurcations and hysteresis. Individual data points have been removed from the plot for the sake of clarity. (c) Relationship between the system's bifurcation frequencies and the amplitude of the excitation voltage. Though slightly problematic from an implementation point-of-view, the strong dependence of the bifurcation frequencies on the amplitude of excitation can be exploited for device tuning and performance metric optimization.

3.1 Nonlinear Frequency Response

The frequency response of the Veeco DMASP probes was studied through the use of a Polytec MSA-400 Laser Doppler Vibrometer and a supplementary LabVIEW interface, which served to control the voltage input and record the steady-state frequency response characteristics of the device. Frequency sweeps were performed at a variety of sweep rates and actuation voltage levels to characterize the nonlinear response of the microcantilevers. Figures 4(a) and 4(b) highlight the recovered amplitude response characteristics. The results depicted in Fig. 4(a) were recovered using a sweep period of 20.48 seconds (Note that this sweep rate is sufficiently fast that the saddle-node bifurcation is captured only in passing). As predicted, the system shows a mild softening behavior and the response increasingly softens with increasing excitation voltage. Similarly, Figs. 4(b) and 4(d) highlight the amplitude and phase response of the system recovered with 20.2 V_{pp} excitation and a 400 second sweep period. As evident, with higher sweep periods, even at a lower voltage, the system exhibits a clear saddle-node bifurcation and significant hysteresis.

3.2 Functionalization

Resonant mass sensors are typically functionalized with a polymer capable of adsorbing/absorbing solvents, vapors, and other analytes of interest [20]. Recently, inkjet technology has been shown to be a viable means of fabrication and sensor functionalization due to its ability to precisely deposit a small volume of liquid onto a substrate [21]. The benefits of this method are its high throughput, low production cost, ability to print a wide variety of liquids onto planar and non-planar surfaces, and environmental cleanliness, compared to conventional fabrication techniques [22].

In the present effort, the beams were functionalized with Poly 4-Vinyl Pyridine to facilitate methanol sensing [23].

Functionalization was carried out using a micro-inkjet printing process with thermal actuation enabled drop formation [24]. In this functionalization process, a resistor is placed within the polymer reservoir. An electrical current is sent through the resistor causing its temperature to rise. The resistor's temperature rises until a vapor bubble forms and subsequently collapses. This induces a wave in the fluid, which, in turn, results in the formation of a drop [25]. The inkjet system used for functionalizing the cantilevers is depicted in Fig. 5(a). It comprises of (a) a two-axis linear stage with an encoder resolution of 0.5 μm, (b) an HP TIPS thermal inkjet drop ejection system with printheads consisting of up to 18 nozzles with the ability to produce drop volumes in the range of 1-220 pL at a frequency of 45 kHz, (c) a CCD imaging system for the visualization of drop deposition, and (d) a laser registration system used for mapping fixed substrate coordinates [22]. Each device was functionalized with 110 pL drops in less than 1 second. The functionalized beams were stored in a nitrogen environment to avoid the interactions with free radicals and to ensure the integrity of the sensor's surface chemistry prior to testing.

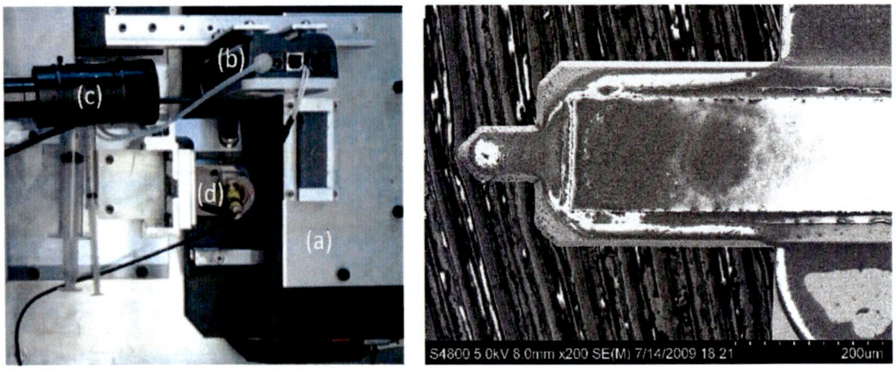

(a) (b)

Figure 5: (a) The micro-inkjet printing setup used for functionalization. (b) A scanning electron micrograph of the functionalized microcantilever.

3.3 Mass Sensing Trials

Once the nonlinear behavior of the resonant microcantilevers had been validated and deemed suitable for sensing, and the probes had been properly functionalized, laboratory-based sensing trials were initiated. The chemical test setup used for this study is depicted in Fig. 6. This test apparatus consists of multiple analyte bubblers, which are connected to precision mass flow controllers and a carrier gas (nitrogen) supply. The temperatures of the bubblers are carefully monitored and controlled through the use of RTDs and adjacent heat pads. By controlling the pressure of the carrier gas at the inlet, choosing appropriate temperature settings, and selecting appropriate flow rates for each of the mass flow controller, any desired analyte/carrier mixture and associated concentration level can be produced. Once fully mixed, this gas is diverted into a isolated test chamber inside of which the functionalized microbeams are mounted. The test chamber is optically accessible from the top, and features a glass window which is specially coated to allow for maximum transmission of the laser signal from the vibrometer. The temperature of the test chamber is also monitored and controlled using RTDs and heat pads to avoid the condensation of analyte onto the walls of the chamber and the optical viewport. Hermetically sealed electrical outlets are located at the base of the chamber to minimize interference with chemomechanical interactions. The exhaust from the chamber is routed into a condensing chamber from which the target analyte is recovered. The entire setup, including the excitation signal to the beam and recovered signal from the vibrometer, is monitored and controlled using a LabVIEW interface.

To validate the performance of the functionalized Veeco DMASP probes and the chemical test chamber itself, the microcantilevers were subjected to a series of mass sensing trials based on their linear frequency response. Specifically, the cantilevers were excited using a 2 V pseudorandom signal. Methanol was then supplied at a pre-determined concentration and rate into the chamber. To achieve successive adsorption and desorption of the gas molecules on the functionalized surface, the methanol/nitrogen mixture was supplied for a fixed period of time and then the system was purged with pure nitrogen. This alternating cycle of gas supply was controlled via LabVIEW. As previously noted, in a linear resonant mass sensor, as the mass is adsorbed on the surface, the natural frequency decreases, and as the mass desorbs, the natural frequency increases. As evident from Fig. 7 the expected behavior was recovered, albeit with comparatively large analyte/carrier concentrations (3.5% by mass). Current experiments are focused on validating the proposed nonlinear

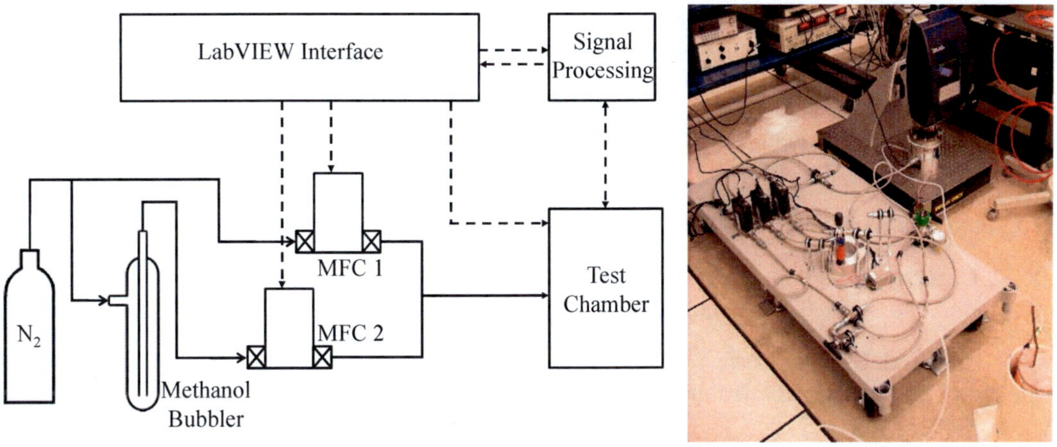

Figure 6: (a) Schematic and (b) pictorial representation of the computer-controlled chemical test apparatus.

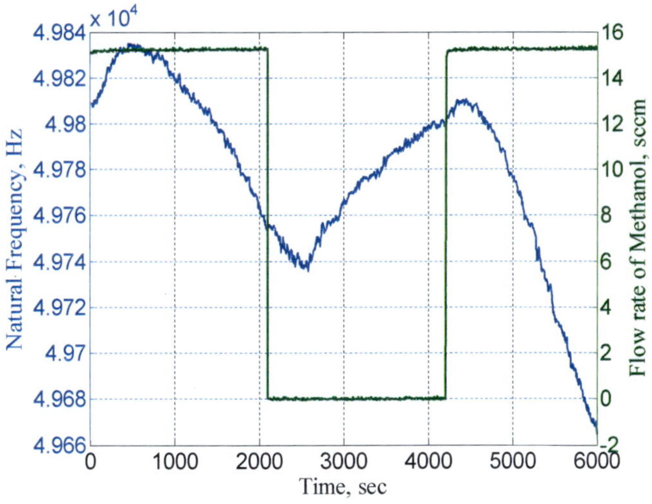

Figure 7: Preliminary linear resonant mass sensing results.

bifurcation-based sensing approach. Initial results have displayed significant promise.

4 Conclusions

The present work details the ongoing development of linear and nonlinear, bifurcation-based mass sensors based on piezoelectrically-actuated microresonators. A detailed nonlinear model for a representative microcantilever has been developed and analyzed to provide a high-level understand of pertinent system dynamics. Nonlinear behaviors deemed suitable for bifurcation-based sensing were experimentally verified using Veeco DMASP probes in conjunction with laser vibrometry. Select probes were functionalized using micro-inkjet techniques and initial experimental trials were completed to validate the integrity of the experimental approach. Current efforts are focused on validating the proposed nonlinear bifurcation-based sensing approach through further experimental investigation. Initial results have been quite promising.

References

[1] Thundat, T., Wachter, E. A. ,Sharp, S. L., and Warmack, R. J., "Detection of mercury vapor using resonating microcantilevers," *Applied Physics Letters*, vol. 66, no. 13, pp. 1695–1697, 1995.

[2] Ono, T., Li, X., Miyashita, H., and Esashi, M., "Mass sensing of adsorbed molecules in sub-picogram sample with ultrathin silicon resonator," *Review of Scientific Instruments*, vol. 74, no. 3, pp. 1240–1243, 2003.

[3] Lavrik, N. V. and Datskos, P. G., "Femtogram mass detection using photothermally actuated nanomechanical resonators," *Applied Physics Letters*, vol. 82, no. 16, pp. 2697–2699, 2003.

[4] Ekinci, K. L., Huang, X. M. H., and Roukes, M. L., "Ultrasensitive nanoelectromechanical mass detection," *Applied Physics Letters*, vol. 84, no. 22, pp. 4469–4471, 2004.

[5] Yang, Y. T., Callegari, C., Feng, X. L., Ekinci, K. L., and Roukes, M. L., "Zeptogram-scale nanomechanical mass sensing," *Nano Letters*, vol. 6, no. 4, pp. 583–586, 2006.

[6] Naik, A. K., Hanay, M. S., Hiebert, W. K., Feng, X. L., and Roukes, M. L., "Towards single-molecule nanomechanical mass spectrometry," *Nature Nanotechnology*, vol. 4, pp. 445–450, 2009.

[7] Lang, H. P., Berger, R., Battiston, F., Ramseyer, J.-P., Meyer, E., Andreoli, C., Brugger, J., Vettiger, P., Despont, M., Mezzacasa, T., Scandella, L., Guntherodt, H.-J., Gerber, C., and Gimzewski, J. K., "A chemical sensor based on a micromechanical cantilever array for the identification of gases and vapors," *Applied Physics A: Material Science & Processing*, vol. 66, supplement 1, pp. S61–S64, 1998.

[8] Nugaeva, N., Gfeller, K. Y. , Backmann, N., Lang, H. P., Duggelin, M., and Hegner, M., "Micromechanical cantilever array sensors for selective fungal immobilization and fast growth detection," *Biosensors and Bioelectronics*, vol. 21, no. 6, pp. 849–856, 2005.

[9] Su, M., Li, S., and Dravid, V. P., "Microcantilever resonance-based DNA detection with nanoparticle probes," *Applied Physics Letters*, vol. 82, no. 20, pp. 3562–3564, 2003.

[10] Boisen, A., Thaysen, J., Jensenius, H., and Hansen, O., "Environmental sensors based on micromachined cantilevers with integrated read-out," *Ultramicroscopy*, vol. 82, no. 1-4, pp. 11–16, 2000.

[11] Senesac, L., and Thundat, T. G., "Nanosensors for trace explosive detection," *Materials Today*, vol. 11, no. 3, pp. 28–36, 2008.

[12] Gurjar, M. and Jalili, N., "Toward ultrasmall mass detection using adaptive self-sensing piezoelectrically driven microcantilevers," *IEEE/ASME Transactions on Mechatronics*, vol. 12, no. 6, pp. 680–688, 2007.

[13] Wolf K., and Gottlieb, O., "Nonlinear dynamics of a noncontacting atomic force microscope cantilever actuated by a piezoelectric layer," *Journal of Applied Physics*, vol. 91, no. 7, pp. 4701–4709, 2002.

[14] Dai, M., Eom, K., and Kim, C.-W., "Nanomechanical mass detection using nonlinear oscillations," *Applied Physics Letters*, vol. 95, no. 203104, 2009.

[15] Younis, M. I. and Alsaleem, F., "Exploration of new concepts for mass detection in electrostatically-actuated structures based on nonlinear phenomena," *Journal of Computational and Nonlinear Dynamics*, vol. 4, no. 2, 021010, 2009.

[16] Postma, H. W. C., Kozinsky, I., Husain, A., and Roukes, M. L., "Dynamic range of nanotube- and nanowire-based electromechanical systems," *Applied Physics Letters*, vol. 86, no. 223105, 2005.

[17] Mahmoodi, S. N., Jalili, N., and Daqaq, M. F., "Modeling, nonlinear dynamics, and identification of a piezoelectrically actuated microcantilever sensor," *IEEE/ASME Transactions on Mechantronics*, vol. 13, no. 1, pp. 58-65, 2008.

[18] Joshi, S. P., "Non-linear constitutive relations for piezoceramic materials," *Smart Materials and Structures*, vol. 1, no. 1, pp. 80–83, 1992.

[19] Doedel, E., Champneys, A., Fairgrieve, T., Kuznetsov, Y., Sandstede, B., and Wang, X., "Auto 97: Continuation and bifurcation software for ordinary differential equations," 1997.

[20] Baller, M. K., Lang, H. P., Fritz, J., Gerber, C., Gimzewski, J. K. , Drechsler, U., Rothuizen, H., Despont, M., Vettiger, P., Battiston, F. M., Ramseyer, J. P., Fornaro, P., Meyer, E., and Guntherodt, H.-J., "A cantilever array-based artificial nose," *Ultramicroscopy*, vol. 82, no. 1, pp. 1–9, 2000.

[21] Calvert, P., "Inkjet printing for materials and devices," *Chemistry of Materials*, vol. 13, no. 10, pp. 3299–3305, 2001.

[22] Boley, W., Bhuvana, T., Hines, B., Sayer, R., Chiu, G.-C., Fisher, T., Bergstrom, D., Reifenberger, R., and Kulkarni, G., "Inkjet printing involving palladium alkanethiolates and carbon nanotubes functionalized with single-strand DNA," in *Proceedings of the DF2009: Digital Fabrication Processes Conference*, (Louisville, Kentucky), pp. 824–827, 2009.

[23] DeMartini, B. E., Rhoads, J. F., Zielke, M. A., Owen, K. G., Shaw, S. W., and Turner, K. L., "A single input-single output coupled microresonator array for the detection and identification of multiple analytes," *Applied Physics Letters*, vol. 93, no. 5, 054102, 2008.

[24] Post, N. J., *Precision Micro-Deposition of Functional Layers Using Inkjet Drop-on-Demand and Applications to the Functionalization of Microcantilever Sensors*. MS thesis, 2007.

[25] Lee, E., *Microdrop Generation*. CRC Press, 1st ed., 2003.

Proceedings of the SEM Annual Conference
June 7-10, 2010 Indianapolis, Indiana USA

Calibrating Force and Displacement in the Face of Property Variation

Jason V. Clark
School of Mechanical Engineering
School of Electrical and Computer Engineering
Purdue University, Discovery Park, BRK-1289, IN 47907
jvclark@purdue.edu

ABSTRACT

We present a theoretical analysis of a packagable method for calibrating the force and displacement of a micro electro mechanical system (MEMS) that is subject to geometric and material property variations. Property variations are usually due to variations in the fabrication processes, packaging, and environmental exposure. Force and displacement are among the most important fundamental mechanical quantities for investigating, discovering, and exploiting micro and nanoscale phenomena. However, finding a practical and traceable method to accurately and precisely measure the minute forces and displacements in MEMS has been elusive. One reason is because the force generated by MEMS is quite often smaller than the force that can be measured by conventional force sensors. Similarly, displacements in MEMS can be as small as a fraction of the diameter of an atom, which is beyond the capabilities of standard displacement sensors. Our present analysis addresses the calibration of force and displacement for MEMS comprising comb drive sensors and actuators. Our approach for calibrating force and displacement is based on the strong and sensitive coupling between mechanical performance and electronic measurands at the microscale. That is, variations in geometry and material properties affect performance, which can be capacitively measured using on-chip or off-the-shelf capacitance meters. A novelty in our analysis is the elimination of unknown properties, which allows us to express mechanical quantities and their uncertainties solely in terms of electrical measurands. We derive analytical expressions for extracting measurements of force, displacement, stiffness, and their uncertainties by electrical probing. And we show how our method is expected to a few orders more precise than convention.

1 INTRODUCTION

Precise measurements of force and displacement are of the most fundamental quantities used to explore, understand, and exploit a wide variety of physical phenomena. For instance, measurements of force and displacement are used in the study of hydrogen bonds of DNA [1], the van der Waals interaction on gecko setae [2], the Casimir effect between conductive materials [3], and a multitude of other phenomena. Unfortunately, for force and displacement magnitudes on the order of nano/piconewtons and nano/picometers, there are currently no commercially available SI traceable reference standards for calibration [4], [5]. Although there are several methods available to calibrate forces (e.g. a review is given in [6]), the accuracies of the methods are unknown. That is, although a minute force or displacement can be sensed by using conventional technology, it has been difficult to precisely know how much force or displacement has taken place [5]. As a consequence, it is not uncommon to see measurements in the MEMS literature expressed with only one significant digit of accuracy, and or no measurements of the uncertainty.

For example, although the atomic force microscope (AFM) is a most popular tool used to explore nanoscale phenomena [6-9], the geometry and material properties of a typical AFM cantilever are often not well known. The stiffness of the AFM cantilever is often measured with a relative error of about 10% [10]. Assuming that that displacement of the AFM cantilever can calibrated with a much smaller relative error, the product of displacement and stiffness yields a force with a relative error of about 10% as well. That is, a measurement of force from this AFM will provide at most one significant digit of accuracy, which can amount to an uncertainty on the order of 100 piconewtons. Not only is there a need to more precisely measure phenomena previously measured by such tools, but there is a need to measure the multitude of phenomena beyond their current resolution. Our capacitive-based technology presented here is expected to lead to a more accurate and practical way to calibrate AFMs, and improved force-displacement probes.

Force-displacement sensors and actuators that can be integrated into a system-on-a-chip typically come in three types: transducers that convert a change in mechanical strain to a change in resistance [11] or change in voltage [12], or transducers that convert a change in mechanical displacement to a change in capacitance [13,14]. Of these, capacitive sensors offer the greatest sensitivity. For example, on-chip capacitance meters have detected changes comb drive capacitance on the order of zeptofarads, which corresponds to a deflection of MEMS flexures on the order of 100 femtometers [15]. Indeed, the MEMS electrical precision is available. What has been lacking is a corresponding precision for MEMS mechanical properties, which we present here. In addition, we also show that the accuracy of force and displacement by our method is independent of the accuracy of capacitance, which is often subject to variations in parasitic capacitance between electrical probed measurements. Our mechanical measurements are functions of changes in capacitance and the precision of capacitance.

Following this introduction, a statement of the problems is discussed in Section 2. A brief overview of electrical probe method is discussed in Section 3. A description of our device and how we calibrate its force and displacement are discussed in Section 4. Descriptions of sensitivity and uncertainty of our method are presented in Section 5. Last, we summarize our finding and subsequent research directions in.

2 PROPERTY VARIATION

It is well known to researchers in the area of micro and nano electro mechanical systems (M/NEMS) that mechanical performance strongly depends on geometric and material properties. These fabricated properties are difficult to predict and difficult to measure. The problem with prediction is that, given any fabrication recipe, the geometric and material properties vary between fabrication facilities, between fabrication runs, and across the wafer itself. A problem with many measurement methods is that they often yield uncertainties that on the order of property variation.

Even with standard commercial fabrication foundries, there are significant variations in geometric and material properties. For example, in Figure 1 we show an optical image of our MEMS that was fabricated using the standard MEMCAP PolyMUMPs process [16]. Two scanning electron microscopy (SEM) insets in Figure 1 show magnified views the tip of a comb drive finger and gap. The insets show significant geometric variations such as fillets instead of sharp 90° corners; coarse sidewalls instead of smooth surfaces; flexure widths which have recessed ~20%; and gaps which have increased ~20%. For clarity, the original layout geometry of the device is superimposed on the SEM insets as dashed lines. Compared to layout geometry, the enlarged gaps decrease capacitance; the narrowed and

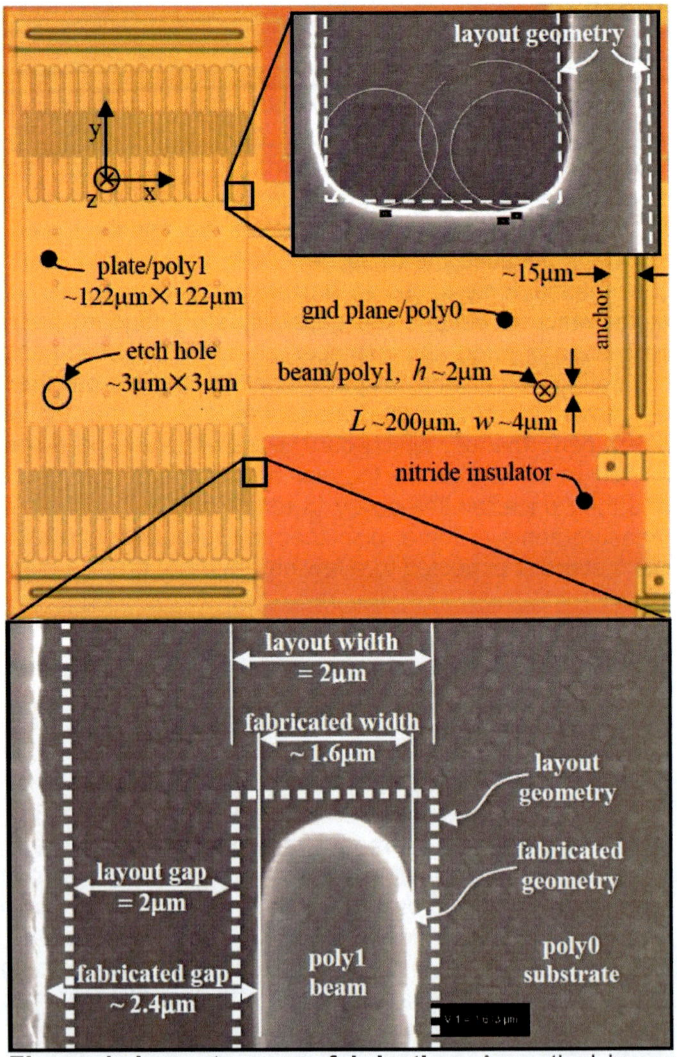

Figure 1: Layout versus fabrication. An optical image shows a MEMS comprising a comb drive supported with a pair of fixed-guided flexures. Two scanning electron microscopy images show magnified views of the fabricated geometry, where layout geometry is depicted as dashed lines. Due to process variation, flexure stiffness have been reduced, capacitance has been reduced, sharp corners have been replaced with fillets and chamfers.

lengthened flexures reduce the effective stiffness; and the fillets increase stiffness.

Regarding material properties, for given a displacement, Young's modulus is often used to determine force in MEMS by Hooke's law. However, the Young's modulus of fabricated MEMS is often unknown. Although many researchers use lookup tables to determine Young's modulus, such values are usually averages of measurements that vary by 10s of percent. For example, Figure 2 shows a scatter plot of disparate measurements of Young's modulus of polysilicon, a most popular MEMS material. Various measurement techniques were used in this data. The plot shows large variations in the reported values and uncertainties of each technique used. Since there is no standard for measuring Young's modulus at the micro and nanoscales, the accuracy of such techniques are unknown.

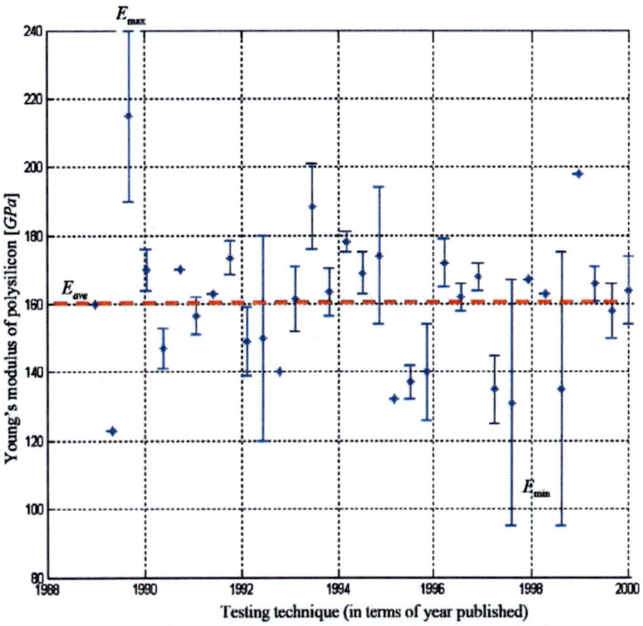

Figure 2: Disparate measurements of Young's modulus of polysilicon. The y-axis represents measurements of Young's modulus. The x-axis represents the year that a research team published the result using a particular technique. The red dashed line at about 160GPa is the average of all reports. The span of modulus values range from 95GPa to 240GPa. This scatter plots is based on data collected by Sharpe [19].

3 ELECTRO MICRO METROLOGY

To address the above metrology issues, we propose the use of a novel metrology method that is expected to facilitate accurate, precise, and practical calibration of force and displacement. Our method is called Electro Micro Metrology (EMM) [17]. The version of EMM presented here is a significant improvement over our previous version reported in [18]. Improvements include a reduction in chip real estate, an increase in robustness, and an improvement in the ease of use.

EMM exploits the coupling between microscale mechanical properties and electrical measurands. That is, in EMM a mechanical property X (e.g. displacement) may be expressed in terms of an electronic measurand e (e.g. capacitance) as $X(\Delta e \pm \delta e) = X(\Delta e) \pm [\partial X/\partial \Delta e]\,\delta e$, where Δe is the change in e, $\partial X/\partial \Delta e$ is the sensitivity, and δe is the uncertainty in e. The uncertainty of measurement using EMM is somewhat controllable as discussed below. And while the sensitivity $\partial X/\partial \Delta e$ can be quite large, the electrical uncertainty can be made to be much, much smaller, which reduces the uncertainty of measured mechanical property.

4 DEVICE

4.1 CONFIGURATION

One configuration of our device is shown in Figures 3, in an unactuated state and two actuated states. The device comprises a pair of asymmetric gap-stops and a pair of comb drives. The drive is supported by fixed-guided flexures; however, fixed-fixed or folded flexures is preferred because they avoid the "beam shortening" effect during large deflections. The central anchor is intentionally offset to produce two unequal gap-stops. If Figure 3, $gap_1 < gap_2$. These two gap-stops allow two measurements to be made within one device, instead of requiring two separate devices as with our previous effort [18]. Since the device only requires a single structural layer and can be fabricated using a single lithography mask, the device can be fabricated in most MEMS processes.

4.2 MEASUREMENT

We show the sequence of EMM measurements of the MEMS in Figure 3. Figure 3a shows the initial, unactuated state. The measured capacitance of the comb drive in this unactuated state has the form

$$C_{initial} = C_0 + C_P \quad (1)$$

where C_P is the parasitic capacitance and C_0 is the pure comb drive capacitance. Although the parasitic capacitance can be different between experiments, it is assumed to be constant during a pair of measurements. For example, when using an off-chip capacitance meter, the meter probes may contact the electrode pad at a slightly different angle if removed and then reapplied; but while conducting a pair of measurements, there should an insignificant change in the amount of C_P. Upon applying a voltage that is greater than the minimum voltage required to close the leftmost gap, gap_1 , a second measurement of capacitance is performed where

$$C_{gap1} = C_1 + C_P \quad (2)$$

is measured (see Figure 3b). Similarly, in closing the rightmost gap, gap_2 , a capacitance

$$C_{gap2} = C_2 + C_P \quad (3)$$

is measured (see Figure 3c). The changes in capacitance between the three states shown in Figure 3 are

$$\Delta C_1 = (C_1 + C_P) - (C_0 + C_P) = C_1 - C_0 \text{ and} \quad (4)$$

$$\Delta C_2 = (C_2 + C_P) - (C_0 + C_P) = C_2 - C_0, \quad (5)$$

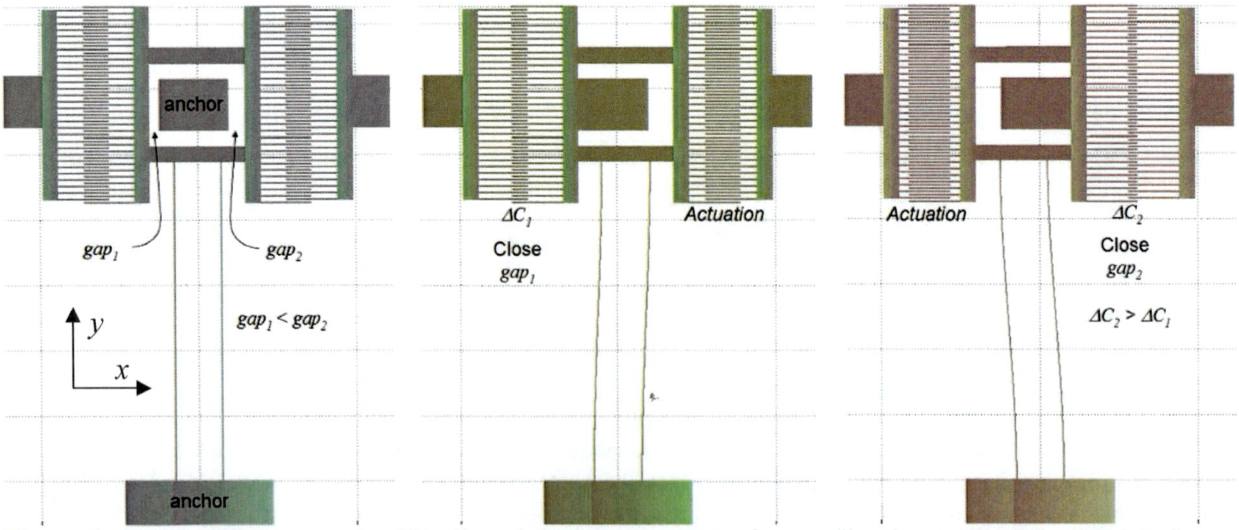

Figure 3: Force-displacement calibration. A 3-step sequence of our calibration methods is simulated. Figure 3a shows are MEMS with an asymmetric gap stop about the central anchor of the comb drive. The capacitance of this zero-state is measured. Figure 3b shows an actuated state where the left gap is closed by applying voltage to the comb drive on the right. The capacitance ΔC_1 is measured. And Figure 3c shows an actuated state where the right gap is closed by applying voltage to the comb drive on the left. The capacitance ΔC_2 is measured.

where C_P cancels in the difference. The elimination of this arbitrary parasitic capacitance C_P allows measurements to be repeatable across different facilities, using different capacitance meters.

4.3 CALIBRATION OF DISPLACEMENT

The measured changes in capacitances in closing the two gaps can be written as

$$\Delta C_1 = 2N\beta\frac{\varepsilon h}{g}gap_1 = 2N\beta\frac{\varepsilon h}{g}\left(gap_{1,layout} + \Delta gap\right) \text{ and} \tag{6}$$

$$\Delta C_2 = 2N\beta\frac{\varepsilon h}{g}gap_2 = 2N\beta\frac{\varepsilon h}{g}\left(n\,gap_{1,layout} + \Delta gap\right) \tag{7}$$

where the unknown parameters ε, h, g, N, and β are the permittivity of the medium, layer thickness, gap between comb fingers, number of comb fingers, and fringing field factor. The unknown quantity Δgap is the difference in gap-stop size between layout and fabrication, which must be found to calibrate displacement. This difference in geometry between layout and fabrication was exemplified in Figure 1. The quantities $gap_{1,layout}$ and $gap_{2,layout} = n\,gap_{1,layout}$ in (6) and (7) are exactly known from layout, where $n \neq 1$ is a layout parameter chosen such that $gap_1 \neq gap_2$. In Equations (6) and (7) we have assumed that the movement of the comb drive shown in Figure 3 is in plane and in the direction of x.

By Equation (6) by (7), we cancel the all unknown quantities except Δgap. We have

$$\frac{\Delta C_1}{\Delta C_2} = \frac{gap_{1,layout} + \Delta gap}{n\,gap_{1,layout} + \Delta gap}. \tag{8}$$

Solving Equation (8) for Δgap yields

$$\Delta gap = gap_{1,layout}\frac{n\,\Delta C_1/\Delta C_2 - 1}{\Delta C_1/\Delta C_2 - 1}. \tag{9}$$

Equation (9) is the geometrical difference between layout and fabrication in terms of measured changes in capacitance and exactly-known layout parameters. Therefore, the EMM-measured gap-stops are $gap_1 = gap_{1,layout} + \Delta gap$ and $gap_2 = n\,gap_{1,layout} + \Delta gap$, where we have assumed that both gap-stops experience the same change in geometry in going from layout to fabrication because the gaps are within close proximity to each other. That is, we have assumed that during fabrication, the two gaps experience the same lithography, deposition, and etching reaction processes.

Now the ratio of the displacement gap_1 to its corresponding change in capacitance ΔC_1 yields a property that we define as the *comb drive constant* Υ,

$$\Upsilon \equiv \frac{\Delta C_1}{gap_1}. \tag{10}$$

where the gap-stop gap_1 is the largest displacement in the positive x-direction for this device. However, for an intermediate displacement of Δx, which produces a change in capacitance ΔC, we have same ratio as (10):

$$\frac{\Delta x}{\Delta C} = \frac{gap_1}{\Delta C_1}. \tag{11}$$

Hence, by accounting for geometric and material process variations, the calibrated displacement of this device is

$$\Delta x = \left(\frac{gap_1}{\Delta C_1}\right)\Delta C = \frac{\Delta C}{\Upsilon}. \tag{12}$$

4.4 CALIBRATION OF FORCE AND STIFFNESS

The electrostatic force of a comb drive can be written as

$$F = \nabla \frac{1}{2} C V^2 = \frac{1}{2}\frac{\Delta C}{\Delta x} V^2, \tag{13}$$

where the partial derivatives can be expressed as differences due to the linearity of our comb drive. Substituting the comb drive constant from Equation (11) into (13) yields the calibrated force measurement

$$F = \frac{1}{2}\left(\frac{\Delta C_1}{gap_1}\right) V^2 = \frac{1}{2}\Upsilon V^2, \tag{14}$$

which accounts for both geometric and material process variations.

In addition, the stiffness of the device can be calibrated by taking the ratio of force (14) and displacement (12)

$$k = \frac{F}{x} = \frac{1}{2}\Upsilon^2 \frac{V^2}{\Delta C}, \tag{15}$$

which is not necessarily constant. That is, the ratio $V^2/\Delta C$ is often nearly constant for small deflection, and often increases nonlinearly for large deflections.

5 SENSITIVITY ANALYSIS

As with any measurement method, there will be some uncertainty associated with measurement. With respect to the above EMM analysis, uncertainties in the measured capacitance δC and voltage δV produce a corresponding uncertainty in the extracted mechanical properties, displacement Δx and force F. We find these mechanical uncertainties by first rewriting all quantities of capacitance and voltage as $\Delta C \to \Delta C + \delta C$ and $\Delta V \to \Delta V + \delta V$ in the previous section, then determining the first order term of the Taylor expansion about the electrical uncertainty. In doing so, the first-order Taylor expansion of displacement (12) about δC gives

$$\Delta x + \delta x = gap_{1,layout}\left(1 + \frac{n\,\Delta C_1/\Delta C_2 - 1}{\Delta C_1/\Delta C_2 - 1}\right)\frac{\Delta C}{\Delta C_1} \pm$$

$$\pm \; gap_{1,layout}\left(\frac{n\dfrac{1-\Delta C_1/\Delta C_2}{\Delta C_2}+\dfrac{(n\Delta C_1-\Delta C_2)(\Delta C_1/\Delta C_2-1)}{(\Delta C_1-\Delta C_2)\Delta C_2}}{\dfrac{\Delta C_1}{\Delta C}\left(\dfrac{\Delta C_1}{\Delta C_2}-1\right)} + \frac{(1-\Delta C/\Delta C_1)}{\Delta C_1}\left(1+\left(n\frac{\Delta C_1}{\Delta C_2}-1\right)\Big/\left(\frac{\Delta C_1}{\Delta C_2}-1\right)\right)\right)\delta C \tag{16}$$

where the second term is the uncertainty in displacement δx. The coefficient of δC is the sensitivity $\partial\Delta x/\partial\Delta C$.

Similarly, the first-order multivariate Taylor expansion of force (14) about δC and δV gives

$$\begin{aligned} F\pm\delta F = &\frac{1}{2}\frac{\Delta C_1 V^2}{gap_{1,layout}\left(1+\dfrac{n\,\Delta C_1/\Delta C_2-1}{\Delta C_1/\Delta C_2-1}\right)} \\ &\pm\frac{1}{2}V^2\left(1-\frac{(\Delta C_1-\Delta C_2)\Delta C_1\left(n\dfrac{1-\Delta C_1/\Delta C_2}{\Delta C_2}+\dfrac{(n\Delta C_1-\Delta C_2)(\Delta C_1/\Delta C_2-1)}{(\Delta C_1-\Delta C_2)\Delta C_2}\right)}{gap_{1,layout}(\Delta C_1+n\Delta C_1-2\Delta C_1)\left(\dfrac{\Delta C_1}{\Delta C_2}-1\right)\left(1+\dfrac{n\,\Delta C_1/\Delta C_2-1}{\Delta C_1/\Delta C_2-1}\right)}\right)\delta C \\ &\pm\frac{\Delta C_1 V}{gap_{1,layout}\left(1+\dfrac{n\,\Delta C_1/\Delta C_2-1}{\Delta C_1/\Delta C_2-1}\right)}\delta V \end{aligned} \tag{17}$$

where the second and third terms combined is the total uncertainty in force δF. The coefficients of δC and δV are the sensitivities $\partial F/\partial\Delta C$ and $\partial F/\partial\Delta V$.

We find the that mechanical uncertainty due to capacitance is typically quite large, while the opposite is true for mechanical uncertainty due to voltage. Fortunately, in practice we see that the precision of capacitance meters are much higher than the precision of voltage source generators. For example, consider our model shown in Figure 3 which consists of 30 comb fingers on each comb drive; a 2 micron gap between each finger; a 10 and 15 micron gap-stop about the central anchor (i.e. the layout parameter n is 1.5); and a layer thickness of 30 microns. Assuming that the change in geometry in going from layout to fabrication is 0.25 microns, then uncertainties in displacement and force from (16) and (17) become

$$\delta x = \pm O(10^8)\delta C, \text{ and} \tag{18}$$

$$\delta F = \pm O(10^7)\delta C \pm O(10^{-8})\delta V. \tag{19}$$

Now with typical values of uncertainties in capacitance δC on the order of attofarads, and voltage δV on the order of millivolts, then from (18) and (19) we find that the uncertainties in displacement and force are expected to be on the order of δC~angstroms and δF~10s of piconewtons. Of course, there are capacitance meters and voltage generators with one to three orders of magnitude better precision than this. In addition, it is important to understand that Equations (16) and (17) are not optimized. That is, by strategic choice of the number of comb fingers, comb finger gaps, gap-stop size, etc., the sensitivities in (16) and (17) can be reduced, which will improve the resulting mechanical uncertainties in displacement and force. We are currently investigating the optimization of such design parameters.

6 CONCLUSION

In this paper we presented a theoretical analysis on a packagable method to traceably calibrate the force and displacement of MEMS that are subject to geometric and material property variations. As a test case we analyzed a MEMS comprised of a comb drive and flexures, and subjected the device to variations in its geometric and material properties. By assuming that geometric errors about the device were locally consistent, we were able to eliminate unknown electromechanical parameters in favor of expressions for calibrating and measuring force and displacement solely in terms of electrical measurands. We showed that our method is insensitive to the accuracy of capacitance probing, but extremely sensitive to changes in capacitance and the precision of capacitance. We found this present method to be a significant improvement over our previous effort by a reduction in chip real estate, an increase in robustness, and an improvement in the ease of use. Through sensitivity analysis we find our method to be more precise than convention. Our analysis explicitly shows the dependence on several key parameters which can be explored for design optimization to further reduce the uncertainty in measurement of force and displacement. In expanding this effort, we are currently relaxing the assumption of locally consistency, i.e. property mismatch due to anisotropic processes; and design optimization for improved sensitivities.

REFERENCES

[1] K. Kim; Z. Lin; P. Shrotriya; S. Sundararajan; Q. Zou, "Iterative Control Approach to High-Speed Force-Distance Curve Measurement Using AFM for Biological Applications," *Automation Science and Engineering.* (2007) pp 219 - 224
[2] M. Reyes, R. S. Fearing, "Macromodel for the mechanics of gecko hair adhesion", *IEEE Conference on Robotics and Automation.* (2008) pp. 1602 - 1607
[3] C. Yamarthy, S. McNamara, "Design of a MEMS sensor to detect the Casimir force", *Nano/Micro Engineered and Molecular Systems. NEMS 2009. 4th IEEE International Conference*, (2009) pp. 645 - 648.
[4] J. Pratt, J. Kramar, D. Newell, and D. Smith, "Review of SI traceable force metrology for instrumented indentation and atomic force microscopy", *Meas. Sci. Technol.* 16, (2005) pp. 2129–37
[5] R. S. Gates, J. Pratt, "Prototype cantilevers for SI-traceable nano-Newton force calibration", *Meas. Sci. Technol.* 17, (2006) pp. 2852–60
[6] M. Tokunaga, T. Aoki, M. Hiroshima, K. Kitamura and T. Yanagida, "Subpiconewton intermolecular force microscopy", *Biochem. Biophys. Res. Commun.* 231, (1997) pp. 566–9
[7] U. Dammer, O. Popescu, P. Wagner, D. Anselmetti, H. Guntherodt, and G. Misevic, "Binding strength between cell adhesion proteoglycans measured by atomic forcemicroscopy *Science* 267, 1997, pp. 1173–5
[8] A. Oberhauser, P. Marszalek, H. Erickson, and J. Fernandez, "The molecular elasticity of the extracellular matrix protein tenascin *Nature* 393, (1998) pp. 181–5
[9] G. Charras, P. Lehenkari, and M. Horton, "Atomic force microscopy can be used to mechanically stimulate osteoblasts and evaluate cellular strain distribution", *Ultramicroscopy* 86, (2001) pp. 85–95
[10] N. Burnham, X. Chen, C. Hodges, G. Matei, E. Thoreson, C. Roberts, M. Davies, and S. Tendler, "Comparison of calibration methods for atomic-force microscopy cantilevers", *Nanotechnology* 14, (2003) pp. 1–6
[11] G. Lin, K. S. J. Pister and K. P. Roos, "Surface micromachined polysilicon heart cell force transducer", *J. Microelectromech. Syst.* 9, (2000) 9–17
[12] R. P´erez, N. Chaillet, K. Domanski, P. Janus, P. Grabiec, "Fabrication, modeling and integration of a silicon technology force sensor in a piezoelectric micro-manipulator," *Sensors and Actuators A,* 128 (2006) 367–375
[13] Y. Sun and B. Nelson, "MEMS capacitive force sensors for cellular and flight biomechanics," Biomedical Materials, 2, (2007) S16-S22
[14] S. Muntwyler, F. Beyeler and B. J. Nelson, "Three-axis micro-force sensor with sub-micro-Newton measurement uncertainty and tunable force range," Journal of Micromechanics and Microengineering 20 (2010)
[15] J. Green, D. Krakauer, "New iMEMS Angular Rate-Sensing Gyroscope," *Analog Devices, Dialogue*, pp. 37-03, (2003).
[16] MEMSCAP Inc. 12 TW Alexander Drive Building 100, Research Triangle Park, NC 27709
[17] J. V. Clark, "*Electro Micro Metrology*," Ph.D. thesis, University of California at Berkeley, Berkeley, California, (2005)
[18] J. V. Clark, D. Garmire, M. Last, J. Demmel, "Practical Techniques for Measuring MEMS Properties", Nanotech2004, Seventh International Conference on Modeling and Simulation of Microsystems, Boston, MA, March 7-11, (2004)
[19] W. Sharpe Jr., "Mechanical Properties of MEMS Materials," *The MEMS Handbook*, pp. 1-33, (2002)

Proceedings of the SEM Annual Conference
June 7-10, 2010 Indianapolis, Indiana USA

Standardization of Nanoscale Interfacial Experiments Using MEMS

Tanil Ozkan[1,*], Qi Chen[1], Ioannis Chasiotis[1]
[1]Aerospace Engineering, University of Illinois at Urbana-Champaign, Urbana, IL 61801, U.S.A.
*tozkan2@illinois.edu

ABSTRACT

A novel experimental method for the interfacial mechanics of nanofibers and nanotubes was developed. The debond force was determined by MEMS devices whose motion was precisely measured from optical images by digital image correlation. Essential elements of this method are the submicron control of nanofiber/nanotube embedded length in a thermoplastic or thermosetting polymer and the application of well controlled pull-out force until terminal debonding. The cross-head displacement resolution is at least 20 nm and the force resolution of the order of nanonewtons. A traceable force calibration technique was integrated to calibrate the MEMS force sensors. The method allows for nanofiber pull-out experiments at time scales varying from microseconds to hours and at hot/cold temperatures. Experiments have been conducted for the first time with ø150-350 nm carbon nanofibers embedded in EPON epoxy to quantify the role of nanofiber surface functionalization in the interfacial shear strength. It was clearly shown than surface functionalization drastically increases interfacial adhesion by a factor of three. The present experiments are the first of their kind both in terms of experimental fidelity and data coherence compared to prior experimental attempts, pointing out the robustness of this new experimental method.

INTRODUCTION

Nanofibers, nanotubes, nanowires and nanorods are expected to play an important role in improved composite media with unique capabilities [1,2]. In mechanical terms, the interface between a structural reinforcement and the surrounding matrix is central to the anticipated property improvement due to a nanoscale agent inclusion. In this regard, a strongly bonded interface is prerequisite for effective mechanical interaction whereas a resilient interface might be preferable from a fracture mechanics point of view. The interfacial shear strength (IFSS) is a important measure of the capability of the matrix to transfer forces to nanoscale reinforcements and vice versa. Until recently, the experimental characterization of the mechanical behavior of interfaces in polymer matrix composites was limited to microscale reinforcements due to experimental limitations in force and strain metrology at the submicron and the nanometer scales. The first experimental inquiries of nanoscale interfacial phenomena became possible with the application of scanning probe microscopy cantilevers [3,4], Figure 1(a). While they represent significant experimental milestones, such cantilever driven carbon nanotube (CNT) drag-out and pull-out techniques have inherent deficiencies, e.g. the applied loading not being uniaxial throughout the course of a pull-out experiment, hence causing mixed mode fracture at the interface, as shown in Figure 1(b). More recently, another study extended the cantilever-based IFSS measurements to thermosetting matrices by pulling protruding high-temperature-heat treated vapor grown carbon nanofibers (VGCNFs) out of an epoxy by using a tungsten cantilever [5]. The force metrology principle was identical to the previous studies, with limited embedded length definition and mixed mode fracture initiation, resulting in exceptionally high IFSS values reaching 632 MPa, which is about 15 times the yield strength of the matrix. The present study developed a novel method to overcome previous experimental limitations and obtain consistent and physically sound experimental measurements of the interfacial adhesion of individual nanoscale fibers embedded in a polymeric matrix. Thorough experiments for the method were conducted for the first time with two different grades of VGCNFs embedded in EPON epoxy to shed light on the role of nanofiber and functionalization in the interfacial shear strength.

EXPERIMENTAL METHOD

Single nanofiber pull-out experiments were conducted by a novel MEMS-based experimental method, which combined the force measurement by on-chip polysilicon loadcells with the use of surface micromachined nanofiber mounts for precise nanofiber mounting and alignment, as shown in Figure 2. In-plane alignment of the

T. Proulx (ed.), *MEMS and Nanotechnology, Volume 2*, Conference Proceedings of the Society for Experimental Mechanics Series 2,
DOI 10.1007/978-1-4419-8825-6_11, © The Society for Experimental Mechanics, Inc. 2011

loadcells with the nanofiber mounts was ensured through the SUMMIT™ ultra-planar, multi-level MEMS manufacturing technology of the Sandia National Laboratories. For nanofiber embedment purposes, 0.8 µm deep trenches were generated by a focused ion beam (FIB), where epoxy was deposited and cured after placement of a VGCNF of desired embedded length.

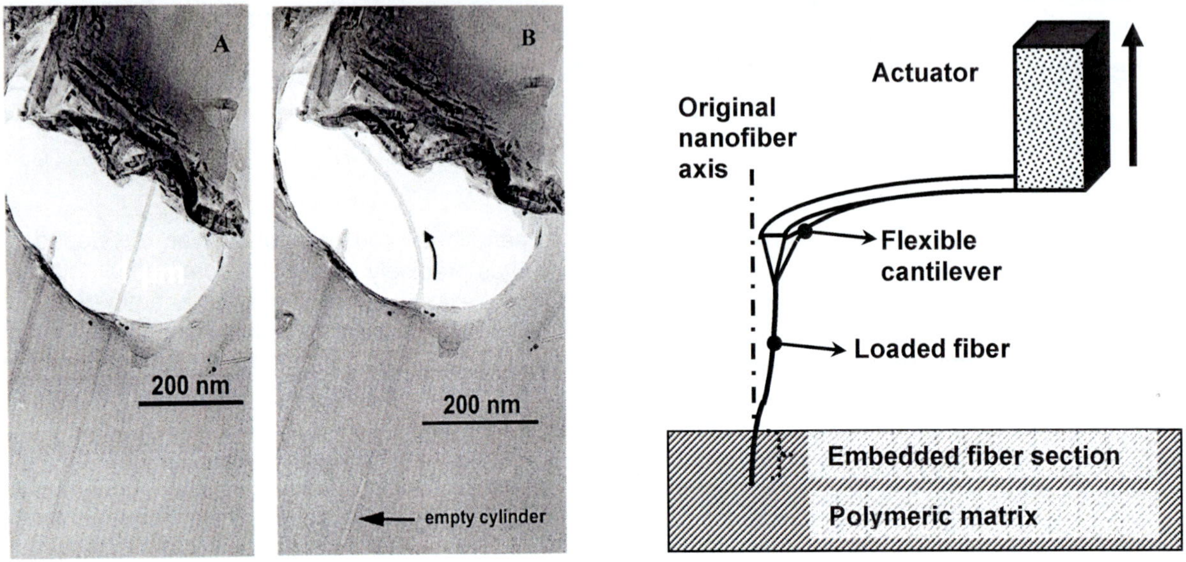

(a) (b)

Figure 1. (a) Nanotube bridging a polymer hole before and after a drag-out experiment by an AFM tip [3]. Figure was reprinted with permission from the American Institute of Physics. **(b)** Schematic of the deflection of a sensitive AFM cantilever during a pull-out experiment. Bending of the flexible cantilever causes the nanofiber to deviate from its original axis resulting in non-uniaxial loading during pull-out.

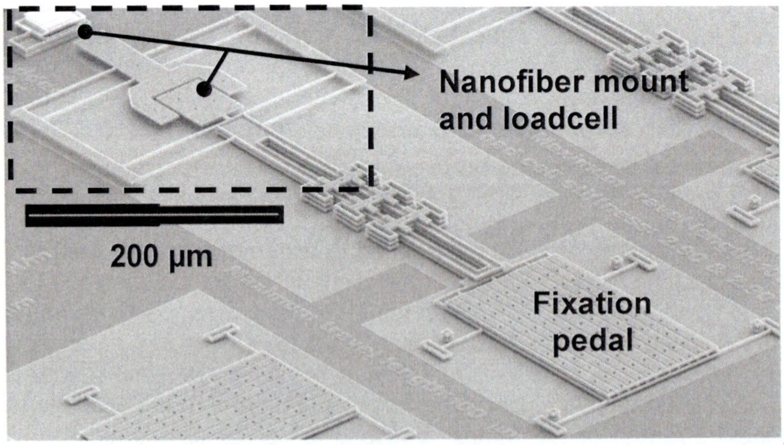

Figure 2. On-chip actuated pull-out device for nanoscale interfacial shear strength experiments.

The embedded lengths were at least as long as the nanofiber diameter or larger. Pull-out experiments were executed with double-column folded beam MEMS-scale loadcells, whose true stiffness was determined by the calibration process discussed in [6]. The actuation of the nanofiber pull-out apparatus and the data acquisition were performed similarly to the methodology described in [7,8]. The ultimate loadcell deflection required to determine the IFSS was measured by digital image correlation (DIC) with an accuracy of ~20 nm which is translated to an accuracy of a few tens of nanonewton in force. Post-mortem SEM images of the remaining pull-out channels inside the polymer containment regions and the debonded VGCNFs ensured that complete nanofiber pull-out took place rather than fiber rupture within or outside the embedded section. Figures 3(a) and

3(b) clearly show the residual opening in the matrix and a high temperature heat treated (but not functionalized) pulled-out nanofiber which cleanly debonded from the matrix.

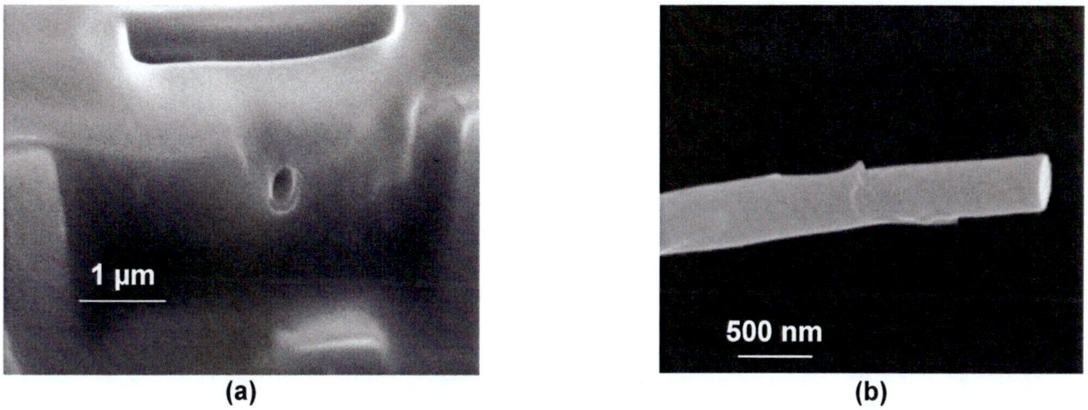

Figure 3. Post-mortem SEM images of **(a)** residual opening after nanofiber pull-out and **(b)** a pulled-out nanofiber end showing clear matrix debonding.

In order to quantify the interfacial shear strength, a uniform yielding criterion over the entire interface was assumed, such that the maximum shear stress τ_{max} at failure was attained and sustained at every point along the interface before complete debonding occurred. This assumption automatically implies constant shear stress distribution along the embedded fiber length and presumes negligible work hardening during the loading phase of the pull-out experiment [9]. Hence, the interfacial shear strength τ_{max} was determined by dividing the debonding force F_d (or the maximum pull-out force) with the total interface area as $\tau_{max} = F_d / (\pi\ d\ L_{emb})$, where d and L_{emb} stand for the nanofiber diameter and the embedded length, respectively.

Two grades of individual VGCNFs with diameters varying from 150-350 nm were tested for their interfacial adhesion to Epon 828/Epikure 3140 epoxy matrix, namely high temperature heat-treated VGCNFs without any functionalization (PR-24-HHT-LD), and high temperature heat treated VGCNFs with oxidative surface functionalization (PR-24-HHT-OX) for improved bonded interactions. IFSS results for the two VGCNF grades are compared to quantify the effect of nanofiber surface functionalization on the mechanical properties of the interface.

EXPERIMENTAL RESULTS AND DISCUSSION

Post-mortem SEM images of the residual pull-out channels in the polymer matrix and the debonded ends of VGCNFs as those shown in Figure 3, pointed out to purely interfacial failure for the PR-24-HHT-LD nanofiber grade and rather mixed interfacial and matrix failure for the PR-24-HHT-OX grade. These observations pave the way to understand the differences in interfacial adhesion and failure of different carbon nanofiber grades embedded in the same epoxy polymer. It is worth noting that in all experiments the slope of the force vs. crosshead displacement curve remained constant until loading, as shown in Figure 4(a), justifying the uniform yielding criterion applied to the data analysis. The IFSS shown in Figure 4(b) for the two VGCNF grades as a function of the nanofiber embedded length exhibited a rather limited small data dispersion, which can be attributed to inherent variability in surface chemistry such as graphitic carbon content and density of the reactive binding sites at the ends of oblique graphitic planes etc. or morphological aspects such as orientation of the graphitic carbon layers with respect to the fiber axis, surface roughness and diameter uniformity. On the other hand, the calculated IFSS were independent of the nanofiber embedded length, which allows the use of the measured values independently of the nanofiber length.

The mean IFSS of non-functionalized PR-24-HHT-LD grade nanofibers to EPON epoxy was 66±10 MPa based on the nominal interfacial area without accounting for the surface roughness of each nanofiber. Much higher IFSS values of 215±12 MPa were measured for oxidatively surface treated nanofibers (PR-24-HHT-OX) that also had larger peak-to-valley surface roughness (~5 nm) compared to the non-functionalized nanofibers. Consequently, surface functionalization and chemical modification of VGCNFs dramatically increased their IFSS by more than a factor of three compared to the non-functionalized primitive (PR-24-HHT-LD grade).

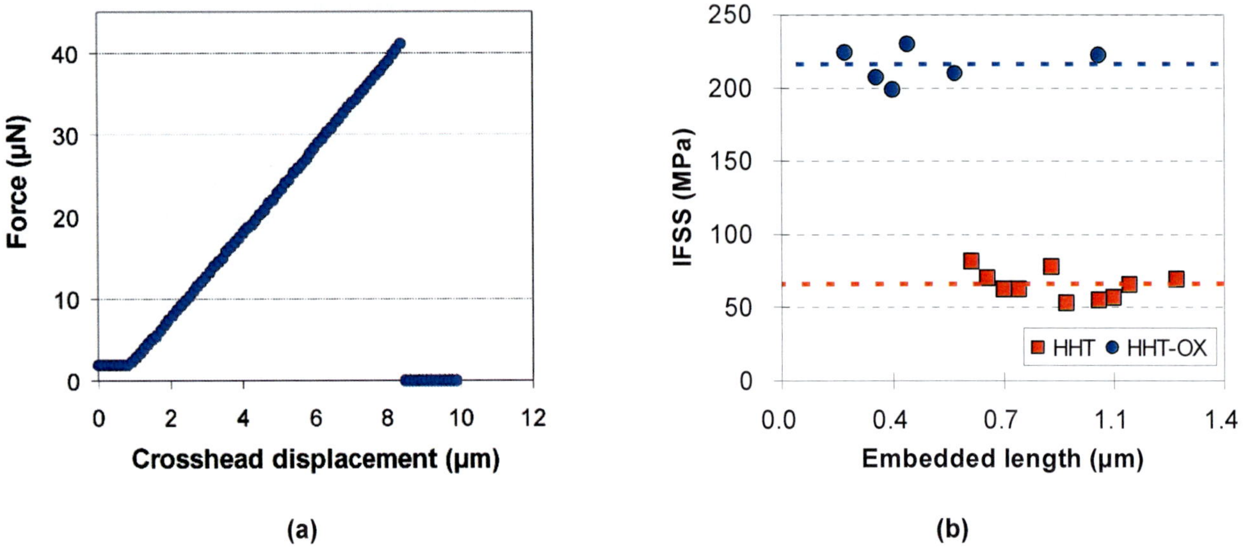

Figure 4. (a) Force vs. crosshead displacement (dominated by the loadcell extension) of a pull-out experiment. **(b)** IFSS of high temperature heat treated (HHT) and oxidized HHT (HHT-OX) VGCNFs.

The average IFSS values for both nanofiber grades indicate that the shear strength of the matrix near the nanofiber is much higher than the typical bulk value of about 40 MPa, which implies that the very existence of the carbon nanofiber could alter the shear properties of the polymer at the interface. The nanometer-level surface roughness of the VGCNFs potentially creates an undulating fracture surface increasing the apparent critical energy release rate. Compared to literature results on the IFSS of CNTs in a polymer matrix obtained in references [3,4], the IFSSs obtained in this study were characterized by much smaller data scatter. Literature data on the IFSS between CNTs and various polymeric matrices varied in the wide range of 20-376 MPa although specimens were taken from the same manufacturing batch [3,4,10].

The IFSS values reported in this study are larger than the corresponding experimental pull-out values for micron sized carbon fibers, which can be as low as 28 MPa for specimens without any surface treatment and can attain values as high as 64.6 MPa following chemical functionalization or surface treatment [9,11]. The average IFSS of around 66 MPa for the PR-24-HHT-LD grade VGCNF-epoxy interface indicates that even the adhesion of the heat-treated, non-functionalized carbon nanofiber grade is almost as good as its chemically functionalized and oxidized microscale counterparts. Such improved adhesion could arise from nanoscale details, such as preferable molecular orientation of polymeric matrix chains along the nanoscale fiber-matrix interface or the presence of denser covalent bonding due to intrinsic defects in the nanofiber exterior surface and its increased surface roughness. Furthermore, such size dependence in IFSS could as well arise from the significantly smaller singular stress region where failure initiates in a nanofiber compared to a microscale fiber. The IFSS values reported in this study also correlate reasonably well with trends in composite fracture toughness measurements performed with the same material systems, indicating the direct relationship between local and effective nanocomposite properties [12,13].

CONCLUSIONS

A novel method for standardized experiments to measure the IFSS of individual nanofibers embedded in a matrix has been reported. The experimental results showed the consistency of measurements conducted by the new method and the good correspondence between theoretical expectations for the effect of surface functionalization on IFSS and the IFSS values measured in this investigation. Furthermore, the trends in composite fracture toughness measurements reported in literature agreed with the single nanofiber IFSS results reported here which provides further validation to the present experimental method for nanoscale interfacial phenomena. From an experimental error analysis point of view, the data scatter arising from methodical deficiencies such as measurement bias and instrument precision errors are dramatically reduced in the present method compared to prior experimental attempts, pointing to the reliability and consistency of the reported data, thus making a definitive step towards a robust experimental method for the interfacial mechanics of nanoscale fibers in soft matrices.

REFERENCES

1. Ramesh K. T., Nanomaterials: Mechanics and Mechanisms, 1st Edition, Springer Science, New York, USA, pp. 1-6, 2009.
2. Klaus F., Stoyko F., Zhang Z., Polymer Composites: From Nano- to Macro-Scale, 1st Edition, Springer Science, New York, USA, pp. 3-22, 2005.
3. Cooper C. A., Cohen S. R., Barber A. H., Wagner H. D., “Detachment of nanotubes from a polymer matrix”, Applied Physics Letters, 81(20), pp. 3873-3875, 2002.
4. Barber A. H., Cohen S. R., Kenig S., Wagner H. D., “Interfacial fracture energy measurements for multi-walled carbon nanotubes pulled from a polymer matrix”, Composites Science and Technology, 64(15), pp. 2283–2289, 2004.
5. Manoharan M. P., Sharma A., Desai A. V., Haque M. A., Bakis C. E., Wang K. W., “The interfacial strength of carbon nanofiber epoxy composite using single fiber pullout experiments”, Nanotechnology, 20, 295701(5pp), 2009.
6. Naraghi M., Chasiotis I., “Optimization of Comb-Driven Devices for Mechanical Testing of Polymeric Nanofibers Subjected to Large Deformations”, Journal of Microelectromechanical Systems, 18(5), pp. 1032-1046, 2009.
7. Naraghi M., Chasiotis I., Kahn H., Wen Y., Dzenis Y., “Novel method for mechanical characterization of polymeric nanofibers”, Review of Scientific Instruments, 78(8), 085108 (7pp), 2007.
8. Ozkan T., Naraghi M., Chasiotis I., “Mechanical properties of vapor grown carbon nanofibers,” Carbon 48, pp. 239-244, 2010.
9. Drzal L. T., Herrera-Franco P.J., “Measurement methods for fiber-matrix adhesion in composite materials”, In: The Mechanics of Adhesion, Dillard D. A. and Pocius A. V. (editors), Elsevier Science B. V., Amsterdam, The Netherlands, pp. 605-660, 2002.
10. Gou J., Minaie B., Wang B., Liang Z., Zhang C., “Computational and experimental study of interfacial bonding of single-walled nanotube reinforced composites”, Computational Materials Science, 31, pp. 225-236, 2004.
11. Jones F. R., “Fiber – matrix adhesion assessment techniques”, In : Handbook of Adhesion, Packham D. E. (editor), 2nd edition, John Wiley & Sons Ltd., West Sussex, U.K., pp. 173-176, 2005.
12. Lafdi K., Fox W., Matzek M., Yildiz E., “Effect of carbon nanofiber-matrix adhesion on polymeric nanocomposite properties—part II”, Journal of Nanomaterials, Article ID 310126, doi:10.1155/2008/310126.
13. Miyagawa H., Drzal L. T., “Effect of oxygen plasma treatment on mechanical properties of vapor grown carbon fiber nanocomposites”, Composites Part A: Applied Science and Manufacturing, 36(10), pp. 1440-1448, 2005.

Proceedings of the SEM Annual Conference
June 7-10, 2010 Indianapolis, Indiana USA

Arrays of Robust Carbon Nanotube-Based NEMS: A Combined Experimental/Computational Investigation

Owen Loh, Xiaoding Wei, Krishanu Nandy, and Horacio D. Espinosa*
Dept. of Mechanical Engineering, Northwestern University, 2145 Sheridan Rd., Evanston, IL 60208-3111
* Corresponding author: espinosa@northwestern.edu

ABSTRACT

We present an investigation of electrostatically-actuated carbon nanotube-based nanoelectromechanical switches. The primary goal of this study is to create a metric for design of robust, high-cycle devices. Methods for fabricating arrays of freestanding carbon nanotubes are discussed. Parametric studies, both experimental and computational, are then conducted to elucidate the failure mechanisms common to this class of carbon nanotube-based nanoelectromechanical systems, and to identify their point of onset within the design space. Experiments are performed *in situ* the scanning electron microscope, enabling direct imaging of device operation and the mode of eventual failure. Complimentary dynamic multiphysics finite element simulations of device operation are also presented to investigate the underlying mechanisms of the experimentally-observed failure modes.

Introduction

Carbon nanotubes are often cited as ideal one-dimensional building blocks to form the next generation of nanoscale electrical, electromechanical, and electrochemical devices. Their immense potential is emergent in theory and preliminary experimental results, which demonstrate, for example, gigahertz switching [1], atomic-scale mass detection [2], and sub-pico Molar virus detection [3]. As such, true CNT-based devices have yet to be realized in statistically-significant numbers.

Nanoscale fabrication is complex and costly. This is compounded by previously insignificant factors, such as van der Waals forces, that arise to create new modes of failure. While issues of robustness and yield are not necessarily the focus of early experiments (and thus rarely reported), it is clear from the fabrication techniques employed and the limited results that these challenges preclude wide-spread integration. Thus, identification and fundamental understanding of the failure modes common to CNTs are critical to their successful implementation in reliable nanoscale devices.

Methodology and Results

A freestanding CNT-based electrostatically-actuated switch [4-6] acts as a platform upon which we study parallel fabrication methods and failure modes. This device encompasses operating principles common to many CNT-based NEMS and thus shares the common failure modes. Accordingly, the results from this research (i.e., identification of failure modes, experimental protocols and model development, and metrics for robust design) are widely applicable.

The device (Figure 1) consists of a CNT fixed at one end to a top electrode and cantilevered over a bottom electrode. For an applied voltage $V > V_{PI}$ (pull-in voltage), the generated electrostatic force causes the CNT to accelerate toward the bottom electrode. When the tip of the CNT approaches the electrode, a tunneling current begins to flow, signifying a closed circuit. This current in turn flows through the feedback resistor R_f, causing a drop in the bias (and thus the electrostatic force). In this way the resistor modulates the bias, stabilizing the switch in the "ON" state.

T. Proulx (ed.), *MEMS and Nanotechnology, Volume 2*, Conference Proceedings of the Society for Experimental Mechanics Series 2,
DOI 10.1007/978-1-4419-8825-6_12, © The Society for Experimental Mechanics, Inc. 2011

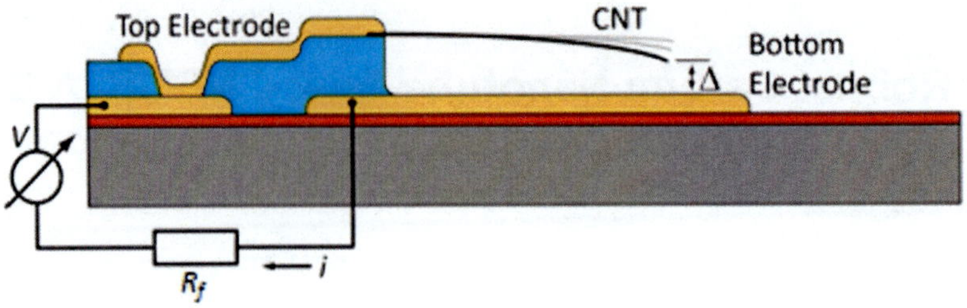

Figure 1 Schematic of the electrostatically-actuated CNT switch. The CNT is fixed at one end by a top electrode and cantilevered over a bottom electrode. The circuit is completed by a feedback resistor and voltage source.

Within this work, our objectives focus on the critical issues of parallel device fabrication and device reliability, both of which currently preclude realization of these CNT-based NEMS in robust sensors and electromechanical logic devices. We first present a novel probe-based nanomanufacturing scheme by which to create large-scale arrays of freestanding CNT-based devices. The method begins with micro fabrication of all necessary electrodes, interconnects, etc, followed by probe-based patterning of catalyst for subsequent chemical vapor deposition growth of CNTs in the desired regions.

Next, using a combined experimental/computational approach to investigate the mechanics and electrical performance of the fabricated devices, we identify the failure modes common to electrostatically-actuated CNT-based NEMS. This is done by characterizing device performance *in situ* the SEM to enable simultaneous electrical measurements and imaging of device operation/failure. We then employ multiphysics modeling to study the dynamic electromechanical response of the device during the pull-in phase and explain the underlying mechanisms for the observed failure modes. Upon validation through comparison to experiments, these models then allow us to predict conditions (device geometry and pull-in voltage) under which device failure is likely, and others in which failure can be avoided. Using this methodology, we perform a comprehensive analysis of the device design space and investigate performance in the context of applications such as NEMS switches, random-access memory elements, and logic devices.

References

1. T. Rueckes, et al. "Carbon nanotube-based nonvolatile random access memory for molecular computing," *Science,* Vol. 289, pp. 94-97, 2000.
2. H.-Y. Chiu, et al. "Atomic-scale mass sensing using carbon nanotube resonators," *Nano Letters,* Vol. 8, No. 12, pp. 4342-4346, 2008.
3. T. Dastagir, et al. "Electrical detection of hepatitis C virus RNA on single wall carbon nanotube-field effect transistors," *Analyst,* Vol. 132, pp. 738-740, 2007.
4. C. Ke and H.D. Espinosa, "In situ electron microscopy electromechanical characterization of a bistable NEMS device," *Small,* Vol. 2, No. 12, pp. 1484-1489, 2006.
5. C. Ke, et al. "Numerical analysis of nanotube based NEMS devices. Part I: Electrostatic charge distribution on multiwalled nanotubes," *J. Appl. Mech.,* Vol. 72, No. 5, pp. 721-725, 2005.
6. C. Ke, et al. "Numerical analysis of nanotube based NEMS devices. Part II: Role of finite kinematics, stretching and charge concentrations," *J. Appl. Mech.,* Vol. 72, No. 5, pp. 726-731, 2005.

Proceedings of the SEM Annual Conference
June 7-10, 2010 Indianapolis, Indiana USA

Electro-Mechanical Response of Carbon Nanotube Reinforced Polymer Composites

Venkat K. Vadlamani[1], Vijaya B. Chalivendra[1*], Arun Shukla[2], Sze Yang[2]
[1]University of Massachusetts, North Dartmouth, MA 02747, USA
[2]University of Rhode Island, Kingston, RI, 02881, USA
*Corresponding author, vchalivendra@umassd.edu, 508-910-6572

ABSTRACT

The effect of nano-deformation, damage and growth of carbon nanotubes (CNTs) reinforced polymer composites is investigated using electro-mechanical response at different loading conditions. Three different polymer systems namely polyurethane, polyurethane reinforced with gas bubbles (that induces porosity) and polyurethane reinforced with Aluminum Silicate hollow microspheres (Cenospheres) in this study. CNTs of different weight percentages are loaded into above three polymer systems and a combination of shear mixing and ultrasonication processes are used to fabricate composites. High-resolution scanning electron microscopy and transmission electron microscopy are used to verify homogenous dispersion of CNTs in the above systems. A four-point probe method is used to measure high resolution electrical response when the above polymer systems are subjected to mechanical loads. The effect of various types of mechanical loading on different stages of deformation of test samples, onset of damage and growth will be discussed using electro-mechanical response of polymer systems.

1. Introduction

Due to their high electrical conductivity, carbon nanotubes are used as sensory network in polymers to understand their deformation and damage in the materials. Wang et al. [1] fabricated continuous carbon fiber reinforced polyphenylene sulfide (PPS) composites and monitored interlaminar thermal damage was monitored using real-time measurement of electrical of interlaminar interface. Recently Thostenson and Chou [2] developed multi-wall CNTs reinforced glass fiber-epoxy laminated composites and they used the distributed sensory network of CNTs to evaluate the onset and the evolution of damage in their composites. In this study, thermoset epoxy was reinforced with two different particulates and using the percolation network generated using multi-walled CNTs in epoxy, extensive experimentation is conducted to understand the deformation, onset of damage and growth in these composites.

1.1 Fabrication Procedure

A two-part thermoset epoxy (Part-A: resin and Part-B: hardener) is used as a matrix. The specific gravity of part-A is 1.1299 and part-B is 0.97 and the mix ratio is 5:1.95. Multi-walled CNTs having a diameter of 30±15 nm and length of 5-20 micron are used to generate percolation network in the system. Two types of particulates namely cenospheres and glass bubbles were used as reinforcement in the epoxy matrix. The cenospheres are a by-product of the fly ash, obtained from coal-fired power plants and glass bubbles are the microscopic spheres of glass which simulate the conditions of porosity.

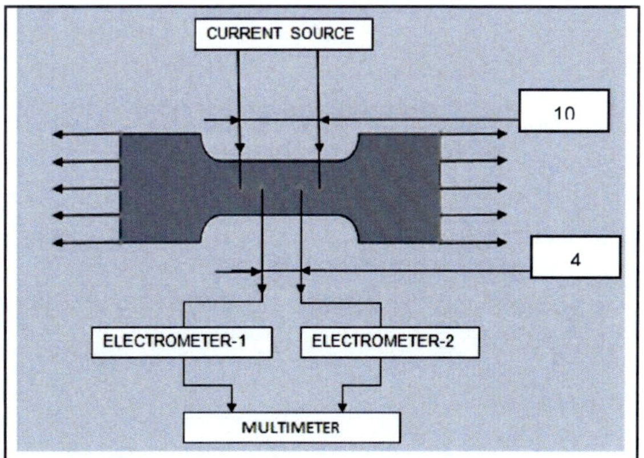

Fig. 1: Four point probe measurements under tensile loading conditions. (All dimensions are in mm)

First, MWCTs of 0.5 wt% are added to epoxy resin and later mix is shear mixed for 30 minutes and ultrasonicated for 60 minutes. In the second step, the particulate materials are introduced and mixed using a shear mixer. The solution is degassed separately to remove the entrapped air. Both the parts are then mixed thoroughly and degassed again to remove any air entrapped during mixing. The mix is poured in suitable molds to make the desired test specimens. Proper care is taken not to induce any air bubbles while pouring the mix into the moulds. The reason

T. Proulx (ed.), *MEMS and Nanotechnology, Volume 2*, Conference Proceedings of the Society for Experimental Mechanics Series 2,
DOI 10.1007/978-1-4419-8825-6_13, © The Society for Experimental Mechanics, Inc. 2011

for degassing before mixing the parts is to reduce the degassing time after mixing, because the epoxy has a short gelation time of 20 min. The specimens are cured at room temperature for 24 hours in the mould.

1.2 Four-point probe system

The Four-point probe technique as in shown in Fig.1 is employed to measure the change in resistance of test specimen when mechanical loads are applied. This system consists of four probes which are attached to the sample, a DC current source (KEITHLEY6220) ranging from 2nA-100mA, a set of electrometers (KEITHLEY6514) and a multimeter (KEITHLEY2000) ranging from 1μV-1000V. The probes are attached along a line and the spacing between the probes is given in Fig.1. Outside probes are used to supply current and the inside probes are used to measure the voltage.

2. Experimental Results

Preliminary tensile experiments are conducted on dog bone specimens of pure epoxy and epoxy composite made out of glass bubbles of 10% volume fraction. The specimens are loaded using Instron materials testing system at a displacement rate of 4mm/min to understand initiation of damage and progressive growth of damage through the variation of percentage change in resistance. Fig. 2(a) shows the variation of percentage change of resistance in pure epoxy and 2(b) shows the percentage change of resistance in composite. It can be noticed from Fig.2 that overall resistance change in glass bubble reinforced epoxy is much higher due to increase in porosity as the load increases. The authors are in the process of conducting several experiments to come up with concrete reasons and conclusions.

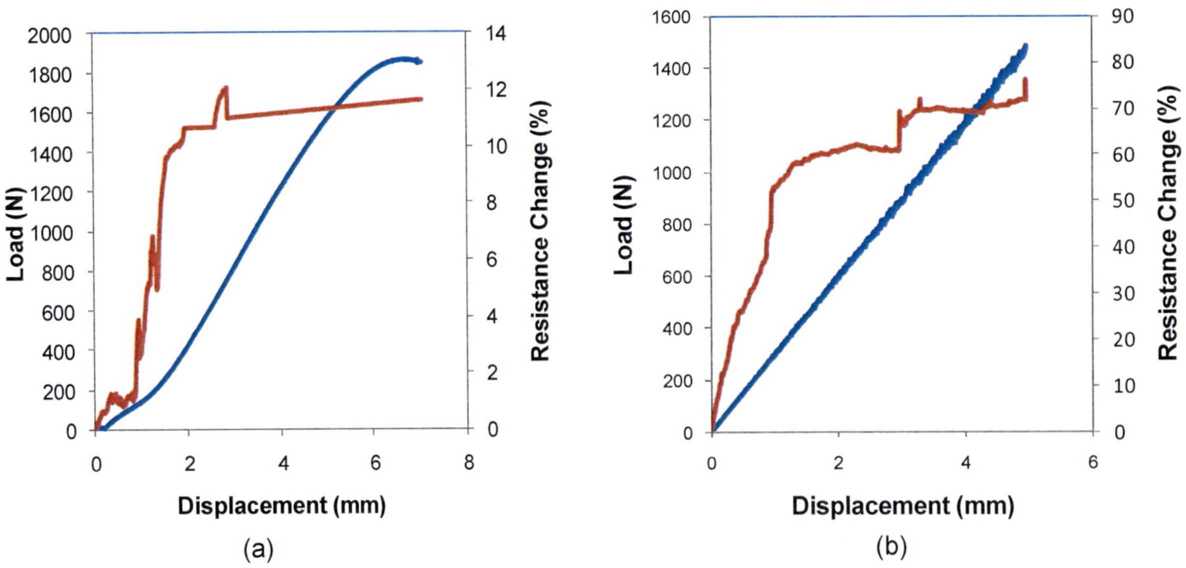

Fig. 2. Percentage resistance change of (a) pure epoxy (b) epoxy reinforced with 10% volume fraction glass bubbles

Acknowledgements

The authors greatly appreciate the financial support from National Science Foundation through the grant number CMMI 0856463.

References

1. Wang S., Mei Z. and Chung D.D.L. "Interlaminar damage in carbon fiber polymer-matrix composites, studied by electrical resistance measurement," *International Journal of Adhesion and Adhesives*, **21**, 465-471, 2001
2. Thostenson E. T. and Chou T-W. "Carbon Nanotube Networks: Sensing of Distributed Strain and Damage for Life Prediction and Self Healing," *Advanced.Materials*,**18**, 2837-2841, 2006.

Proceedings of the SEM Annual Conference
June 7-10, 2010 Indianapolis, Indiana USA

A Test Platform for Systematic Investigation of Tribology in MEMS

N. Ansari[a,*] and W. R. Ashurst[a]
[a]Department of Chemical Engineering, Auburn University, Auburn, AL 36849 USA
*Corresponding author: ansarna@auburn.edu (N. Ansari)

Abstract

Tribology remains an active area of research for the Micro-electromechanical Systems (MEMS) community. At the micron scale, which is the scale relevant to commercial MEMS, a number of factors namely roughness, apparent contact area, surface topography, surface chemistry, etc. are known to have a significant impact on the tribological properties. Historically, researchers have found it difficult to study the effects of these factors individually. We report on a test platform designed and fabricated using a single mask scheme within a relatively smooth SOI wafer. The test platform includes several different micromechanisms on the same chip so that a systematic investigation of the factors that influence tribology in MEMS can be carried out. The test platform is therefore an ideal stage for testing and comparing the various strategies that can be used to address the tribological issues that are presently plaguing the MEMS community.

1. Introduction

In the micro-domain, tribology is influenced by several factors, only some of which are known. Taller asperities being the real points of contact between two surfaces, the real area of contact between them is determined by their roughness, material properties and the shapes of asperities on them. Roughness and topography of contacting surfaces therefore have a strong influence on tribology [1–3]. Both of these factors have their origins in grain orientation and processes used in the fabrication of MEMS, as a result of which, we have very limited control over them [4]. The chemical compositions of the contacting surfaces are also known to have a significant influence on tribology. For example, a hydrophobic surface with a relatively low surface energy is less prone to adhesion [5–7]. Also, environmental factors like high relative humidity can lead to capillary condensation in narrow gaps between hydrophilic surfaces, thereby, making them more prone to adhesion [3]. Studying the effects of each of these factors individually is therefore critical for a precise understanding of tribology.

Historically, the test platforms used to investigate tribology are fabricated using polysilicon structural layers. The inherent roughness of the in-plane (rms roughness is 4.91 nm) and sidewall (rms roughness is 13.8 nm) surfaces of polysilicon structural layers is evident in the AFM images shown in Fig. 1a and Fig. 1b. Further, linescans across these AFM images, shown in Fig. 1a and Fig. 1b indicate that both the in-plane as well as sidewall polysilicon surfaces have significantly high peak to valley ratio. As a result, researchers have found it difficult to study the effects of factors influencing tribology individually since the effects of inherent roughness of the polysilicon structural layers is always coupled with the effects of other factors. Additionally, the material properties of the polysilicon structural layers vary from fabrication to fabrication and sometimes even from chip to chip. Properties like residual stress, Young's modulus and stress gradient are extremely sensitive to the deposition and annealing conditions of polysilicon structural layers and hence these properties sometimes even vary on the same chip [8, 9].

2. The Test Platform

In order to address these issues of inherent roughness and inconsistent material properties of polysilicon structural layers used in conventional test platforms, we have designed a single mask scheme to fabricate a test platform within an SOI wafer. The structural layer of our test platform is a 2 μm thin film of Si(100) on top of a 500 μm thick Si(100) substrate. A 2 μm thin sacrificial BOX oxide is sandwiched between the substrate and the structural layer. Since the substrate and the device layer are both single crystal silicon surfaces, they are relatively much smoother than the polysilicn structural layers used in conventional test platforms. This is clearly evident from the AFM image of the in-plane surface (rms roughness is 0.097 nm) of the test platform reported in this paper, shown in Fig. 2a. A linescan

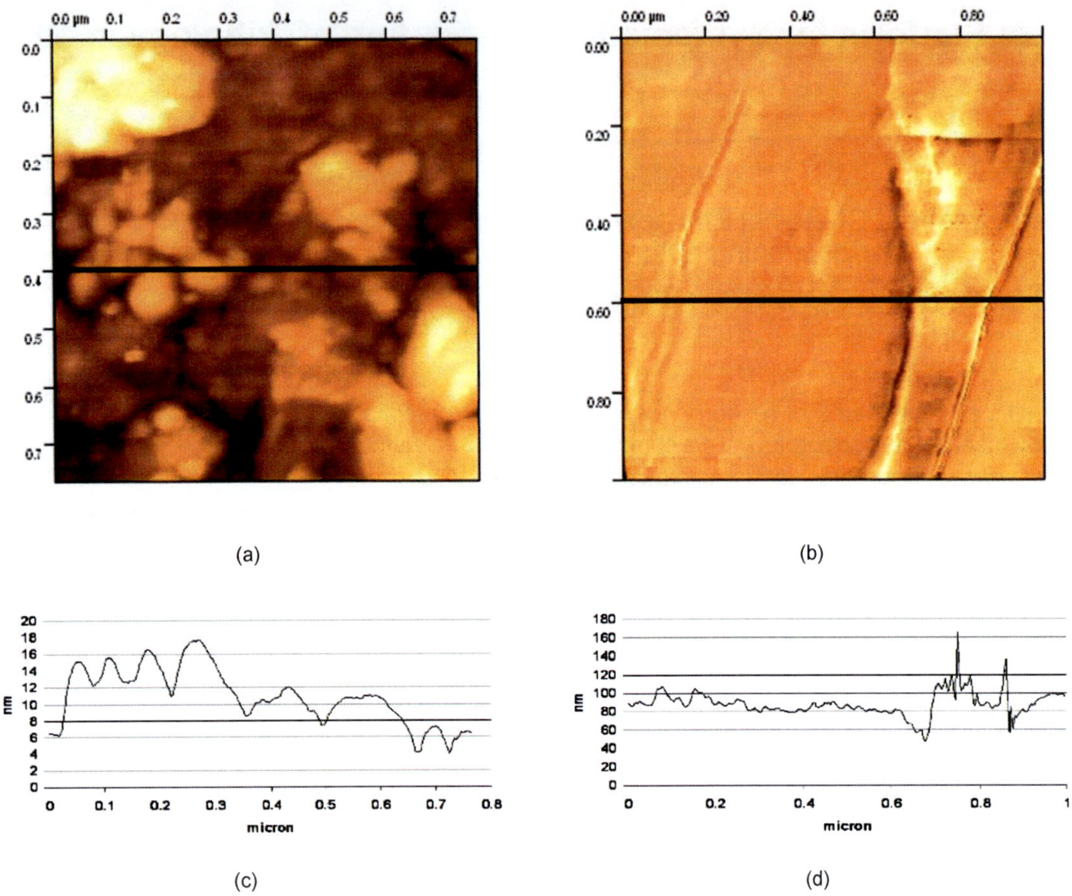

Figure 1: (a) AFM image of an in-plane polysilicon surface (b) AFM image of a sidewall polysilicon surface (c) Linescan across image a (d) Linescan across image b)

across this AFM image of the in-plane surface shows significantly reduced peak to valley ratio. It must be noted that a large fraction of this already small peak to valley ratio is due to the inherent noise of the AFM. The sidewall surface (rms roughness is 6.52 nm) of the test platform reported in this paper has a peak to valley ratio higher than that of its in-plane surface, but it is still lesser than that of a polysilicon sidewall surface. Also, the material properties of single crystal silicon surfaces do not vary much. Furthermore, to eliminate the inconsistencies resulting from the fabrication process variations, the test platform includes several different micromechanisms so that a complete systematic investigation of all the tribological properties can be carried out on the same chip.

The Test Platform (TP) reported in this paper is fabricated using a single mask scheme and hence is facile and inexpensive to fabricate. Also, the time required to fabricate TP is significantly lesser than that required to fabricate most of the conventional test platforms. In essence, TP is an extremely useful and versatile tool, which on one hand saves a lot of time and money while on the other hand facilitates a systematic investigation of several tribological properties critical to MEMS. TP has micromechanisms that can be used to determine several material properties like the residual stress in a released structural layer. The fracture strength and the Young's modulus of the structural layer can also be determined.

Conventionally, most of the adhesion studies are conducted on in-plane surfaces although, it is the sidewall surfaces that often come into contact in most useful MEMS. Additionally, it is clearly evident from the AFM images shown in Fig. 2a and Fig. 2b that the in-plane and sidewall surfaces are significantly different topographically. They are also expected to be different chemically. These differences are a result of different processing environments to which these surfaces have been exposed during fabrication. Since the in-plane and sidewall surfaces are topographically

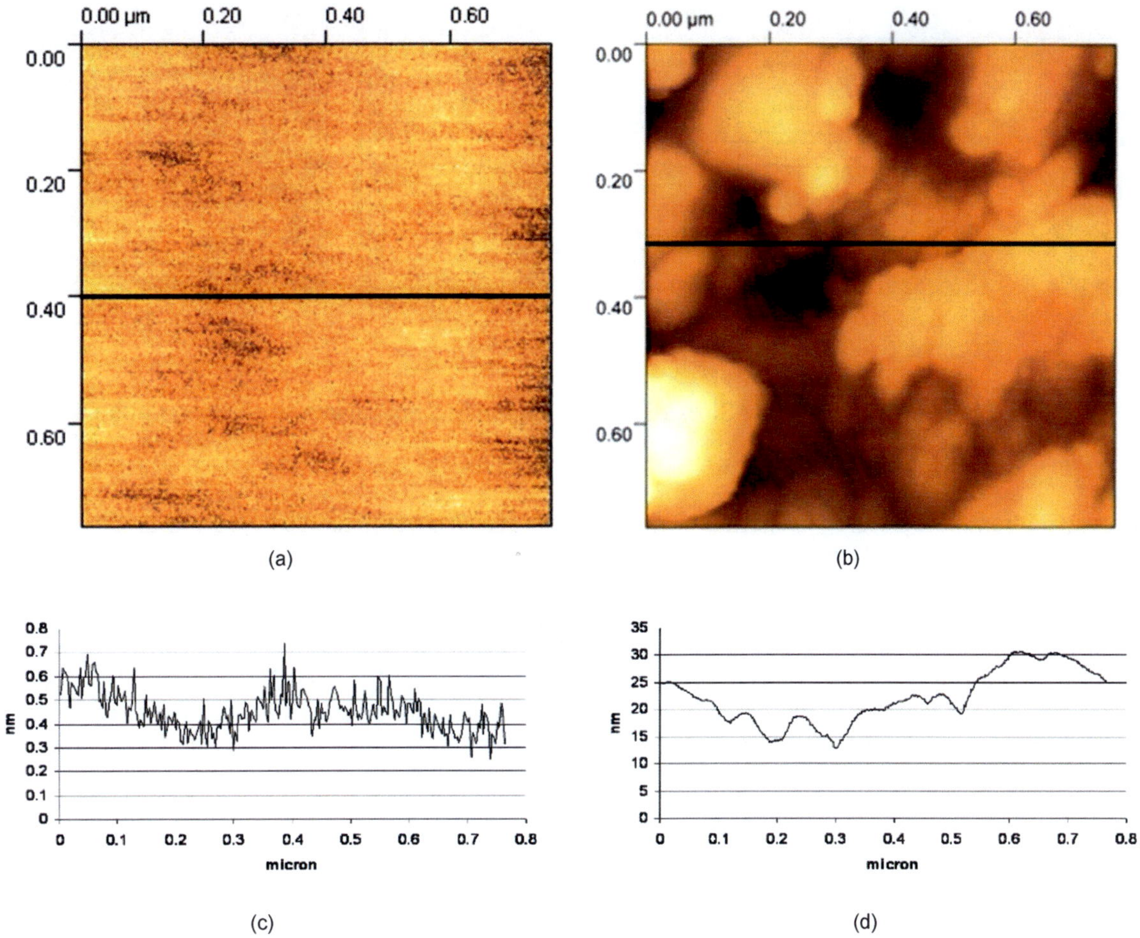

Figure 2: (a) AFM image of an in-plane surface of the test platform(b) AFM image of a sidewall surface of the test platform (c) Linescan across image a (d) Linescan across image b

and chemically different, different microinstruments are needed to study tribology between them. TP conveniently addresses this need. It has microinstruments to study adhesion between both in-plane as well as sidewall surfaces. It also has three different microinstruments that can be used to study and compare friction and wear between sidewall surfaces.

3. Microinstruments

3.1 Material Properties

TP has microinstruments that can be used to determine certain material properties of the structural film. Figure 3a is an optical image of a Residual Stress Tester (RST), which when released can be used to determine the residual stress in the structural film. Depending on the nature of residual stress in the structural film (i.e., tensile or compressive), the suspended beam shown in Fig. 3a either rotates clockwise or anti-clockwise respectively on getting released. By optically matching the patterns shown in Fig. 3b, the rotation (θ) of the released suspended beam can be accurately determined. The residual stress (σ_R) in the released structural film is determined using Eq. 1, where z is a constant that depends on the material properties of the structural film and the physical dimensions of the suspended beam, w is the width of the supporting suspension beams that connect the suspended beam to the anchors, t is the thickness of the structural film and d as shown in Fig. 3a is the distance between the two supporting suspension beams.

$$\sigma_R = z(\frac{\theta}{wtd}) \tag{1}$$

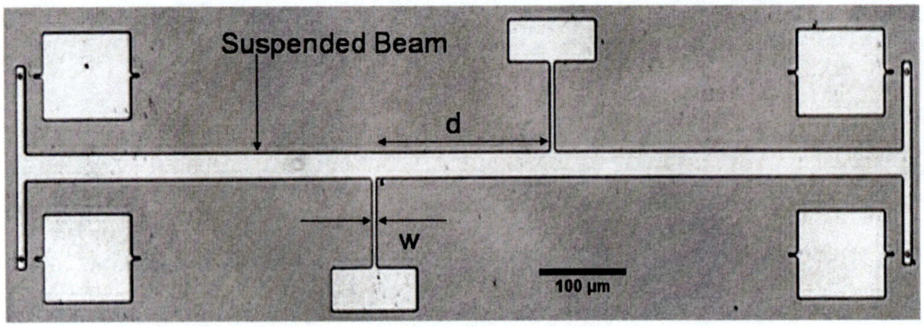

(a)

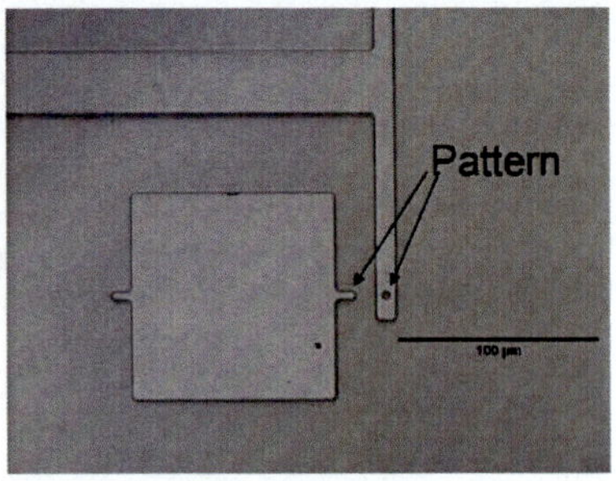

(b)

Figure 3: (a) Optical image of the Residual Stress Tester (b) Magnified optical image showing patterns that are used for pattern matching

The Mechanical Strength Tester (MST) shown in Fig. 4a can be used to determine the fracture strength of the structural film. The shuttle, which is suspended using four suppporting beams is pushed in the direction indicated by the arrow in Fig. 4a using a Signatone SE-20T probe tip, which is controlled using an open loop peizo. As the displacement of the shuttle increases, the beams that are protruding out of the shuttle are pushed against the anchored stopper blocks until they break-off from the shuttle. Optical pattern matching is used to determine the displacement of the shuttle. The fracture strength (σ_{max}) of the structural film is calculated using Eq. 2, where E is the Young's modulus of the structural film, w is the width of the fractured beam, δ is the distance by which the shuttle is displaced when the beam gets fractured and L_c is the distance between the suttle and the anchored stoppers. Each MST has three sets of two beams each with each set of beams having a different length than the other two sets.

$$\sigma_{max} = \frac{3Ew\delta}{2L_c^2} \tag{2}$$

Young's modulus (E), which is one of the most critical material properties of the structural film, can be determined using the resonators. The resonators are actuated elecrostatically by applying an AC voltage across one set of comb fingers and grounding the other set of comb fingers as well as the suspended resonating structure. Resonance is determined optically with an accuracy of $\pm$ 10 Hz. The resonance frequency (f_R) of a resonator is given by the Rayleigh equation (i.e., Eq. 3 given below), where k_x is the stiffness of the folded suspension beams supporting the resonating structure and it depends on E of the structural film and the true dimensions of the folded supporting suspension beams and M_{eff} is the effective mass of the resonating structure. The rearranged form of the Rayleigh

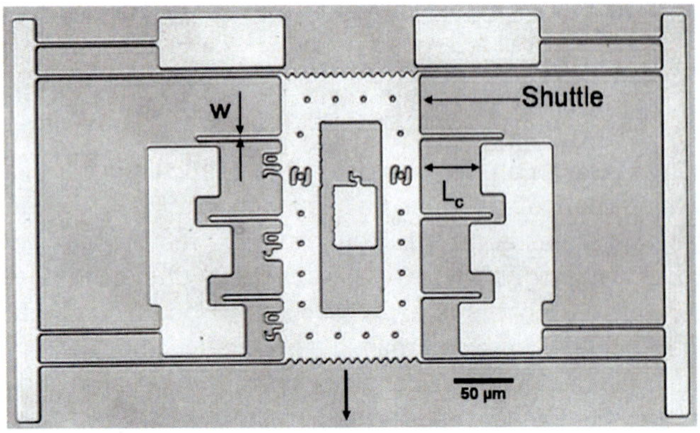

(a)

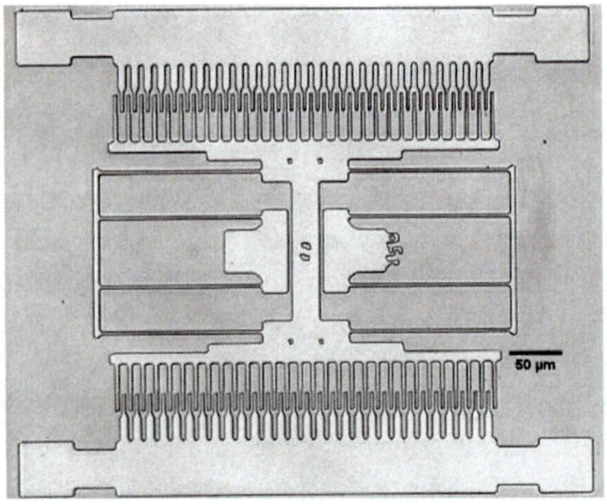

(b)

Figure 4: (a) Optical image of the Mechanical Strength Tester (b) Optical image of a Resonator equation given by Eq. 4 can be used to extract the Young's modulus of the structural film, where t is the thickness of the structural film and w and L are width and length of the folded suspension supporting beams respectively. TP has resonators with seven different folded suspension beam lengths and there are multiple copies of each one of them.

$$f_R = \frac{1}{2\pi}\sqrt{\frac{k_x}{M_{eff}}} \tag{3}$$

$$f_R^2 = (\frac{tw^3}{2\pi^2 M_{eff} L^3})E \tag{4}$$

3.2 Adhesion

Apparent work of adhesion ($\mathcal{W}_{ip}$) between in-plane surfaces can be accurately determined using the Cantilever Beam Array (CBA) shown in Fig. 5a. Each beam in the array is actuated manually by pushing it with a sharp tungsten probe tip until it contacts the substrate, after which the probe tip is retracted carefully and slowly, which causes the beam to peel apart up to a certain characteristic length called as the crack length (s). The crack length depends on the interfacial properties of the two contacting surfaces. The crack length of each actuated beam is determined optically

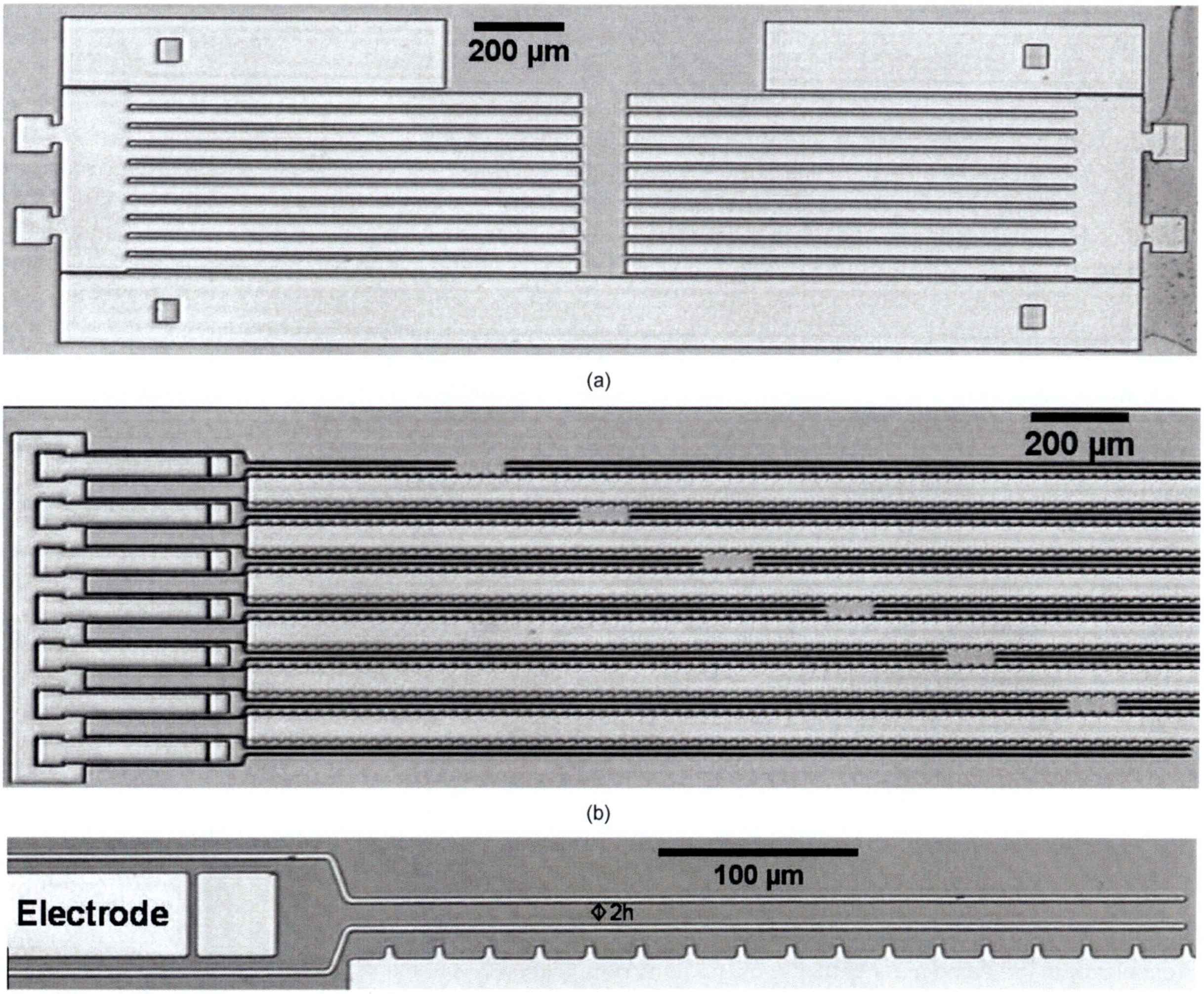

(a)

(b)

(c)

Figure 5: (a) Optical image of a Cantilever Beam Array (b) Optical image of a Sidewall Beam Array (c) Magnified optical image of a single pair of sidewall beams using interferometry. Apparent work of adhesion between the substrate and the bottom surface of the structural film can be determined using Eq. 5 suggested by Hurst et. al., where E and t are the Young's modulus and thickness of the structural film respectively, s is the crack length and a_1 and a_2 are parameters extracted from the fitted height profile of the actuated cantilever beam, which is obtained eperimentally using phase shifting interferometry [10]. Instead of cantilever beams with increasing lengths, the CBA's in TP have 10 cantilever beams each, all of the same length. This allows estimation of 10 distinct data points from each CBA, which not only saves a lot of space on TP but also enables a statistical analysis of the local adhesion. To preferentially use S-shaped cantilever beams, which provide a more accurate estimation of $\mathcal{W}_{ip}$ for determining $\mathcal{W}_{ip}$ between most of the available in-plane surfaces, TP has three sets of eight CBA's each. The lengths of cantilever beams in those three sets are 500 μm, 1000 μm and 2000 μm respectively.

$$\mathcal{W}_{ip} = \frac{3Et^2}{2s^4}\left(a_1^2 + a_1 a_2 + \frac{1}{3}a_2^2\right) \tag{5}$$

The Sidewall Beam Array (SBA) shown in Fig. 5b can be used to determine the apparent worh of adhesion between sidewall surfaces ($\mathcal{W}_s$). There are 8 SBA's in TP and each one of them has seven sidewall beam pairs. The shortest sidewall beam pair has a length of 750 μm and each successive sidewall beam pair is 250 μm longer.

$$\mathcal{W}_s = \frac{3Et^3h^2}{s^4} \tag{6}$$

Each sidewall beam pair is actuated electrostatically. A DC voltage is applied to the electrode shown in Fig. 5c, while the sidewall beams and the substrate are grounded. As the voltage applied to the electrode is increased from 0 V at a rate of 1 V/s, the two sidewall beams are drawn together like tweezers. At 90 V, which is the maximum voltage applied, the beams are in contact over much of their lengths, leaving a length of about 150 μm near the electrode. The actuation voltage is then ramped down to 0 V at 1 V/s and the beams are observed to peel apart up to a certain characteristic length called as the crack length (s), which depends on the interfacial properties of the two contacting surfaces. Aapparent work of adhesion between sidewall surfaces is determined using Eq. 6, where E is the Young's modulus of the structural film, t is the width of a sidewall beam, h is half the distance between the unactuated sidewall beam pair and s is the crack length which is determined optically [7]. It should be noted that both the CBA and SBA only determine the apparent work of adhesion between contacting surfaces since all the contacting surfaces have a finite roughness and topography.

3.3 Friction

The device shown in Fig. 6a is a Sidewall Adhesion and Friction Tester (SFAT), which is a modified version of the device used by Timpe et. al. [11]. The SFAT ia a very versatile tool, which can be used to study static as well as dynamic friction between sidewall surfaces. It can also be used to study adhesion as well as wear between sidewall surfaces. It is actuated electrostatically. The substrate, suspended structures in both the arms and the unused set of comb fingers are grounded. By applying a DC voltage across one of the two sets of comb fingers in the normal arm, the normal arm is either loaded onto the contact block that protrudes out of the tangential arm or is pulled away from it. Similarly, by applying a DC voltage to one of the two sets of comb fingers in the tangential arm, the tangential can be rubbed against the normal arm.

$$\mu_s = \frac{F_f}{F_n} \tag{7}$$

Figure 6b is an optical image of another type of friction tester called as the Sidewall Friction Tester (SFT), which can also be used to study static friction, dynamic friction and wear between sidewall surfaces. The SFT is also actuated electrostatically in the same manner as the SFAT. However, unlike the contacting sidewall surfaces of SFAT, which are plane surfaces, the contacting sidewall surfaces in SFT are a pair of a plane and a cylindrical surface. The normal arm in the SFT loads a suspended beam structure onto an anchored post structure and the tangential arm slides the beam against the post. By appropriately pre-caliberating the normal and tangential arms in both the SFAT and SFT, the normal force F_n as well as the tangential force F_f applied by the two arms respectively at the onset of slip between

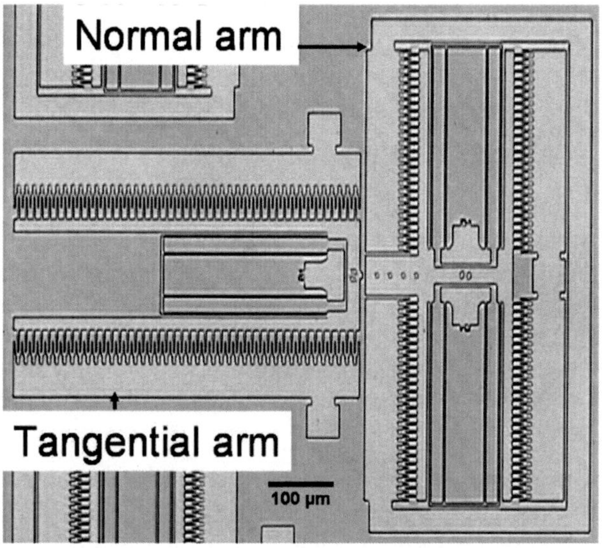

(a)

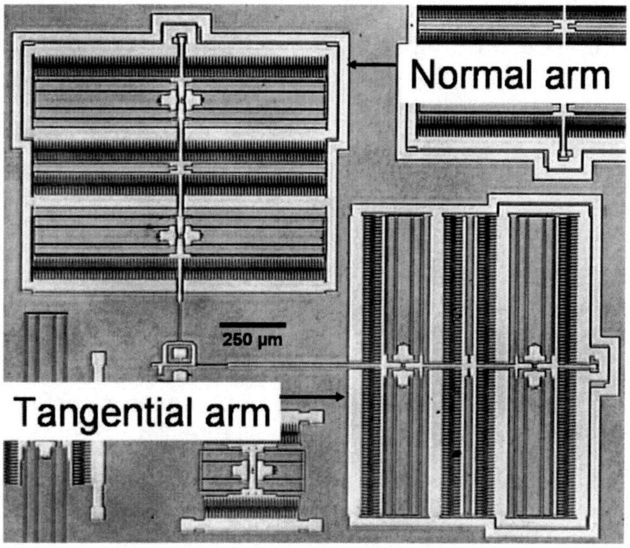

(b)

Figure 6: (a) Optical image of a Sidewall Adhesion and Friction Tester (b) Optical image of a Sidewall Friction Tester

the contacting sidewall surfaces can be accurately determined. The onset of slip is detected optically using a pattern matching technique and the coefficient of static friction (μ_s) is calculated using Eq. 7.

4. Fabrication and Testing procedure of Microinstruments

Standard surface micromachining technology is used to fabricate the TP. Since the TP is fabricated within an SOI wafer, which has only one structural layer, it can be fabricated using a single mask scheme. In addition to avoiding the errors associated with mask misallignment and making the fabrication procedure extremely facile, the single mask scheme designed by us also makes it extremely inexpensive. The fabricated TP is released by etching in a solution of conc. HF (49 wt.%). After all the microinstruments on the TP are released, the etchant is completely rinsed away with deionized (DI) water. The TP is then placed in hot H_2O_2 (75-80 °C) for 10 min. for oxidizing the test surfaces after which it is rinsed with DI water again.

Two types of surfaces are studied in this paper, namely hydrophobic surfaces coated with OTS SAM coating and hydrophilic surfaces having only native oxide on them. To obtain test surfaces having only native oxide on them, DI water is rinsed away with isopropanol (IPA) and the TP is supercritically dried in a Tousimis Autosamdri 815-B CPD system. Samples dried in this CPD system are known to have contaminated surfaces. In order to clean the test surfaces before testing, the TP is treated with a low power (25 W), capacitively coupled RF (13.56 Mhz) plasma of oxygen (O_2 pressure used is 300 mtorr). This treatment results in clean and oxidized test surfaces, including surfaces underneath the microinstruments. Water Contact angle measured on a simultaneously processed Si(100) test chip after this treatment is found to be $<5°$, indicating that the treated surfaces are hydrophilic. The TP is actuated immediately after plasma cleaning to minimize contamination from the ambient. To obtain test surfaces coated with OTS SAM coating, IPA is further rinsed away with anhydrous hexane. The TP is then transferred into a conditioned OTS coating solution in a glass dish which had been previously treated with OTS. After 45 mins, the TP is transferred back to pure anhydrous hexane, which is subsequently displaced with IPA and the IPA is similarly displaced with DI water. The TP is then removed from DI water in one, slow but continuous motion in such a way that its in-plane surfaces are perpendicular to the liquid surface. All the solvent displacements are carried out using a fill/drain approach suggested by Ashurst et. al. [7]. After this treatment, the water contact angle on a simultaneously processed Si(100) test chip is measured to be 108°, indicating that the treated surfaces are hydrophobic.

Electrical contacts to the microinstruments are made by touching the actuation pads (fabricated in the microinstruments) with sharp tungsten probe tips. Interferometry using monochromatic green light (532 nm) is employed to determine the out-of-plane height profile.

5. Results and Discussion

In order to test the validity of the performance of the TP, the performance is tested on two different chips. The test surfaces on one of the chips (referred as chip A) are coated with OTS SAM coating and the test surfaces on the other chip (referred as chip B) have only native oxide on them. The tribological properties of these two types of surfaces are considered as benchmarks by the MEMS community. The following sections will individually discuss the performance of the different microinstruments on the test platform.

5.1 Material Properties

For both chip A and chip B, optical images of the RST are collected before and after releasing it. By applying the pattern matching procedure to these optical images, it is found that the suspended beams do not exhibit any rotation. This result implies that the structural films on both the chips A and B do not have any residual stress in them, as expected. Also, as expected, the OTS SAM coating does not induces any stress in the Si(100) structural film, which is an indication that either the OTS SAM coating is conformal and uniform or it does not have any significant influence on the stress state of the structural Si(100) film.

On each of the two chips A and B, 16 MST's are tested. The width of each of the tested beams is determined optically to be 3.76 μm (design width is 4 μm). Table 1 lists the displacement at fracture of the tested beams along with the strain at fracture in the tested beams. Each reported value is an average of 16 data points. The mean values reported in Table 1 indicate that consistent results are obtained irrespective of the length of the tested beam. Also, as expected, the OTS SAM coating does not change the fracture strength of the Si(100) structural layer. The results obtained from both chip A and chip B are in good agreement with the values of fracture strength reported for Si(100) films in literature [12–15].

Seven resonators, each having a different folded suspension beam length are used to determine the Young's modulus (E) of the structural film on chip B. Table 2 lists the folded suspension beam lengths along with the corresponding observed resonance frequencies of the seven resonators used. Each resonance frequency reported in Table 2 is an average of three data points collected using three distinct resonators with same folded suspension beam lengths. By fitting the rearranged form of the Rayleigh equation given by Eq. 4 to the values reported in Table 2, the value of E for the structural film on chip B can be extracted. The slope of the fit shown in Fig. 7 (i.e., 129 GPa), which is the experimentally obtained value of E for the structural film on chip B, is in good agreement with that of a Si(100) film published in literature (i.e., 130 GPa) [9].

5.2 Adhesion

Four CBA's consisting of ten beams each are actuated on both chips A and B to determine the apparent work of adhesion between in-plane surfaces. The experiments are conducted in laboratory air at room temperature (i.e. 22°C) and a relative humidity (RH) of 13%. In-plane apparent work of adhesion reported in Table 3 for both the surface treatments are higher than those reported by Ashurst et. al. [6, 7]. Even though their experiments were conducted at a higher RH of 50%, they reported a lower in-plane apparent work of adhesion of 0.012 mJ/m^2 and 13.1 mJ/m^2 for OTS coated polysilicon surfaces and polysilicon surfaces with only native oxide on them respectively. The high $\mathcal{W}_{ip}$ exhibited by in-plane surfaces of both chip A and chip B even at a lower RH can be attributed to the smoother topography of the test surfaces of the TP reported in this paper as compared to the test surfaces used by Ashurst et. al. Two CBA's actuated on chip B under the same conditions except at a higher RH of 58% exhibited $\mathcal{W}_{ip}$ of 139 mJ/m^2. The apparent work of adhesion exhibited by the in-plane surfaces on chip B at high RH is very close to the theoretical limiting value of 144.96 mJ/m^2, which corresponds to 2γ, where γ is the surface tension of water (72.48 dynes/cm at 22 °C). This further substantiates the extremely smooth topography of the in-plane test surfaces of the TP reported

Beam Length (μm)	Displacement at Fracture $\delta_f \pm$ std.dev. (μm)		Strain at Fracture $\epsilon_f \pm$ std.dev. (%)	
	Chip A	Chip B	Chip A	Chip B
50	8.51 ± 0.49	8.48 ± 0.63	1.92 ± 0.11	1.91 ± 0.14
60	12.24 ± 1.24	12.30 ± 0.82	1.92 ± 0.19	1.93 ± 0.13
70	17.49 ± 0.39	17.48 ± 0.35	2.01 ± 0.05	2.01 ± 0.04

Table 1: Displacement at fracture of tested beams and strain at fracture tested beams on chip A and chip B

Supporting Beam Length (μm)	Resonance Frequency $f_R \pm$ std.dev. (KHz)
200	13.50 $\pm$ 0.01
250	9.15 $\pm$ 0.01
300	8.34 $\pm$ 0.01
350	6.14 $\pm$ 0.01
400	4.72 $\pm$ 0.01
450	3.92 $\pm$ 0.01
500	3.64 $\pm$ 0.01

Table 2: Resonance frequencies of resonators with different folded suspension beam lengths.

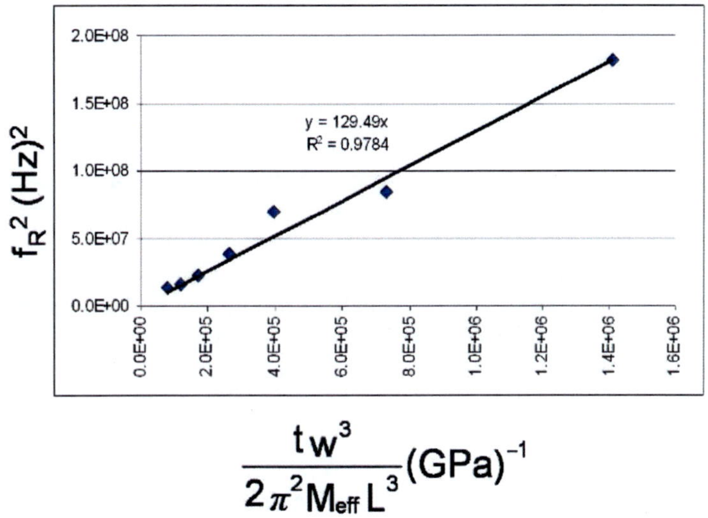

Figure 7: Plot exhibiting the fitting of Eq. 4 to the values reported in Table 2. The slope of the fitted line is the experimentally obtained value of E for the structural film on chip B.

in this paper. To determine the apparent work of adhesion between sidewall surfaces, two SBA's are actuated on both chip A and chip B in laboratory air at room temperature and a RH of 13%. Apparent work of adhesion between sidewall surfaces on both chip A and chip B as reported in Table 3 is less than that between the corresponding in-plane surfaces. This substantiates the role of roughness in reducing adhesion, since the sidewall surfaces (rms roughness is 6.52 nm) on both the chips have a rougher topography than the corresponding in-plane surfaces (rms roughness is 0.097 nm).

Chip	Apparent In-plane Work of Adhesion $\mathcal{W}_{ip} \pm$ std.dev. (mJ/m^2) (RH = 13%)	Apparent Sidewall Work of Adhesion $\mathcal{W}_s \pm$ std.dev. (mJ/m^2) (RH = 13%)
A	0.02 $\pm$ 0.001	$<$0.04
B	22.92 $\pm$ 2.5	1.065 $\pm$ 0.07

Table 3: Adhesion energies between in-plane and sidewall surfaces on chip A and chip B.

5.3 Friction

All the experiments to determine the coefficients of static friction (μ_s) between sidewall surfaces are conducted in laboratory air at room temperature and a RH of 13%. Three SFAT's and three SFT's are actuated on both chip A and chip B. As seen in Table 4, for sidewall surfaces on both chip A and chip B, the static friction coefficients determined using both the SFAT and SFT are in good agreement. However, Ashurst et. al. reported static friction coeffcients of 0.07 $\pm$ 0.005 and 1.1 $\pm$ 0.1 for OTS coated surfaces and surfaces with only native oxide on them respectively [6]. The lower μ_s reported by Ashurst et. al could be attributed to the rougher topography of the polysilicon sidewall surfaces (rms roughness is 13.8 nm) of their test devices in comparison to the sidewall surfaces (rms roughness is 6.52 nm) of the TP reported in this paper. On the contrary, in-plane static friction coefficients of 0.09 $\pm$ 0.01 and 2.3 $\pm$ 0.8 for OTS coated polysilicon surfaces (rms roughness is 5 nm) and polysilicon surfaces having only native oxide on them (rms roughness is 5 nm) respectively, reported by Srinivasan et. al. are in good agreement with the sidewall static friction coefficients determined using the TP reported in this paper [5]. These results clearly emphasize the role of roughness in friction between contacting surfaces.

Chip	$\mu_s \pm$ std.dev.	
	SFAT	SFT
A	0.17 $\pm$ 0.03	0.15 $\pm$ 0.02
B	1.98 $\pm$ 0.4	1.69 $\pm$ 0.3

Table 4: Coefficients of static friction between sidewall surfaces (rms roughness is 6.52 nm) of chip A and chip B determined using both SFAT and SFT.

6. Conclusions

We designed a single mask scheme for facile and inexpensive fabrication of a micromachine test platform, which enables the test platform to be built in different types of structural layers. The test platform is successfully fabricated and shown to be a versatile tool that enables systematic investigations of several tribological properties. The micromachined surfaces of the test platform are the smoothest test surfaces used to-date. This attribute of the test platform can be used to systematically investigate the effects of roughness and topography on tribological properties by tailoring the roughness and topography of the test surfaces. The test platform is an ideal stage for testing and comparing the various strategies that can be used to address the tribological issues that are presently plaguing the MEMS community.

7. Acknowledgments

The authors wish to thank the Electrical & Computer Engineering department at Auburn university for allowing them to use the AMSTC facilities to fabricate the test platform. The authors also gratefully acknowledge the financial support of Auburn University.

References

[1] L. Phinney, G. Lin, J. Wellman, A. Garcia, Surface roughness measurements of micromachined polycrystalline silicon films, Journal of Micromechanics and Microengineering, 14, 927–931, 2004.

[2] R. Alley, G. Cuan, R. Howe, K. Komvopoulos, The effect of release etch processing on surface microstructure stiction, In Proc. of IEEE Solid State Sensor and Actuator Workshop, Hilton Head '92, 202–207, 1992.

[3] M. P. de Boer, T. A. Michalske, Accurate method for determining adhesion of cantilever beams, Journal of Applied Physics, 86 (2), 817–827, 1999.

[4] W. R. Ashurst, M. P. de Boer, C. Carraro, R. Maboudian, An investigation of sidewall adhesion in MEMS, Applied Surface Science, 212, 735–741, 2003.

[5] U. Srinivasan, J. Foster, U. Habib, R. T. Howe, R. Maboudian, D. Senft, M. Dugger, Lubrication of polysilicon micomechanisms with self-assembled monolayers, In Proc. of the 1998 Solid - State Sensor and Actuator Workshop, Hilton Head '98, 156–161, 1998.

[6] W. R. Ashurst, C. Yau, C. Carraro, R. Maboudian, M. T. Dugger, Dichlorodimethylsilane as an anti-stiction monolayer for MEMS: A comparison to the octadecyltrichlosilane self-assembled monolayer, Journal of Microelectromechanical Systems, 10 (1), 41–49, 2001.

[7] W. R. Ashurst, C. Carraro, R. Maboudian, W. Frey, Wafer level anti-stiction coatings for MEMS, Sensors and Actuators A-physical, 104 (3), 213–221, 2003.

[8] M. de Boer, T. Mayer, Tribology of MEMS, MRS Bulletin, 302–304, 2001.

[9] B. D. Jensen, M. P. de Boer, N. D. Masters, F. Bitsie, D. A. LaVan, Interferometry of actuated microcantilevers to determine material properties and test structure nonidealities in mems, Journal of Microelectromechanical Systems, 10 (3), 336–346, 2001.

[10] K. M. Hurst, C. B. Roberts, W. Ashurst, A New Method to Determine Adhesion of Cantilever Beams Using Beam Height Experimental Data, Tribology Letters, 35, 9–15, 2009.

[11] S. J. Timpe, K. Komvopoulos, Microdevice for measuring friction and adhesion properties of sidewall contact interfaces of Micro-electromechanical Systems, Review of Scientific Instruments, 78 (6), 065106, 2007.

[12] C. Chen, M. Leipold, Fracture toughness of silicon, Ceramic Bulletin, 59 (4), 1980.

[13] J. F. Nye, Physical Properties of Crystals, Oxford University Press, Oxford, 1985.

[14] R. Ballarini, R. L. Mullen, Y. Yin, H. Kahn, S. Stemmer, A. H. Heuer, The fracture toughness of polysilicon microdevices: A first report, Journal Of Materials Research, 12 (4), 915–922, 1997.

[15] H. Kahn, N. Tayebi, R. Ballarini, R. L. Mullen, A. H. Heuer, Fracture toughness of polysilicon mems devices, Sensors And Actuators A-Physical, 82 (1-3), 274–280, 2000.

Proceedings of the SEM Annual Conference
June 7-10, 2010 Indianapolis, Indiana USA

Full Optical Scatter Analysis for Novel Photonic and Infrared Metamaterials

Thomas M. Fitzgerald[1] and Michael A. Marciniak[2]
Department of Engineering Physics
Air Force Institute of Technology
Wright-Patterson AFB OH, USA
[1]thomas.fitzgerald@afit.edu
[2]michael.marciniak@afit.edu

ABSTRACT

Artificial structures with sub-optical wavelength features are engineered to demonstrate material properties for optical and infrared permeability and permittivity not otherwise found in nature. Such artificial structures are referred to as optical and infrared metamaterials. The application space of electromagnetic metamaterials includes novel sub-wavelength waveguides and antennas, true time delay devices, optical filters, plasmonic electronic-optical interfaces, optical limiters and thermal management systems.

In this paper, we present an optical diagnostic technique adapted for measuring and analyzing bidirectional polarimetric scatter from novel optical and infrared metamaterials of interest. This optical diagnostic technique is also broadly applicable to other optical/infrared metamaterial structures as well as other nanostructures such as plasmonic devices or photonic crystals that may be proposed or developed in the future.

The specific project goals are

a) Demonstrate a novel metamaterial characterization full-polarimetric diffuse ellipsometry technique suitable to measure desired material properties with stated uncertainty limits for novel optical and infrared metamaterials of interest.

b) Demonstrate incorporation of predictive computational codes that estimate the electro-magnetic property values for metamaterial designs and concepts of interest.

I. INTRODUCTION

Artificial structures with sub-optical wavelength features can have engineered non-conventional values for material properties such as optical and infrared permeability and permittivity. Such artificial structures are referred to as optical metamaterials.[1] Engineering macroscopic optical/infrared metamaterial property values such as permittivity and permeability is facilitated by access to mature semiconductor processing techniques and advanced computer processing. Manipulation of permittivity and permeability values through novel structure geometry throws open wide the design space for electro-optical devices and enables tailored values of material optical impedance and refractive index, and even negative index material development.

The application space of engineered metamaterials includes sub-wavelength waveguides and antennas, true time delay devices, optical filters, plasmonic electronic-optical interfaces, optical limiters and thermal management systems.[1] Metamaterials development is an exciting research area and is deservedly getting a lot of attention. However, much work remains to be accomplished in the area of optical metamaterial performance characterization.

This research effort is focusing on the development of a novel optical diagnostic technique for measuring and analyzing full-angle optical scatter from metamaterials of interest. The approach is an evolutionary continuation of the work by Germer et al at NIST. Germer has published significant work on the employment of the model based SCATMECH scatter library for characterization of surface roughness.[2-5]

T. Proulx (ed.), *MEMS and Nanotechnology, Volume 2*, Conference Proceedings of the Society for Experimental Mechanics Series 2,
DOI 10.1007/978-1-4419-8825-6_15, © The Society for Experimental Mechanics, Inc. 2011

We are extending the previous body of work to metamaterials research. We are incorporating appropriate numerical scatter models for the as built optical metamaterial architecture to include the effects of fabrication defects. We anticipate improving the understanding of loss mechanisms that effect optical metamaterial performance in order to bridge the gap between metamaterial design performance and metamaterial "as-built" performance and to verify that simulated metamaterials are behaving like fabricated metamaterials.

The specific goals of the AFIT team are to:

a) Demonstrate a novel metamaterial characterization full-Polarimetric diffuse ellipsometry technique suitable to measure desired material properties with stated uncertainty limits for novel optical and infrared metamaterials of interest.

b) Demonstrate incorporation of predictive computational codes that estimate the electro-magnetic property values for metamaterial designs and concepts of interest.

Metamaterial property values for permittivity and permeability are used to derive macro property values such as impedance and index of refraction. Full scatter angle Mueller Matrix measurements capture the complete interaction of a material with incident energy. The measured Mueller matrix stores the degree of attenuation, de-polarization, and retardance for all angles of interest. The currently accepted technique for measuring metamaterial permittivity and permeability typically uses normal incidence polarized light to determine the reflection and transmission coefficients of the metamaterial. Then the permittivity and permeability are estimated.[6] [7][8]

We have modified a Schmitt Measurement Systems (SMS) Complete Angle Scatter Instrument (CASI) to a Dual Rotating Retarder (DRR) configuration in order to measure all Mueller scatter coefficients for out-of-plane transmission and reflection of a metamaterial of interest. We use the polarized angle scatter reflectance and transmission coefficients to determine the effective permeability and permittivity of a metamaterial. These coefficients are compared with prediction using standard measurement uncertainty techniques.

Mueller matrix ellipsometry requires strong predictive models in order to properly interpret the results. We have incorporated the Direct Dipole Approximation (DDA) based DDSCAT computational model to develop scatter predictions for the metamaterial geometry of interest. We are developing these models to understand the CASI-DRR measured perturbations and deficiencies in material properties due to imperfect fabrication and layer defects. The finite element model will provide absorption and scattering coefficients for the desired geometry based on a selected input $E(k,\lambda)$ where k gives the incident k-vector and λ is the optical wavelength in air. Permittivity and permeability can then be derived from the calculated reflection and transmission coefficients.[6] [7]

Our approach to metamaterials property prediction, structure fabrication, and optical measurement will be broadly applicable to other optical/infrared metamaterial structures that are being developed now and in the future.

II. DESCRIPTION OF THE APPROACH

a. DESCRIPTION OF THE CASI-DRR INSTRUMENT

We have modified an available in-house SMS CASI Bi-directional Reflectance Distribution Function (BRDF) measurement system to the DRR configuration proposed by Azzam and developed by Chipman.[9, 10] To achieve complete angle DRR configuration, we have added four additional motion control channels to our CASI instrument. These channels automate the rotational motion of two linear polarizers and two quarter wave plates. The representative optical layout for our instrument is shown in Figure 1.

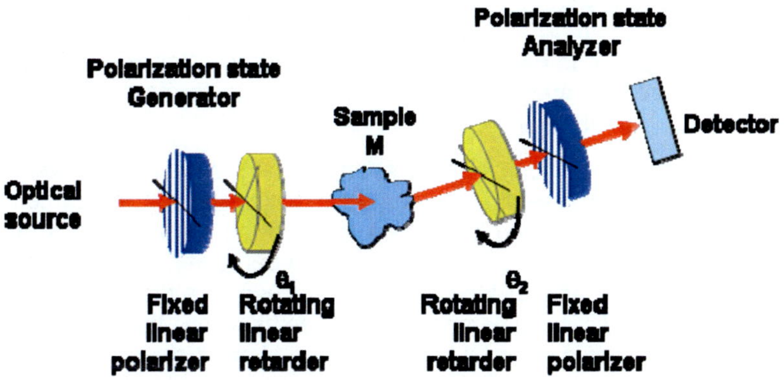

Figure 1. Optical configuration for AFIT DRR

Our DRR addition consists of an input Polarization State Generator and an output Polarization State Analyzer. The Generator and Analyzer stages each feature a linear polarizer and a linear retarder. The linear retarders are rotated to produce and analyze complete polarization states. The Analyzer stage features rotational motion in a horizontal plane about the sample. Sample M has full 6-Degree of Freedom mobility (DOF) necessary for full scatter characterization.

We have rigorously applied the condition number DRR instrument analysis approach proposed by Smith to determine instrument operating mode.[11] In the Mueller algebra a DRR instrument can be described by

$$I = (1\;0\;0\;0)\Pi_a \Delta_a M_s \Delta_g \Pi_g S^T \qquad (1)$$

Where *I* is the measured intensity or Stokes S_0 parameter of the output polarization state mathematically selected by $(1\;0\;0\;0)$, Π is the Mueller representation for a polarizer at angle θ with respect to the polarization axis of interest, Δ is the Mueller representation for a waveplate with δ phase retardation, M is the sample of interest and S is the Stokes vector of the input light. Equation 1 can be recast into the W-Matrix form as

$$I = WM_s \qquad (2)$$

W is an [N x16] matrix that fully describes the instrument configuration for any desired Generator-retarder/Analyzer retarder combination. N is the number of independent measurements necessary to return a full Mueller matrix. Smith's condition number approach to W-Matrix analysis allows determination of both N and suitable angle-increment pairs for rotation of the linear retarders. By searching the entire space of possible angle pairs Smith determined that (34,26) degree increments for the Generator and Analyzer retarders respectively are optimal for inversion of the W-matrix.

$$M_s = \left(W^T W\right)^{-1} W^T I \qquad (3)$$

Smith's W-Matrix analysis did not include the effects of instrument scatter or optical component misalignment. These are areas of active research and discussion for our team. We are using the Compain DRR instrument characterization and error correction technique to correct for systematic alignment errors in our instrument.[12]

Finally we are using the optical component characterization technique described by Chenault to determine characteristic values of Diattenuation and retardance for the optical components.[13] Chenault's technique is particularly well suited as it features spinning the item to be characterized between fixed linear polarizers. The technique exploits our existing DRR configuration.

b. DESCRIPTION OF THE DDSCAT MODEL

The method proposed for use in this research is the Discrete Dipole Approximation (DDA) developed by Purcell and implemented by Draine as DDSCAT.[14, 15] DDSCAT uses a hybrid FE/CDM approach. The finite elements are point polarized dipoles on a cubic lattice. The dipoles are coupled as in CDM, and the resulting geometry and material dependent spectral scatter and absorption coefficients are calculated.[16]

DDSCAT has been successfully applied to determining localized surface plasmon resonance for nano-particle bio- and chem-sensor development, interstellar spectroscopy of microscopic graphite, and is starting to be used for optical metamaterial design and analysis. [17-29]

DDSCAT takes into account much higher localized field strengths due to near-field interactions, local surface plasmon resonances, and interaction between the incident field and the structure geometry.[30-33] DDSCAT is used to determine the extinction (scattering and absorption) coefficients for arbitrarily shaped nano-particles. The nano-particles can either be free or placed on or in a substrate. In particular, DDSCAT is used to determine the optimal nano-particle shape, geometry, and surface density for its use as a plasmonic bio-sensor. Noble metal (gold and silver) nano-particles have distinct plasmon resonances based on their geometry. When other substances such as DNA or other organic material bind to the metal nano-particle the plasmon resonance shifts in a measurable way. The relative shift can be measured for even one molecule of the foreign substance. Plasmon resonances shifts are the leading edge of molecular detection and the participants use DDSCAT for calculating the shift distances. The subtle shifts in plasmonic resonance due to the addition of foreign agents or chemicals can be measured and reported.[28, 29]

III. INSTRUMENT CHARACTERIZATION

The AFIT CASI-DRR instrument was assembled in January 2010. We are now going through characterization trials and optical alignment in order to optimize the performance of the device. We have developed a strong DRR Mueller Calculus model in the Python language that allows us to model a measurement and compare the measurement to the ideal. The residual difference between the measurement and ideal states the sensitivity of the instrument. The residual is given by

$$\varepsilon = \frac{I_{Ideal} - I_{Meas}}{I_{Ideal}} \tag{4}$$

The Mueller distance between Mueller matrices M_1 and M_2 is another useful tool for measurement analysis. The Mueller difference is given by Equation 4 below. The Mueller distance between our measurement result and the desired Identity matrix is 0.064.[34] For comparison sake the Mueller distance for the NIST team led by Dr. Germer is 0.006.[35]

$$\varepsilon_{1,2} = \sqrt{\frac{\sum_{i,j}^{4,4}\left(M_{1_{i,j}} - M_{2_{i,j}}\right)^2}{16}} \tag{5}$$

The Mueller difference for several canonical Mueller forms is b=given in Table 1.

Table 1. Mueller differences for several canonical Mueller forms

	Blank	Depolarizer	Lin Horizontal	Lin Vertical	QWP Vertical	QWP Horizontal
Blank	0.00	0.43	0.43	0.43	0.50	0.50
Depolarizer	0.43	0.00	0.25	0.25	0.43	0.43
Lin Horizontal	0.43	0.25	0.00	0.35	0.43	0.43
Lin Vertical	0.43	0.25	0.35	0.00	0.43	0.43
QWP Vertical	0.50	0.43	0.43	0.43	0.00	0.71
QWP Horizontal	0.50	0.43	0.43	0.43	0.71	0.00

We measured canonical Mueller matrices such as "No Sample", Depolarizer, and Linear Polarizer to both characterize and correct the instrument for misalignment.[12] The "No Sample" case has the [4 x 4] Identity matrix as its Mueller representation. A typical "No Sample" measurement compared to (a) an ideal measurement and (b) a measurement limited only by random rotational variation at the 0.02 deg limit of our rotation stages is shown in Figure 2. The residual difference for Figure 2 is shown in Figure 3.

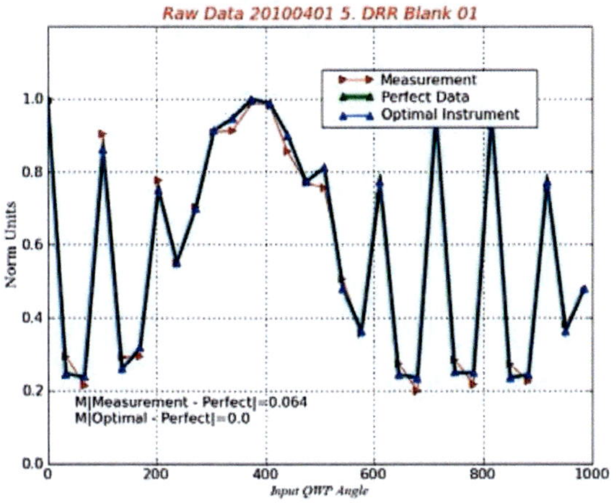

Figure 2. Intensity trace for "No Sample" case over 30 measurements at (34,26) degree increment pairs for Generator and Analyzer retarder rotation stages.

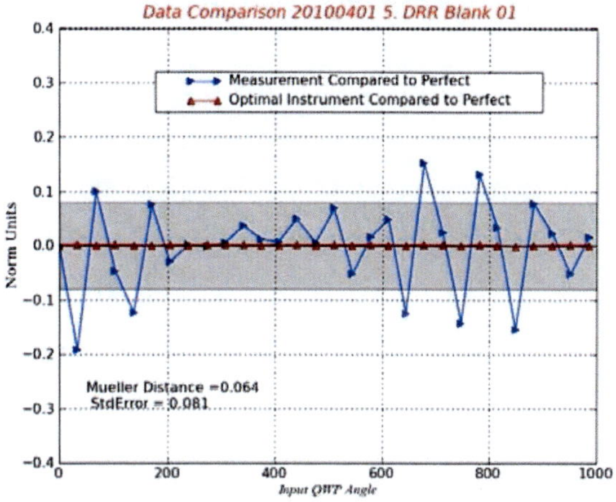

Figure 3. Residual comparison for "No Sample" case of current configuration to rotation stage limited instrument.

Our typical "Depolarizer" measurement compared to (a) an ideal measurement and (b) a measurement limited only by random rotational variation at the 0.02 deg limit of our rotation stages is shown in Figure 4. The residual difference for Figure 4 is shown in Figure 5. Our Mueller difference is 0.044 for measurement compared to ideal.

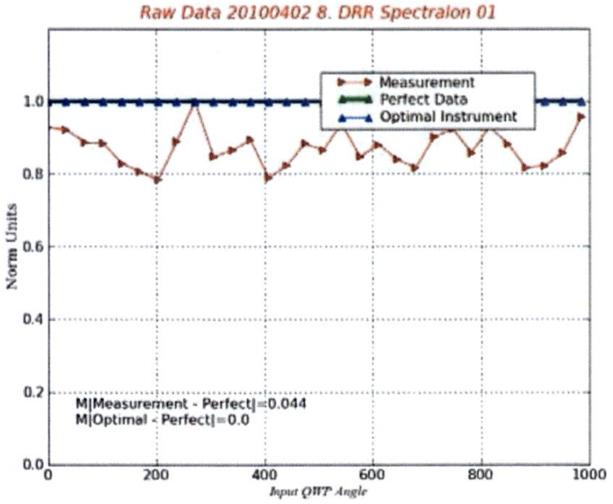

Figure 4. Intensity trace for Spectralon™ depolarizing surface.

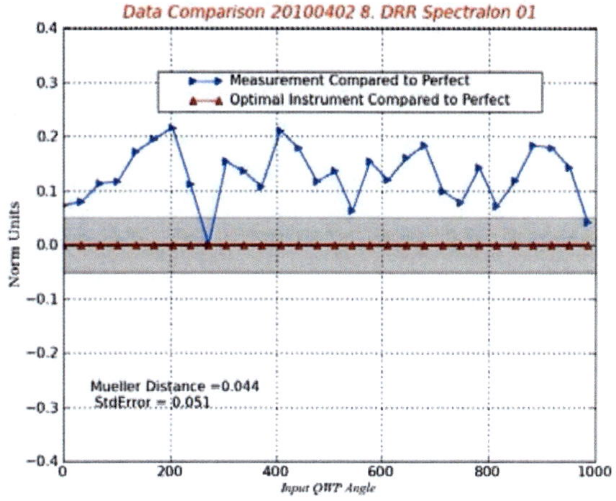

Figure 5. Residual comparison for Spectralon

We are preparing to apply our technique to the measurement and characterization of a layer of 50nm silver nanocubes on a glass slide. The nanocubes are randomly oriented in rotation but seem well aligned to the substrate face. An SEM photograph of the nanocubes is shown in Figure 6. The DDSCAT derived spectral extinction coefficients of the nanocubes is shown in Figure 7.

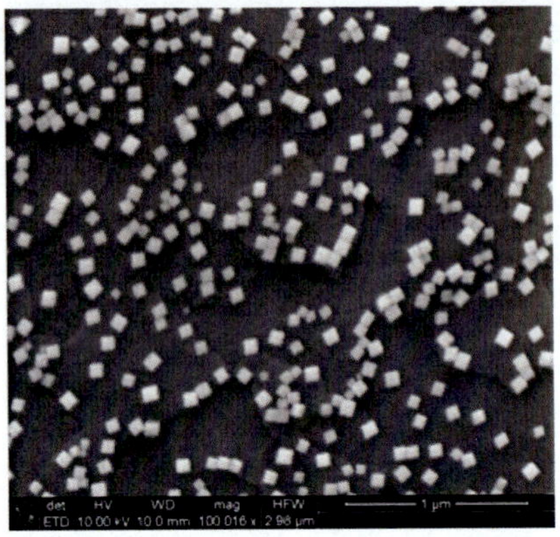

Figure 6. Silver nanocubes on ITO coated glass substrate

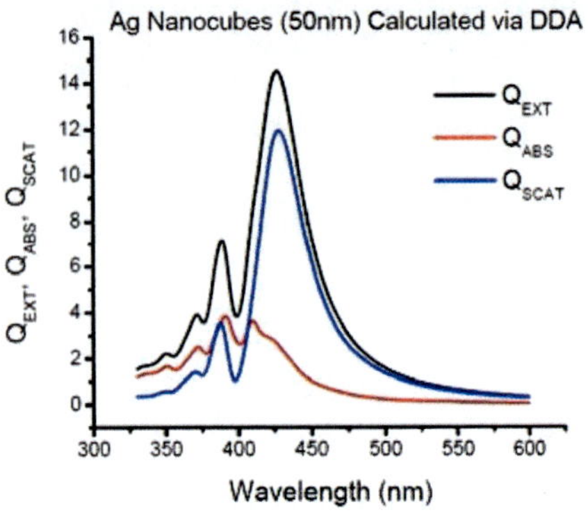

Figure 7. DDSCAT scattering coefficients for silver nanocube

Figure 8 shows initial measurements for the silver nanocubes on an ITO coated glass substrate. The data is presented for each Mueller element as Value versus log(angle in degrees) for positive angles. Using log(angle) stretches the x-axis to more clearly show specular features. The non-physical values of some elements > 1 are due to greater uncertainty still present in the instrument for those elements.

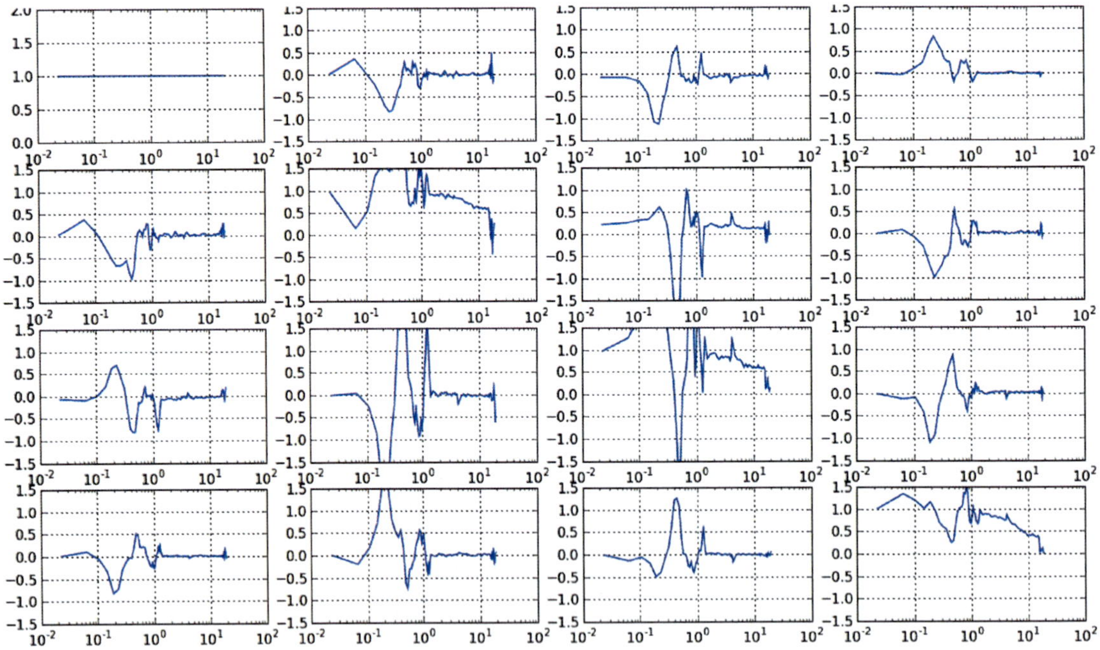

Figure 8. Initial Mueller elements vs log(angle) for silver nanocubes on glass substrate illuminated with 544 nm light incident at +10 degrees. Non-physical values > 1 are being corrected with additional instrument characterization.

IV. CONCLUSION

We have completed initial development of a full-scatter DRR configuration ellipsometer suitable for novel metamaterial characterization. The technique is well suited to measure desired material properties with stated uncertainty limits for novel photonic and infrared metamaterials of interest. We have incorporated predictive computational codes that estimate the electro-magnetic property values for metamaterial designs and concepts of interest. We are anxious to apply our resources and instruments to targets of interest to the community. The authors can be contacted at the email addresses given.

References

1. E. Ozbay, "Plasmonics: Merging Photonics and Electronics at Nanoscale Dimensions," Science **311,** 189-193 (2006).

2. T. A. Germer, C. Wolters, and D. Brayton, "Calibration of wafer surface inspection systems using spherical silica nanoparticles," Optics Express **16,** 4698-4705 (2008).

3. T. A. Germer and M. J. Fasolka, "Characterizing surface roughness of thin films by polarized light scattering," in *Proceedings of SPIE,* Anonymous (, 2003), pp. 264-275.

4. J. H. Kim, S. H. Ehrman, G. W. Mulholland, and T. A. Germer, "Polarized light scattering by dielectric and metallic spheres on silicon wafers," Appl. Opt. **41,** 5405-5412 (2002).

5. T. A. Germer, "Characterizing Interfacial Roughness by Light Scattering Ellipsometry," in *AIP CONFERENCE PROCEEDINGS,* Anonymous (IOP INSTITUTE OF PHYSICS PUBLISHING LTD, 2001), pp. 186-190.

6. D. R. Smith, S. Schultz, P. Markoš, and C. M. Soukoulis, "Determination of effective permittivity and permeability of metamaterials from reflection and transmission coefficients," Physical Review B **65,** 195104 (2002).

7. C. Menzel, C. Rockstuhl, T. Paul, F. Lederer, and T. Pertsch, "Retrieving effective parameters for metamaterials at oblique incidence," Physical Review B **77,** 195328 (2008).

8. C. Menzel, T. Paul, C. Rockstuhl, T. Pertsch, S. Tretyakov, and F. Lederer, "Validity of effective material parameters for optical fishnet metamaterials," Physical Review B **81,** 35320 (2010).

9. R. M. A. Azzam and N. M. Bashara, *Ellipsometry and polarized light* (North-Holland, 1977).

10. D. H. Goldstein and R. A. Chipman, "Error analysis of a Mueller matrix polarimeter," Journal of the Optical Society of America A **7,** 693-700 (1990).

11. M. H. Smith, "Optimization of a dual-rotating-retarder Mueller matrix polarimeter," Appl. Opt. **41,** 2488-2493 (2002).

12. E. Compain, S. Poirier, and B. Drevillon, "General and self-consistent method for the calibration of polarization modulators, polarimeters, and Mueller-matrix ellipsometers," Appl. Opt. **38,** 3490-3502 (1999).

13. D. B. Chenault and R. A. Chipman, "Measurements of linear diattenuation and linear retardance spectra with a rotating sample spectropolarimeter," Appl. Opt. **32,** 3513-3519 (1993).

14. B. T. Draine, "The discrete-dipole approximation and its application to interstellar graphite grains," Astrophys. J. **333,** (1988).

15. B. T. Draine and P. J. Flatau, "Discrete-dipole approximation for scattering calculations," JOURNAL-OPTICAL SOCIETY OF AMERICA A **11,** 1491-1491 (1994).

16. B. T. Draine and P. Flatau, "User Guide for the Discrete Dipole Approximation Code DDSCAT 7.0," eprint arXiv: 0809.0337 (2008).

17. C. Noguez, "Surface plasmons on metal nanoparticles: the influence of shape and physical environment," J.Phys.Chem.C **111,** 3806-3819 (2007).

18. C. Noguez, C. E. Roman-Velazquez, R. Esquivel-Sirvent, and C. Villarreal, "High-multipolar effects on the Casimir force: The non-retarded limit," Europhys. Lett. **67,** 191-197 (2004).

19. C. Noguez, I. O. Sosa, R. G. Barrera, and M. DF, "Light Scattering by Isolated Nanoparticles With Arbitrary Shapes," in *MATERIALS RESEARCH SOCIETY SYMPOSIUM PROCEEDINGS,* Anonymous (Warrendale, Pa.; Materials Research Society; 1999, 2001), pp. 275-280.

20. I. O. Sosa, C. Noguez, and R. G. Barrera, "Optical Properties of Metal Nanoparticles with Arbitrary Shapes," J Phys Chem B **107,** 6269-6275 (2003).

21. D. Cho, F. Wang, X. Zhang, and Y. R. Shen, "Contribution of electric quadrupole resonance in optical metamaterials," in *American Physical Society, 2008 APS March Meeting, March 10-14, 2008, abstract# D35. 012,* Anonymous (, 2008).

22. W. Cai, U. K. Chettiar, A. V. Kildishev, and V. M. Shalaev, "Optical Cloaking with Non-Magnetic Metamaterials," Arxiv preprint physics/0611242 (2006).

23. W. Cai, U. K. Chettiar, A. V. Kildishev, and V. M. Shalaev, "Optical cloaking with metamaterials," Nature Photonics **1,** 224–226 (2007).

24. U. K. Chettiar, A. V. Kildishev, H. K. Yuan, W. Cai, S. Xiao, V. P. Drachev, and V. M. Shalaev, "Dual-Band Negative Index Metamaterial: Double-Negative at 813 nm and Single-Negative at 772 nm," (2008).

25. E. M. Hicks, O. Lyandres, W. P. Hall, S. Zou, M. R. Glucksberg, and R. P. Van Duyne, "Plasmonic Properties of Anchored Nanoparticles Fabricated by Reactive Ion Etching and Nanosphere Lithography," Journal of physical chemistry.C **111,** 4116-4124 (2007).

26. A. Penttilä, E. Zubko, K. Lumme, K. Muinonen, M. A. Yurkin, B. Draine, J. Rahola, A. G. Hoekstra, and Y. Shkuratov, "Comparison between discrete dipole implementations and exact techniques," Journal of Quantitative Spectroscopy and Radiative Transfer **106,** 417-436 (2007).

27. F. S. Diana, A. David, I. Meinel, R. Sharma, C. Weisbuch, S. Nakamura, and P. M. Petroff, "Photonic crystal-assisted light extraction from a colloidal quantum Dot/GaN hybrid structure," Nano Letters **6,** 1116-1120 (2006).

28. L. J. Sherry, R. Jin, C. A. Mirkin, G. C. Schatz, and R. P. Van Duyne, "Localized surface plasmon resonance spectroscopy of single silver triangular nanoprisms," Nano Lett **6,** 2061 (2006).

29. L. J. Sherry, S. H. Chang, G. C. Schatz, R. P. Van Duyne, B. J. Wiley, and Y. Xia, "Localized surface plasmon resonance spectroscopy of single silver nanocubes," Nano Lett **5,** 2034–2038 (2005).

30. K. Byun, S. Kim, and D. Kim, "Design study of highly sensitive nanowire-enhanced surface plasmon resonance biosensors using rigorous coupled wave analysis," Optics Express **13,** 3737-3742 (2005).

31. J. Lu, C. Petre, E. Yablonovitch, and J. Conway, "Numerical optimization of a grating coupler for the efficient excitation of surface plasmons at an Ag-SiO_2 interface," Journal of the Optical Society of America B **24,** 2268-2272 (2007).

32. M. G. Moharam and T. K. Gaylord, *Rigorous Coupled-Wave Analysis of Planar-Grating Diffraction* (OSA, 1980).

33. J. Spiegel, J. de la Torre, M. Darques, L. Piraux, and I. Huynen, "Permittivity Model for Ferromagnetic Nanowired Substrates," IEEE Microwave and Wireless Components Letters **17,** 492 (2007).

34. R. A. Chipman, "Handbook of Optics, vol. II, ch. 22," Optical Society of America, (2000).

35. T. A. Germer and C. C. Asmail, "Goniometric optical scatter instrument for out-of-plane ellipsometry measurements," Rev. Sci. Instrum. **70,** 3688 (1999).

Proceedings of the SEM Annual Conference
June 7-10, 2010 Indianapolis, Indiana USA

Thermal Management and Metamaterials

Calvin T. Roman, Ronald A. Coutu, Jr. and LaVern A. Starman

Department of Electrical and Computer Engineering,
Air Force Institute of Technology
2950 Hobson Way, WPAFB, OH 45433, USA

E-mail: calvin.roman@us.af.mil

Abstract

Thermal metamaterials are materials composed of engineered, microscopic structures that exhibit unique thermal performance characteristics based primarily on their physical structures and patterning, rather than just their chemical composition or bulk material properties. In many cases, the heat transfer performance attributes of the thermal metamaterial are such that similar performance cannot be obtained using conventional materials or compounds. Thermal metamaterials are an emerging technology, and are just now beginning to be acknowledged and developed by the microelectronics and material sciences community. This paper presents a series of ten proof-of-concept thermal metamaterial devices. Modeling and testing of these microelectromechanical systems (MEMS) based thermal metamaterial prototypes showed that the electrical and thermal conductivity of the material can be switched or tuned within a certain operational range, and that this switching is a function of passive or active actuation of the metamaterial's structural elements, not just its chemical composition.

1. Introduction

Today's scientists and engineers have a plethora of materials, both natural and synthetic, encompassing a wide spectrum of physical properties at their disposal. Metamaterials, which are fabricated structures with response functions that do not occur in natural materials or chemical compounds, are a next step in the ongoing quest for new and novel material properties. "Meta" is Greek for "after" or "beyond" [1]. Metamaterial engineering facilitates obtaining theoretically possible, but often unavailable, material properties as the limits of natural and chemically-synthesized materials are approached.

Although the general definition of metamaterials encompasses all material properties, the vast majority of metamaterial-related research conducted to date has focused on manipulating the manner in which a material responds to an applied electromagnetic energy or optical wave. Commonly engineered metamaterial properties include negative dielectric permittivity ($-\varepsilon$) and negative magnetic permeability ($-\mu$). Materials with these properties are commonly referred to as double-negative (DNG) materials, left-handed media (LHM), or backward-wave media (BW media), and can have a negative index of refraction when evaluating their response to RF energy of a certain wavelength [2].

While the bulk of metamaterial research performed has involved engineering and tuning a material's ε and μ response – the broad definition of metamaterials does not limit them to only unusual electromagnetic properties. This paper examines thermal metamaterials, an emerging technology just now beginning to be acknowledged by the microelectronics and material sciences community. Thermal metamaterials are materials composed of engineered, microscopic structures that exhibit unique thermal performance characteristics based primarily on their physical structures and patterning, rather than just their chemical composition or bulk material properties. In many cases, the heat transfer performance attributes of the thermal metamaterial are such that similar performance cannot be obtained using conventional materials or compounds.

Metamaterials engineered for their electromagnetic responses focus on ε and μ as the primary material properties of interest. Metamaterials engineered for their thermal properties focus on different material properties, including, but not limited to, thermal conductivity and electrical conductivity (κ and σ) [3]. Figure 1 shows a few example materials covering a wide range of κ and σ characteristics. Note that κ and σ are not zero or negative in any quadrant, as is the case when evaluating ε and μ. Existing chemical compounds and natural materials provide for many combinations of κ and σ. However, the allure of a thermally-tuned metamaterial is not only that the designer may be able to independently select the value of κ and σ needed for the material application, it is that the material's κ and σ may be controlled within a certain range via a mechanical actuation scheme.

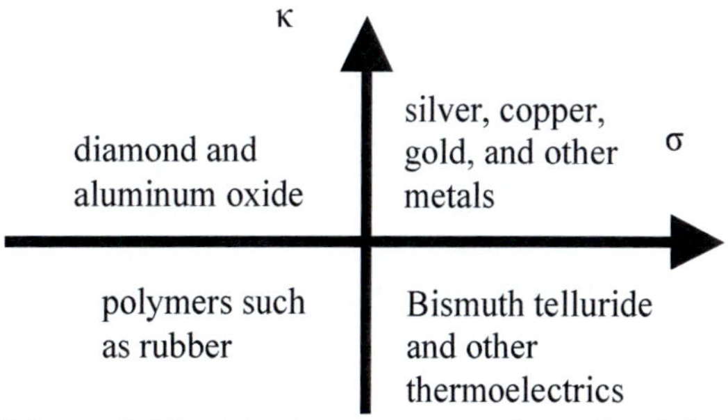

Figure 1. Electrical verses thermal conductivity.

T. Proulx (ed.), *MEMS and Nanotechnology, Volume 2*, Conference Proceedings of the Society for Experimental Mechanics Series 2,
DOI 10.1007/978-1-4419-8825-6_16, © The Society for Experimental Mechanics, Inc. 2011

The objective of this paper was to present a series of MEMS-based thermal metamaterial designs. Modeling and characterization of these proof-of-concept designs showed that materials composed of nano or microscopic structures can realize tuned thermal performance characteristics based primarily on actuation of their physical structures rather than just their chemical composition.

2. Design

The goal of our thermal metamaterial design was to control and exploit the thermal properties of a thin film, specifically κ and σ, in a way that is not possible using existing materials and chemical compounds. It is executed on a scale that historically would have precluded a mechanical structure solution.

2.1. Concept of Device Operation

The designs are based upon a metamaterial composed of dense arrays of MEMS cantilever beams 150-300 μm long. These beam arrays, designed in thermally actuated and electrostatic actuated variants, create a metasurface (a 2-D metamaterial). This metasurface has variable electrical and thermal conductivity via switching (opening/closing) of the arrays of cantilever beams.

Application of this metasurface includes use as a thin film (less than 5 μm thick) which exhibits a somewhat tunable, but not decoupled, σ and κ when considered from an input side to an output side. Figure 2 (a) and (b) show unit cells of a conceptual drawing of this idea, in thermally-actuated and electrostatic-actuated versions, respectively.

In Figure 2 (a), a "thin film" of MEMS structures facilitating a user and/or temperature controlled σ and κ is shown. This concept can operate in two ways. The first is when the user or application manages the control voltage "*V*". This control signal generates resistive $I^2 \cdot R$ heating in the heating element shown in red. This heating causes the bimorph cantilever beams to close and make contact with the I/O terminals shown in yellow. As the beams begin to close, the σ and κ of the system, when considered from the I/O, will begin to dramatically increase as more conductive "paths" through the system appear. Therefore, the user, through manipulation of the control signal "*V*", can vary, switch, or tune σ and κ within a certain range with respect to the system's I/O.

The second way to operate this concept is simply through system temperature dependence. In this scenario, the bimorph cantilever beams close and make contact with the I/O terminals as a function of temperature and bimorph beam length only, with no user input to control "*V*". The result is a thin film which, when evaluated from the I/O terminals, passively switches from a low/poor σ and κ at lower system temperatures to a higher σ and κ at a selected higher temperature.

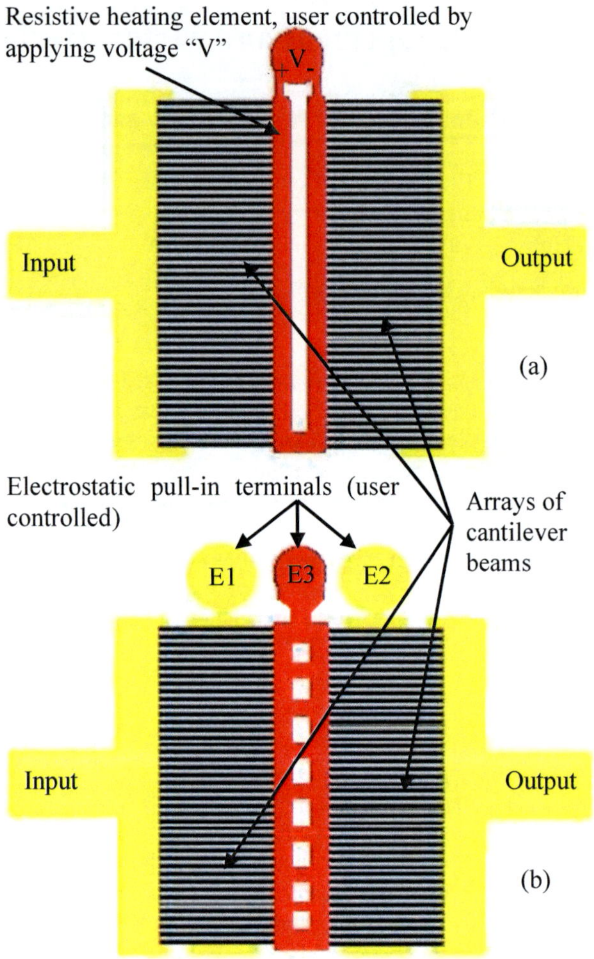

Figure 2. Conceptual drawings of unit cells of a thermally-actuated thermal metamaterial (a), and a electrostatic-actuated thermal metamaterial (b).

Another variation of this same idea involved actuating the arrays of beams via electrostatic attraction instead of through heating of a bimorph device, as shown in Figure 2 (b). This is advantageous as it draws less power than resistive heating, has a reduced IR signature, and allows for faster, more active switching control of the beam arrays. In Figure 2 (b), the user or application manages electrostatic pull-in control voltages "*E1*", "*E2*", and "*E3*". Manipulation of these control signals causes the beam arrays to snap downward due to electrostatic attraction and pull-in, completing a conductive path between each I/O terminal and the middle structural element shown in red. "*E1*" controls all of the beams to the left of center, "*E2*" all those on the right, and "*E3*" holds all beams at a common potential to facilitate pull-in. As the beams snap down, the σ and κ of the system, when considered from the I/O, will dramatically increase. Therefore, the user, through manipulation of the control signals, can vary or tune κ and σ within a certain range with respect to the system's I/O.

In addition to the switchable κ and σ thin film application just presented, this same concept has a potential application as a "heat steering" device. In this application, the metamaterial could provide tailored, highly localized thermal management to a hot spot of microelectronics/VLSI logic on a substrate. More specifically, as part of a thermal management system, in which this metamaterial could be used to "steer heat" in the X-Y plane via switchable thermal conductivity. This could allow heat to be retained by the logic or circuitry being managed during periods of inactivity or extreme cold, and conducted away to a heat exchanger system during periods of elevated operating temperatures when the logic can benefit from ridding itself of excess thermal energy.

A conceptual drawing of this application is shown in Figure 3. Microelectronics logic under metamaterial-based thermal management is surrounded by a dense array of MEMS cantilever beams. The MEMS beams are anchored to the substrate, possibly directly above the logic being managed, as a back-end-of-line (BEOL) process. When actuated, the arrays of cantilever beams make conductive contact with a heat sink material layer, effectively steering some of the logic's thermal energy via conduction. From the logic's perspective, the metamaterial surrounding it in its X-Y plane has variable thermal and electrical conductivity. This is of course because, for this structural arrangement, κ and σ properties are a function of which and how many cantilever beams are open or closed.

As was seen with the tuned κ and σ thin film application, this "heat steering" application could be implemented in either a bimorph actuated or electrostatic actuated version. However, as cooling is a three-dimensional discipline, this concept would be used in conjunction with, not as a complete replacement of, a traditional heat sink setup in medium-high power devices. In lower power applications, this metasurface could be used effectively if paired with a low-κ layer beneath the logic being managed. This would minimize thermal conduction downward into the substrate and allow the metasurface to steer much of the logic's excess thermal energy in the X-Y plane.

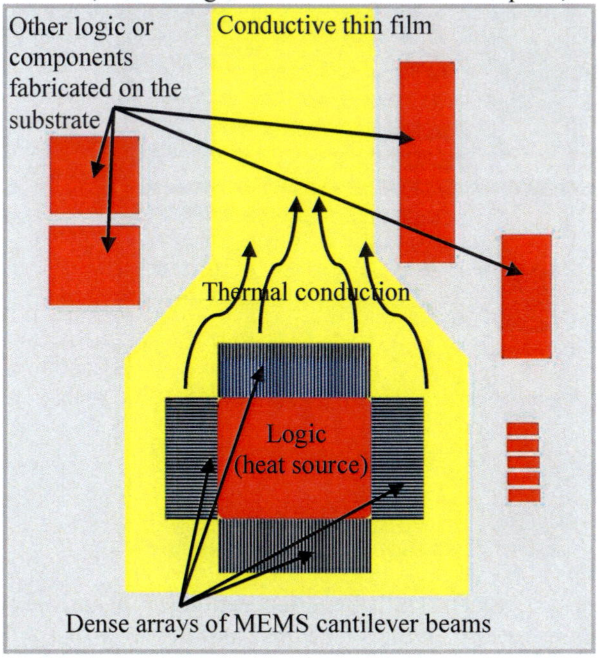

Figure 3. Conceptual drawing of thermal metamaterial in heat steering application. Heat generated by the logic or circuitry is conducted outward in the X-Y plane through an array of MEMS beams, and into a conductive thin film.

2.2. Fabrication

The MEMPCAP polysilicon deposition multi-user MEMS process, or PolyMUMPs™, was selected as the fabrication process for this research effort. PolyMUMPs™ is a three-layer polysilicon surface and bulk micromachining process that utilizes two sacrificial layers of silicon dioxide to provide separation between three mechanical layers of polysilicon [4]. Eight mask levels result in seven material layers [4]. The minimum feature size is 2 μm. Figure 4 shows a cross section of the PolyMUMPs process layers.

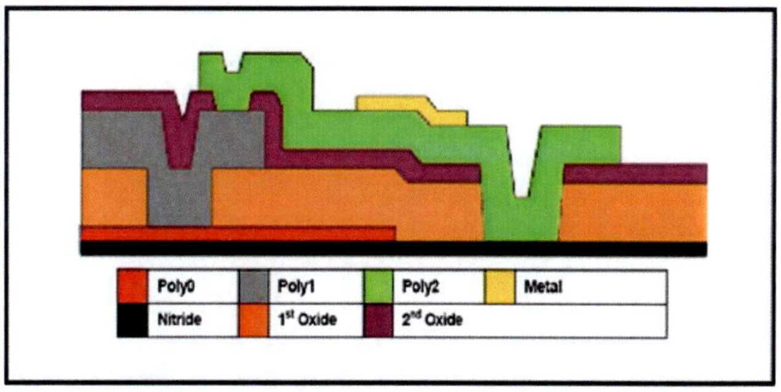

Figure 4. Cross sectional view of all seven layers of PolyMUMPs process [4].

The fabricated devices are shown in Figure 5. All 10 metamaterial versions were produced on each 3 mm X 3mm die sample.

2.3. Design Description

Building upon the tuned κ and σ thin film and heat steering concepts, a series of ten design layouts were fabricated. Four versions featuring passively actuated, bimorph cantilever beam arrays as the main structural

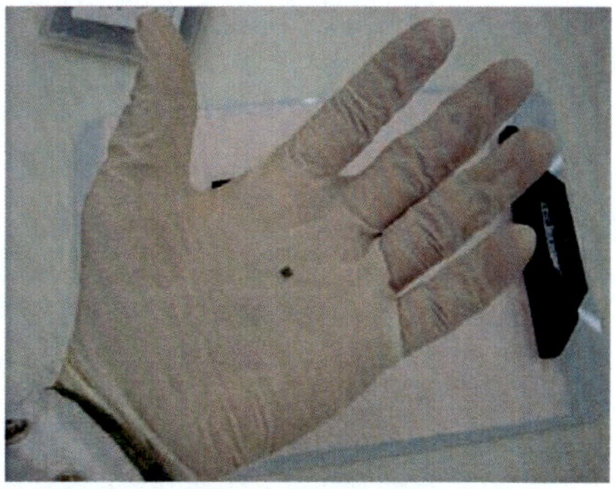

Figure 5. PolyMUMPs™ fabricated device in hand to show scale.

feature of the metamaterial are presented first, as shown in Figure 6. The designs incorporate aspects of both concepts presented in Section 2.1. A polysilicon resistive heating element fills the role of the heating element shown in Figure 2 (a), as well as that of the "logic" of Figure 3. Polysilicon sheets on the top and bottom of Figure 6 perform as either I/O terminals (Figure 2 concept) or as a heat sink sheet (Figure 3 concept).

Four variations of this design were developed, with cantilever beam length being the only variable. The temperature at which the bimorph beams bend down and make conductive contact with the polysilicon layer below is a function of beam length. Lengths of 150, 200, 250, and 300 μm were made, corresponding to closing temperatures of 100, 90, 80, and 70 °C, respectively. Beam width is 30 μm in all cases. Bimorph beam thickness, which was dictated by the PolyMUMPS fabrication process, is 2 μm (1.5 μm of polysilicon and 0.5 μm of gold).

In addition to the bimorph cantilever beam array based metamaterials just discussed, several versions featuring electrostatic actuated cantilever beam arrays as the main structural feature of the metamaterial were designed. These versions spotlight a significantly different beam actuation technique, which provides somewhat different overall capabilities.

Figure 7 shows an electrostatic actuated variant. There are several significant differences between this design and the bimorph beam based design. Here, since the opening/closing of the cantilever beam contacts is actively controlled via electrostatic attraction, two contact pads are added (electrostatic pull-in terminals in Figure 7). Each pad is connected to polysilicon sheets which run along the length of each cantilever beam, 2 μm beneath them. A sufficient potential difference between the bottom electrode of polysilicon and the cantilever beam above will cause the beam to bend downward and pull-in, making contact with the polysilicon I/O terminal or heat sink. Note that this design uses cantilever beams consisting of only a single polysilicon layer (no metal layer on top).

Six variations of this design were developed, with cantilever beam material (Poly1 verses Poly2) and resistive heater grid layout/material being the variables. The first variation featured a Poly1 resistive heater element and Poly1 cantilever beams. Second, a design using Poly2 for both the heater element and cantilever beams was drawn. The third design had a heater element consisting of Poly1

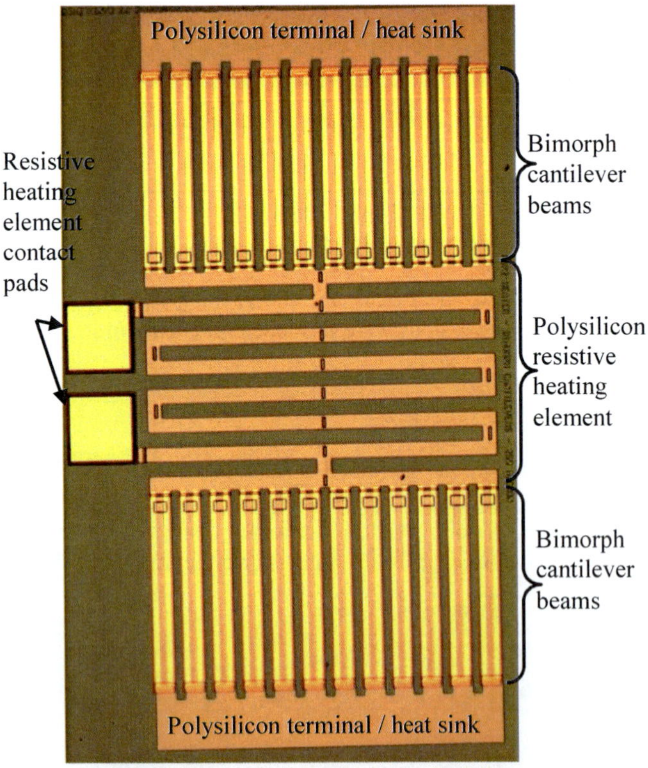

Figure 6. Bimorph cantilever beam array metamaterial. Bimorph cantilever beams curl upward and bend downward in response to ambient temperature and/or $I^2 \cdot R$ heating from the resistive heating element, creating conductive paths through the system.

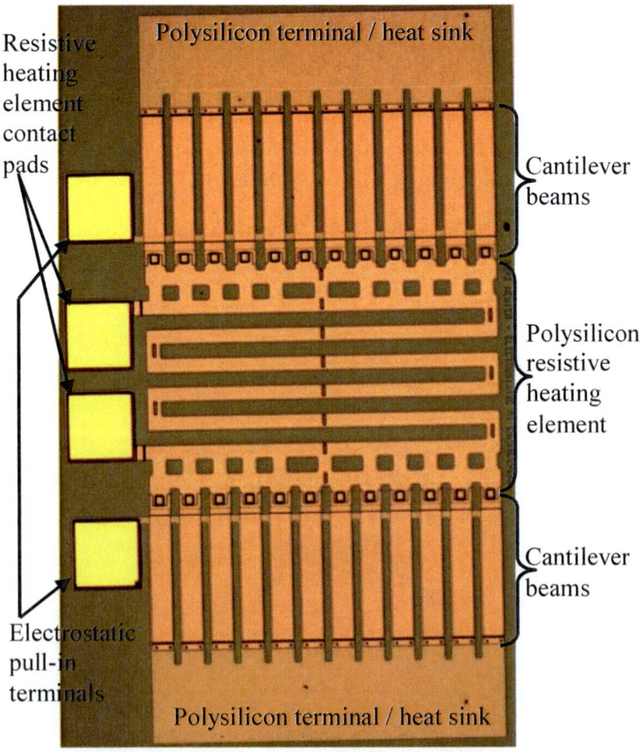

Figure 7. Electrostatic-actuated cantilever beam array metamaterial. Cantilever beams snap downwards (pull-in) in response to electrostatic attraction caused by a difference in potential between the polysilicon cantilever beams and the polysilicon electrodes attached to each electrostatic pull-in terminal.

and Poly2 stacked with Poly1 cantilever beams. The fourth version also had a heater element consisting of Poly1 and Poly2 stacked, but with Poly2 cantilever beams. The fifth and sixth variants featured a modified heater arrangement with Poly1 (fifth version) and Poly2 (sixth version) cantilever beams.

Unlike the previous design in which each beam actuated individually, these cantilever beams make contact in sets of 12 when the appropriate potential is applied to the pull-in terminals. This results in trading highly localized passive thermal management for more active control. Electrode structure sizing for pull-in voltage was determine by:

$$V_P = \sqrt{\frac{8 \cdot k \cdot g^3}{27 \cdot \varepsilon_0 \cdot W \cdot W_e}} \quad (1)$$

where V_p is the pull-in voltage (V), k is a spring force constant (N/m), and is a function of the beam material and dimensions, g is the air gap between the Poly0 electrode and cantilever beam (μm), ε_0 is the permittivity of free space (F/m), W is the cantilever beam width (μm), and W_e is the width of the electrode (μm). The electrode was sized to accommodate a pull-in voltage of approximately 8 V and 12 V, for the variants with Poly2 beams and Poly1 beams, respectively.

CAD drawings of the ten layouts were imported into CoventorWare MEMS 2008 for finite element modeling (FEM). Modeling verified the devices structural operation, including correct bimorph cantilever beam curling verses temperature and as-designed electrostatic pull-in voltage.

As no provision was made to hermetically seal or package these prototypes, the devices were stored in a nitrogen overpressure drybox within a cleanroom environment. This minimized native oxide growth on the cantilever beam tips and I/O terminal or heat sink sheets – which would have resulted in increased resistivity and thermal contact resistance and thereby decreased prototype performance.

3. Experiment and Results

3.1. Interferometer-based beam displacement testing

The mechanical motion of the prototype metamaterial devices is nearly entirely in the Z-axis. For each of the four bimorph cantilever beam array based prototypes, beam displacement verses system temperature was measured. For each of the six electrostatic-actuated cantilever beam array based prototypes, beam pull-in voltage verses system temperature was recorded. System temperature was swept using a calibrated thermoelectric heating/cooling device. Test results for the bimorph cantilever beam array based metamaterials are plotted in Figure 8. Note that this test uniformly heated the entire device. Because of this uniform heating, the temperature at which the array of cantilever beams "closed" is higher than when the devices are heated via Joule heating from the adjacent resistive heating element. When heated via conduction from the heater grid, the gold layer becomes significantly hotter than the polysilicon – resulting in a lower closing temperature than when uniformly heating the device.

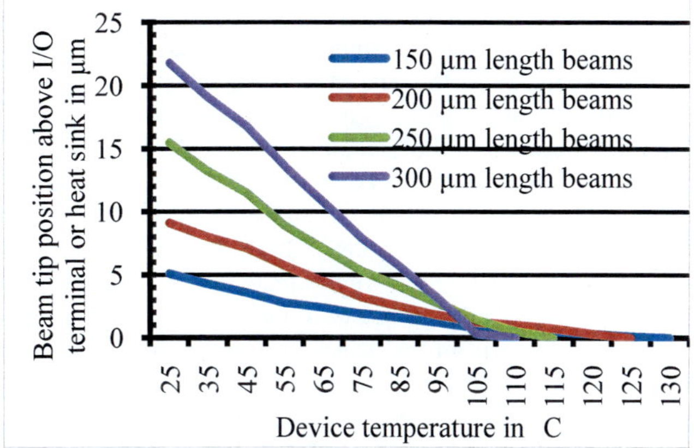

Figure 8. Bimorph cantilever beam tip position verses temperature. When the position equals zero, the beam tip has made contact with the poly layer beneath it. 150 μm beams make contact at 127 °C, 200 μm beams make contact at 123 °C, 250 μm beams make contact at 114 °C, and 300 μm beams make contact at 110 °C.

Measured pull-in voltages for each electrostatic actuated thermal metamaterial variant very closely correlated with the mathematically calculated (using Equation 1) pull-in voltages (8V and 12V). None of the samples evaluated exhibited any temperature dependency in their pull-in performance, and all measurements were within 8% of the calculated pull-in.

3.2. Thermal imaging temperature mapping test

In a metamaterial engineered to provide a unique or tailored thermal response, characterization of heat transfer throughout the device is a critical performance metric. An accurate way to evaluate heat transfer on a MEMS-scale device is by using a thermal imaging microscope system.

Each of the four bimorph cantilever beam array based prototypes were temperature mapped over the course of an operational sequence. This sequence began with each metamaterial device at a uniform temperature (no power applied to the heater grid). Then, at time T=0 seconds, 24 VDC was applied across the polysilicon resistive heating element, causing ≈ 300 mW of power consumption by the element. Over the next several seconds, the device would heat up, causing the bimorph cantilever beam arrays to bend downwards and close. At approximately T=10 seconds, the device had reached a steady state thermal condition. The potential was then removed from the heater grid – causing the device to rapidly cool and the bimorph beams to curl upwards. Once all 24 of the bimorph beams had recovered to their open position, the operational sequence was complete.

Figure 9 illustrates thermal characterization of the bimorph beam array thermal metamaterial from initial power application through steady state operation, with particular emphasis placed on thermal conduction from the bimorph beams into the heat sink. At time T=0, a 24 VDC potential is applied across the heater grid. Figure 9 shows how the device structures conduct heat and move in response to this Joule heating, which is occurring off the right side of the figure. Approximately 1.5 seconds after power was applied, the beams fully close – completing a conductive path from the heater grid, through the bimorph beams, and into the polysilicon sheet. The device reaches its thermal steady state at≈ T=10 seconds. At steady state, thermal conduction from the heater grid, through the 12 bent down bimorph cantilever beams, and into the heat sink layer heats the sink to≈ 10 °C hotter than the background wafer. The metamaterial has switched from a low-κ, low-σ state into a high-κ, high-σ state.

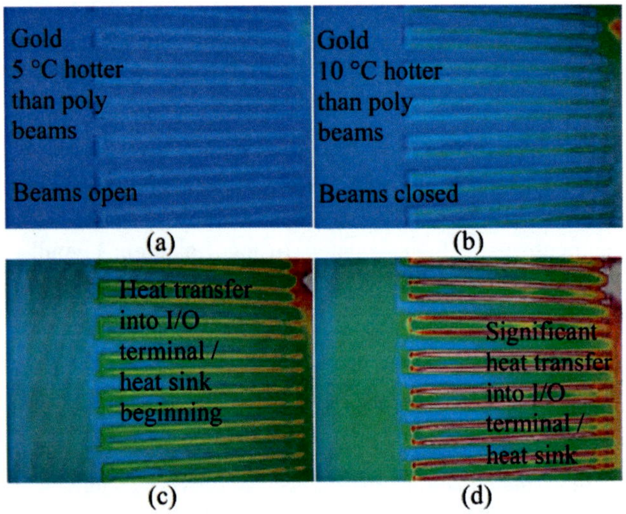

Figure 9. Bimorph beam array heat transfer into the heat sink. The 250 μm variant is shown under a 15X optical and 4X digital zoom, with emphasis on the beam array and heat sink layers. In (a), the beam array and heat sink are shown ½ second after voltage was applied across the heating element. The gold begins to heat, but the beams are still curled upwards with no thermal conduction into the heat sink. One second later in (b), the beams have all closed and conduction into the heat sink can now begin. At 2.5 seconds, a complete outline of the Poly0 heat sink layer can be seen, because it's temperature has been raised (heated) via conduction from the bimorph beams (c). Lastly, in (d), the system has reached steady state at 10 seconds. The outline of the gold layer in each beam is clearly seen, and is 15 °C hotter than the Poly2 structural layer of each beam. The heat sink's outline is clearly visible, and is nearly the same temperature as the Poly2 in the bimorph beams which are directly conducting heat into the sink.

Bimorph cantilever beam based variants closing or switching temperatures were within 2% of the values predicted during FEM, as was presented in Section 2.3. Unfortunately, due to extended thermal imaging equipment failures, the electrostatic actuated variants were not thermally characterized.

3.3. Electrical resistance test

The ability to provide variable electrical conductivity is fundamental to the functionality of the thermal metamaterial prototypes. For each of the four bimorph cantilever beam array based prototypes, electrical resistance across the device was measured. The resultant data product was electrical resistance verses device temperature, and is shown in Figure 10.

Each of the four bimorph designs performed similarly, with "off scale high" (> 1 GΩ) resistance at temperatures less than their beam closing temperature (from Section 2.3) minus ≈ 4 °C. This high resistance is due to a several micron "open circuit" at each of the 24 beam tip to polysilicon sheet interfaces. As the temperature was increased to within≈ 4 °C of the beam array closing temperature, some of the beams began to make conductive contact with the sheet below. At this point, the resistance dropped from "off scale high" to between 100-150 kΩ. Further increasing the system temperature resulted in complete beam array closure, with the resistance reaching a minimum value of≈ 3 kΩ.

This test quantified the temperature dependent electrical resistance of these four bimorph cantilever beam array based thermal metamaterials. The resistance of the bimorph beam arrays is effectively a thermostat type "on/off switch",

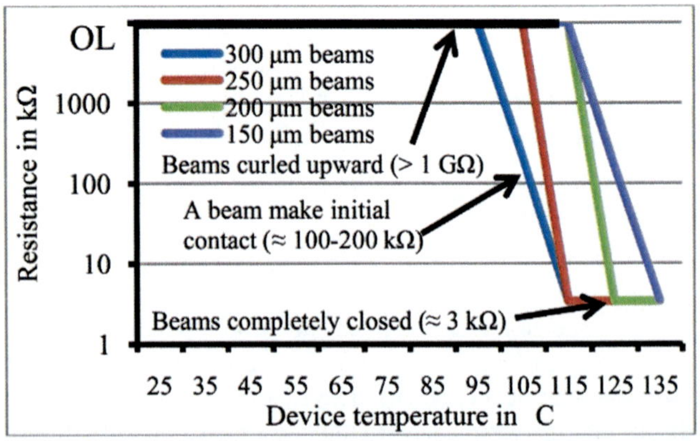

Figure 10. Electrical resistance across the bimorph cantilever beam array metamaterial (across I/O terminals). All designs exhibited off scale high (GΩ's) resistance at temperatures less than approximately their closing temperature (from Section 2.3) minus 4 °C. As each design's closing temperature was approached, the resistance would drop from "overload" to approximately 100-200 kΩ. Increasing the temperature further resulted in the resistance decreasing to ≈ 3 kΩ.

where "off" is > 1 GΩ and "on" is ≈3 kΩ. The contact resistances present at each of the 24 beam tip to polysilicon sheet interfaces was the dominant factor in the systems overall resistance. This data showed that the σ of the bimorph beam based metamaterial can be switched based on passive motion of its structural features, not simply reliant on its material composition.

Similar measurements were recorded for each of the six electrostatic actuated cantilever beam array based prototypes. However, for each test point, a measurement was made with the beams "open" and also with them "pulled-in" or "closed". The resultant data product was electrical resistance verses device temperature verses beam array position, and is show in Figure 11.

All six designs exhibited "off scale high" (> 1 GΩ) resistance when their arrays of beams were in the open position (not pulled-in). This high resistance is due to an approximately 1 μm "open circuit" at each of the beam tip to polysilicon sheet interfaces. As seen in Figure 11, the resistance across each design increased ≈ 2-11% when heated from 25-135 °C. This increase is expected as the polysilicon used in this process experiences a resistivity increase of ≈ 24% when heated from 25-135 °C. The reason that the resistance of each sample increased only 2-11% over the same range is that the dominant factor in the samples resistance is the contact resistance at each of the 24 beam tips, not the polysilicon structural resistance. As expected, versions constructed of materials with greater sheet resistance generally exhibited higher resistance.

This test quantified the pull-in and temperature dependent electrical resistance of these six electrostatic actuated cantilever beam array based thermal metamaterials. The resistance of the electrostatic actuated beam arrays is

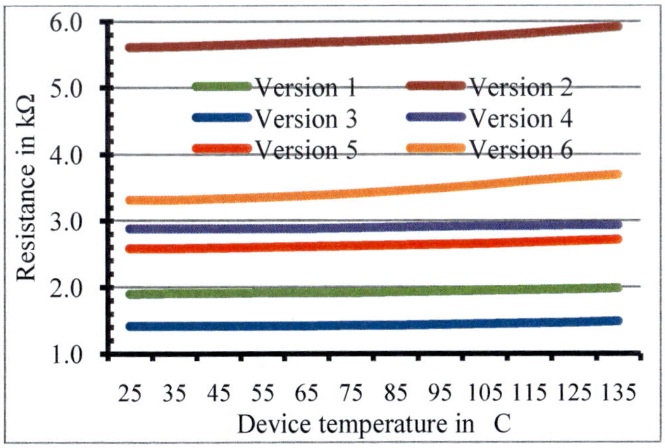

Figure 11. Electrical resistance of electrostatic actuated cantilever beam array metamaterials, across I/O terminals, with beams pulled-in or closed. Variations in resistance between versions is due to sheet resistance differences in beam material (Poly1 or Poly2), heater grid material (Poly1, Poly2, or Poly1 & 2 stacked), and heater grid type. Versions were described in Section 2.3. Electrical resistance increased ≈ 2-11% when heated from 25-135 °C.

effectively an "on/off switch", where "off" is > 1 GΩ and "on" is between 1-6 kΩ, depending on which materials were used in the design. This data supports that the electrical resistivity and conductivity of the electrostatic actuated beam based metamaterial can be switched based on actuation of its structural features, not simply its material composition.

4. Conclusion

This paper developed, modeled, fabricated, and characterized a series of proof-of-concept thermal metamaterial devices. Characterization of these prototypes showed that the σ and κ of the material can be switched within a certain operational range, and that this switching is a function of actuation of the metamaterial's structural elements, not just its chemical composition. The designs presented, discussed, and tested here are just a few examples of the countless material possibilities that can be realized by exploiting thermal metamaterials.

Disclaimer:
The views expressed in this document are those of the author and do not reflect the official policy or position of the United States Air Force, Department of Defense, or the U.S. Government.

References

[1] "Electromagnetic Metamaterials." Meta Group, Duke University. <http://people.ee.duke.edu/~drsmith/about_metamaterials.html>. 2009.

[2] J. Engheta, N., Ziolkowski, R. "Metamaterials, Physics and Engineering Explorations."Wiley Interscience. 2006.

[3] Dudis, D. "Metamaterials for Thermo-electrical Control in Laminated Architectures." Air Force Research Lab, Nonstructural Materials Branch presentation. Aug 2009.

[4] Carter, J., Cowen, A, Busbee, H., *et al*, "PolyMUMPs Design Handbook," Revision 11.0. MEMSCAP Inc. 2005.

Proceedings of the SEM Annual Conference
June 7-10, 2010 Indianapolis, Indiana USA

MEMS integrated metamaterial structure having variable resonance for RF applications

Derrick Langley, Ronald A. Coutu Jr. Dr.,
LaVern A. Starman, Dr., Peter J. Collins, Dr.,
Air Force Institute of Technology, 2950 Hobson Way, Bldg 641, Wright-Patterson AFB OH 45433, Ronald.Coutu@afit.edu

ABSTRACT

Metamaterial structures for RF applications are becoming essential in the race to reduce the footprint of antenna and components necessary for RF systems. Metamaterials provide a viable option to engineer structures from commonly used materials and processes to reduce the weight and size requirements for systems that normally operate at ¼ wavelength or greater in size for optimal performance. The Split ring resonators (SRR) first developed by Pendry, et al., has proven to be a viable component necessary to create negative index material structures. A fabricated SRR has a specific resonant frequency brought on by its permanent geometry. By incorporating a MEMS electrostatic cantilever with the SRR, an investigation into the ability to vary the resonant frequency of the SRR was completed. This paper reports on the modeling, design, fabrication and testing for this integrated component.

Keywords: Metamaterial, Split Ring Resonator, MEMS,

Background

Metamaterials developed from the original analytical work of Veselago, who in 1968 presented a paper explaining the concept behind "gyrotopic substances possessing both plasma and magnetic properties.[1]" Based on the analytical results he presented, a high degree of research goes into metamaterial structures. A majority of the investigations focus into research of material structures that do not exist in nature. One of the early breakthroughs in metamaterial research came from the work of Pendry, et al, who developed the Split Ring Resonator (SRR) [2].

The SRR design is a solenoid structure with a capacitive gap. The SRR helps tailor the resonant frequency for a planer structure based on the geometry, capacitive gap and materials. Based on these parameters, the SRR reacts to propagating electromagnetic waves based on the resonant frequency of the material and the ability to geometrically define the resonant frequency of the structure. Figure 1 shows two interlacing SRRs that due to the layout and size define the resonant frequency at RF wavelengths. Interlacing the structures helps to enhance the resonance and improves the figure of merit used as a quality factor in design requirements.

For this investigation, cantilevers incorporated into the design of the SRR affect the overall resonance of the structure. Using an electrostatic cantilever, the SRR's resonant frequency shifts allowing a different resonant frequency based on the proximity of the cantilevers. Isolating the tip of cantilevers from the other side of SRR gap is a dielectric layer which serves as a barrier to direct contact between the cantilever tip and SRR eliminating a short circuit of the structure. Combining the SRR and cantilevers allows for a novel tuning of the resonant frequency based on the pull-in activation of the cantilevers.

Disclaimer: The views expressed in this article are those of the authors and do not reflect the official policy or position of the United States Air Force, Department of Defense, or the U.S. Government.

DOI 10.1007/978-1-4419-8825-6_17, © The Society for Experimental Mechanics, Inc. 2011

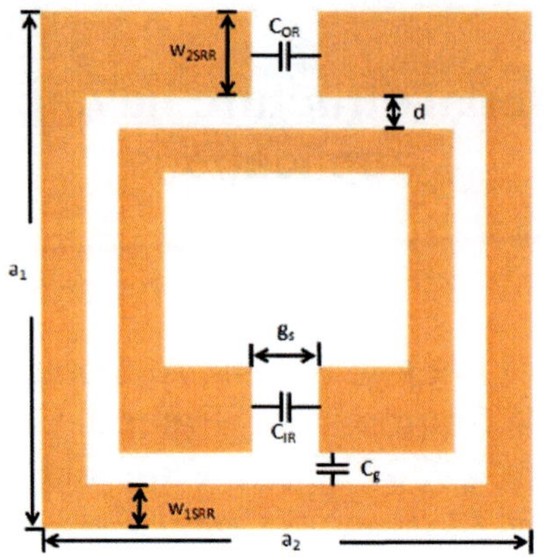

Figure 1. SRR unit cell with labeled ring dimensions

Analytical modeling for SRR

The SRR design based on initial work conducted by Pendry, et al., considers the interaction of plasma and magnetic properties of structures interacting with an electric field. The structure arranged like a solenoid with a capacitance gap gives the SRR inductive and capacitive properties effecting the overall permeability and permittivity of the structure. Based on the resonant frequency of interest, the SRR dimensions, material and gap determine the resonance of the structure. This investigation focuses on SRR structures that have a resonance in the range of 1.0 to 10 GHz built using microelectronic fabrication techniques. Based on the frequency range, the size of the SRR is limited to 3.0 to 30 mm in order to be considered a metamaterial structure.

To begin structural designs, we use the resonant frequency of interest to help determine the SRR dimensions. Equation (1) defines the resonant frequency.

$$\omega_0 = \frac{1}{\sqrt{LC}} \tag{1}$$

Where *L* is the inductance and *C* is the capacitance of the SRR. Dimensions and material characteristics determine the *L* based on the inductance of a solenoid given by Equation (2) using the length (*l*), thickness (*t*) and free-space magnetic permeability (μ_0). The dielectric materials and capacitive gap determine the overall capacitance of the SRR. The change in capacitance determines the cantilever beam dimensions based on the design requirements. Equation (3) calculates the capacitance based on the *t*, width (*w*), capacitive gap (*d*), free-space permittivity (ε_0) and relative permittivity (ε_r) of the material in the gap of the capacitor.

$$L = {\mu_0 l^2}/{t} \tag{2}$$

$$C = {\varepsilon_0 \varepsilon_r w t}/{d} \tag{3}$$

Using ω_0, Equation (4) models the effective response of magnetic permeability ($\mu(\omega)$) for an SRR structure. Equation (4) includes the fill factor (*f*) and the damping factor (Γ_m) associated with losses in the material. Assuming there are no losses for frequencies less than the plasma frequencies ($\omega < \omega_{pm}$), the constitutive parameter of $\mu(\omega)$ is negative.

$$\mu(\omega) = 1 + \frac{f\omega^2}{\omega_0^2 - \omega^2 - j\Gamma\omega} \qquad (4)^{[4]}$$

Equations (1) – (4) model a single SRR. To model two interlacing SRR as shown in Figure 1, the model must account for the interaction of the inner and outer SRR. When using two SRR, Equation (5) calculates the resonant frequency. Notice in the model, L_{av} is the average inductance of the two rings; C_g is the capacitance due to the gap of the concentric rings; C_{IR} and C_{OR} are capacitances due to the inner and outer splits, respectively [3].

$$\omega_0 = \sqrt{2\pi L_{av}\left(\left(\frac{\pi C_g}{2}\right) + C_{OR} + C_{IR}\right)}\ \text{(Hz)} \tag{5}$$

Figure 1 shows the location of the parameters used in calculating the ω_0 for the paired SRRs. Using the design, Equation (6) calculates the induction for the SRR using the free space wavelength (side length $a=\lambda_0/10$), and the width of the SRR, w_{SRR}, at the gap of each SRR [3].

$$L \approx \frac{\lambda_0 \mu_0}{5\pi}\left(2.3\log_{10}\left(\frac{8\lambda_0}{5w_{SRR}}\right) - 2.85\right)\ \text{(H)} \tag{6}$$

The capacitance of the split rings takes into account the gap capacitance between rings as shown in Figure 1. Equation (7) calculates the capacitance which feeds into the ω_0.

$$C_1 = C_{OR} + \frac{C_g}{2}\ \text{(F)}, \qquad C_2 = C_{IR} + \frac{C_g}{2}\ \text{(F)} \tag{7}$$

Cantilevers are beams fixed at one end and free at the other end (fixed-free beam). In MEMS, Cantilevers work as both DC and RF contact switches. For DC contacts, cantilevers open and close a circuit. Cantilevers used in RF circuits vary the capacitance gap for MEMS structures. For the SRR, the cantilevers change the capacitance between the gaps. To vary capacitance, the SRR is integrated with five cantilevers at the gap. Important modeling parameters of the cantilever are the spring constant and deflection. The spring constant (k_m) is the ratio of the applied force (F) and the resultant displacement (x) given by Equation (8) [5].

$$k_m = \frac{F}{x} \tag{8}$$

The deflection (d) occurring at the free-end of the beam is based on the applied force, beam length, Young's modulus and inertia. Equation (9) calculates the maximum deflection of the cantilever due to a force at the free end. The force deflects the cantilevers of the SRRs with electrostatic force which is created by applying a DC bias to pull-down pads underneath the cantilever.

$$d = \frac{Fl^3}{3EI}$$ (9)[5]

Equation (10) defines the magnitude of the electrostatic force necessary to move the cantilever based on applied voltage.

$$F_{electric} = \frac{1}{2}\frac{CV^2}{(d_{gap}+x)}$$ (10)[5]

Where C is the capacitance, V is the applied voltage, d_{gap} is the original distance between the cantilever and the biased plate and x is the resulting equilibrium displacement.

The well established pull-in voltage equation calculates the voltage necessary to have the cantilever contact the lower electrode. Equation (11) is the pull-in voltage equation for a simple cantilever beam. Varying the beam length changes the required pull-in voltage and allows variation in the capacitance for the SRR. A dominating parameter in the pull-in voltage is the spring constant based on Equation (8).

$$V_p = \frac{2x_0}{3}\sqrt{\frac{k_m}{1.5C_0}}$$ (11)[5]

Having the analytical background for the SRR and the cantilever, the integration of the metamaterial with MEMS becomes possible for structures operating at RF frequency range. The total design consists of arrays of SRR arranged to optimize the response at a range in the low GHz.

Design and fabrication

Designing metamaterial structures operating at 1.0 to 10 GHz makes fabrication, as an in-house effort, possible on 75 mm Quartz wafers. The structural dimensions in this RF range make it feasible to build structures using microelectronic fabrication techniques. We completed the design layout using L-Edit a circuit layout software tool available with the MEMS Pro software package. Figure 3 shows the isometric view of the SRR integrated with cantilevers.

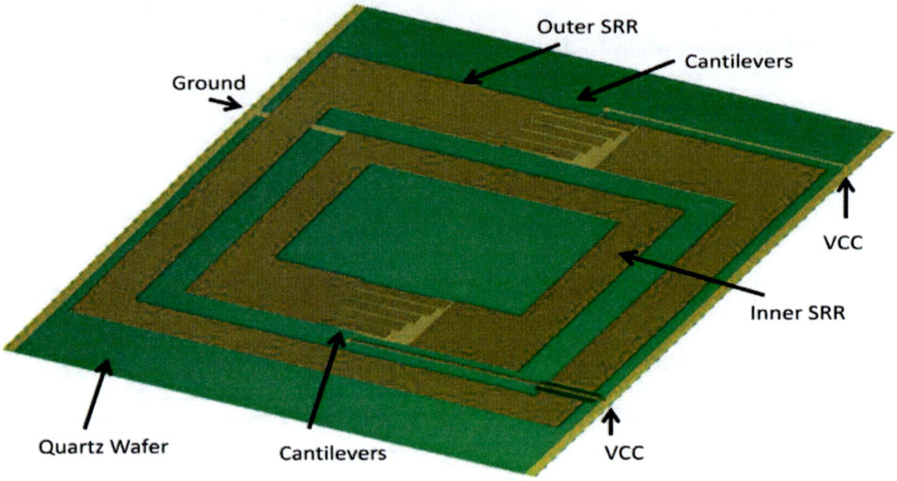

Figure 3 showing the SRR integrated with cantilever beams

Figure 4 shows the fabrication process flow for creating the structural layers. The wafer composed of quartz serves as the foundation of the structure helping isolate the SRR from adjacent structures. The wafer allows the transmission of GHz frequencies thus allowing only the SRR to affect the propagating signal. Next, the SRR structure is defined with an evaporated Ti/Au layer. The second process step defines the Ti/Au electrode used to apply biasing voltage and pull-in the cantilever. The third process step calls for depositing and defining the silicon nitride for the dielectric covering the electrode and gap of the SRR. The next two process steps consist of defining a sacrificial photoresist layer and patterning to protect the electrode and gap while opening anchor areas of the cantilever. Reflowing the photoresist helps to curve the cantilever reducing deflection due to stress. A seed layer of sputtered Au is deposited on the wafer helping with electroplating. After the seed layer, regions of the SRR and cantilever are defined with a 10 μm thick photoresist preparing the wafer for the Au plating bath. Using the Au plating bath, a 3-5 μm thick structure is plated for the SRR and cantilevers. The photoresist and seed layer are removed after completing the electroplating process. Dicing the wafers is the next process step to fit arrays into testing strip-line or waveguide. The release process, which incorporates a photoresist removal process and a critical carbon dioxide drying process, is the final step for removing the sacrificial layer allowing free standing cantilevers.

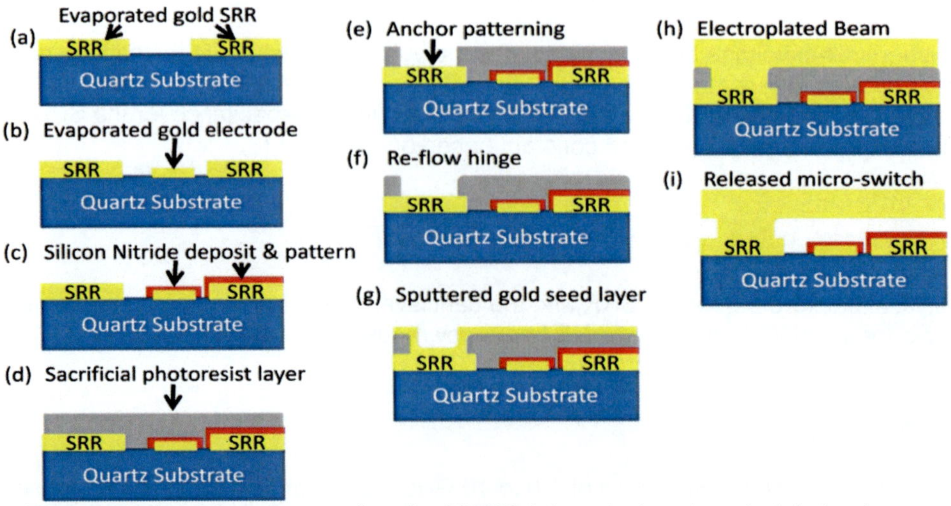

Figure 4 Fabrication process flow for MEMS integrated metamaterial structure.

Testing

Testing consists of two phases which involve DC and RF testing phases. In the first phase, cantilevers are tested for DC actuation. Each cantilever has a different length to vary the capacitive area within the SRR gap. Applying a DC voltage to the electrode pad within the gap draws the cantilevers towards the other side of SRR gap. The testing method used to observe the pull-in involves a white light interferometer and DC actuation using a voltage regulator. Figure 5 shows results from the white light interferometer indicating the open and close stages of the two longest cantilevers. The voltage required to close the 400 μm cantilever is 34 V, which verifies the analytical and FEM modeling for the cantilevers.

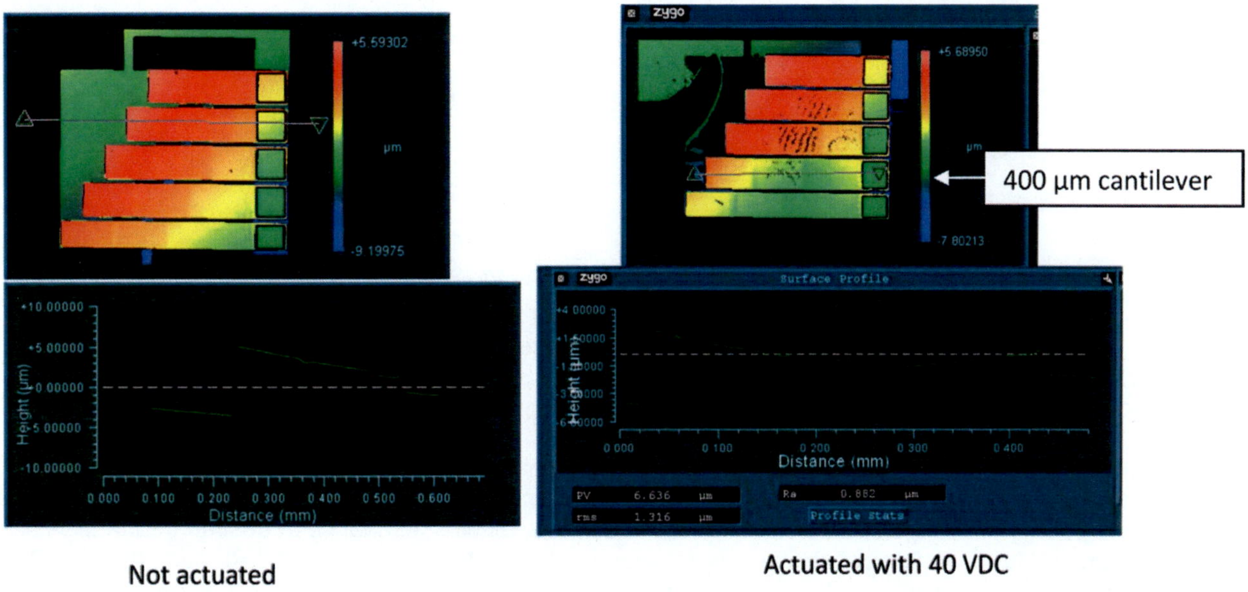

Figure 5. Pictures showing open and close cantilevers captured with Zygo Interferometer system

The second test phase involves inserting the metamaterial array into a strip-line to measure the change in refractive index at various frequencies. To perform the testing, an RF signal is transmitted through the strip-line with and without the metamaterial inserted in the plane of the propagation EM field/ wave. The test compares response and helps to verify the analytical modeling for the metamaterial array. Phase two testing will be completed after characterization and testing of the cantilevers is established.

Results

Using the Zygo Interferometer system, we observed the following results for the cantilever structures. Various cantilevers were tested using test structures. The length of the cantilevers varied from 400 μm down to 300 μm while the pull-in voltage varied from 34 V to 55 V with the initial test structures only tested for pull-in voltage. The next iteration of test structures will include capabilities to test the structures with RF signals similar to RF MEMS switches. The following table gives values for the pull-in voltage based on applied DC voltage.

Table 1. Pull-In Voltage Measured for various cantilever

Cantilever	Measured pull-in voltage
300	54
325	50
350	45
375	36
400	34

Similar to RF MEMS switches, the cantilever were tested for capacitance values in the open and closed states. The control voltage for the cantilevers was biased from 0 V to 55 V which pulled in the cantilevers making contact with the silicon nitride layer. As each cantilever contacts the nitride and lower electrode, the measurements of the capacitance is obtained using a Agilent Technologies LCR meter set up to measure capacitance at 1 MHz. Figure 6 shows the capacitance values obtained for various pull-in voltages on the cantilevers beginning with 0.144 pF measurement without biasing and 0.53 pF with biasing at 54.6 V.

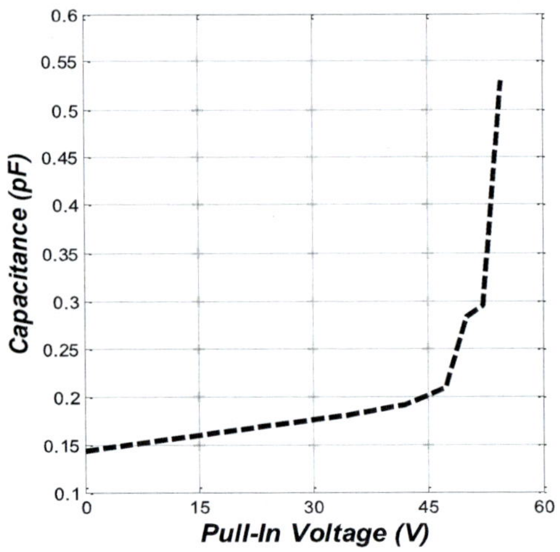

Figure 6. Capacitance Measurements versus Pull-In Voltage for Cantilever test structure

Conclusion

The design, fabrication and testing of the integrated RF metamaterial structure shows good agreement with experimental results. Based on analytical modeling and experimental testing, the integrated metamaterial design provides a capability to change structural features and achieve specific resonance at desired frequencies along the RF spectrum. Having tuning capabilities opens avenues for modifying antenna and RF components with metamaterial structures. Using results of our in-house effort, modifications will be made to investigate possibilities to reduce size, weight and power of RF systems for military applications.

Future Work

Future projects will focus on integrating RF cantilevers not only between the resonant gap of the SRR but the integrating cantilevers between the inner and outer SRR. Modification to the foundational SRR structure developed by Pendry, et al., opens lots of design opportunities for RF metamaterials. A goal of the future work will be to provide solutions to reducing size, weight and power requirements for RF antenna and system components.

Acknowledgements: The financial support and sponsorship of this project were provided by Drs. Katie Thorp and Augustine Urbas from the Air Force Research Laboratory (AFRL), Materials and Manufacturing Directorate. The authors are also thankful to the AFRL, Sensors Directorate for assistance and advice during device fabrication.

References

[1] Veselago, V.G., "The Electrodynamics of substances with simultaneously Negative Values of ε and μ," *Uspekhi Fizicheskikh Nauk*, 10(4), pp. 509-514, (1968).
[2] Pendry, J. B., Holden, A. J., Robbins, D. J., and Stewart, W. J., "Magnetism from Conductors and Enhanced Nonlinear Phenomena," *IEEE Transactions on Microwave Theory and Techniques*, Vol. 47, No. 11, pp. 2075-2084, 1999.
[3] Luke, R. A., "A MEMS Multi-Cantilever Variable Capacitor On Metamaterial," Master's thesis, Air Force Institute of Technology, 2008.
[4] Ramakrishna, S. A., Grzegorczyk, T. M., *Physics and Applications of Negative Refractive Index Materials.* Washington: SPIE PRESS, 2009.
[5] Liu, C., Foundations of MEMS. New Jersey: Pearson Education, Inc., 2006.

Proceedings of the SEM Annual Conference
June 7-10, 2010 Indianapolis, Indiana USA

Characterization and Testing of Adaptive RF Metamaterial Structure Using MEMS

Christopher A. Lundell, Peter J. Collins, Lavern A. Starman, Ronald A. Coutu, Jr.
Air Force Institute of Technology
2950 Hobson Way
Wright-Patterson AFB, OH 45433
christopher.lundell@us.af.mil

Abstract

Metamaterials are materials with periodic, sub-wavelength inclusions believed by some to generate responses that behave electromagnetically like effective mediums. This behavior provides the capability to synthesize materials with special parameters that will allow for size reduction of existing devices as well as the creation of new devices with unique capabilities. A key limiting factor for metamaterials, however, is the restriction of the special parameters to a small frequency band. To overcome this limitation, a microelectromechanical systems (MEMS) cantilever-beam device is placed over the gaps of a pair of split ring resonator (SRR) particles designed to operate in the radio frequency (RF) regime. Along with a wire lattice, the SRR particles are believed by many to generate an effective medium with a negative index of refraction over a small frequency band. The variable capacitors change the capacitance of the SRRs and shift the resonant frequency of the device. Efforts to predict and measure the behavior of this device are critical to the design effort. This paper demonstrates the techniques employed to model and test the electromagnetic properties of this device.

Introduction

Metamaterial devices have gained notoriety over the past decade. Government and industry expenditures have grown as researchers investigate the many different application areas. Some of the application areas that have been prominently mentioned include device miniturization, subwavelength focusing, and even electromagnetic cloaks.

Metamaterials are constructed using periodic inclusions that, when sized appropriately, give the structure unique properties. Many believe metamaterials create an effective medium with its own electric permittivity and magnetic permeability that are not found in the constituent materials. A class of metamaterials, called double negative (DNG) metamaterials, is of particular interest since they are believed by some to have a frequency band where the effective permittivity and permeability are simultaneously negative. This feature is not found in nature and leads to some of the more controversial and interesting applications. The negative permittivity and permeability are the result of the resonances of the periodic inclusions which are typically a wire lattice and concentric SRR pair for a DNG metamaterial.

DNG metamaterial designs are constrained by the fact that the negative permittivity and permeability are confined to a resonance frequency band. In traditional passive metamaterial devices, this resonance band is fixed. Many researchers have described ways to create an adaptive metamaterial that can change resonant frequencies. Most adaptive metamaterial designs focus on manipulating the resonance of the SRR elements. One common method to introduce frequency adaptability is to add an additional magnetic field that passes through the gap of the SRR. In one such design, described in [1], an external magnetostatic field is applied to the structure that shifts the resonant frequency. By adding a magnetic layer into the structure, the design in [2] also applies an additional magnetic field and shifts the resonant frequency.

Another common method to make an adaptive metamaterial is to add a circuit component, typically a capacitor, to the SRR particles. Gil et al. [3] and Shadrivov et al. [4] propose metamaterial devices with additional varactor diodes that adjust the resonant frequency. The design analyzed in this paper is very similar to those devices.

T. Proulx (ed.), *MEMS and Nanotechnology, Volume 2*, Conference Proceedings of the Society for Experimental Mechanics Series 2,
DOI 10.1007/978-1-4419-8825-6_18, © The Society for Experimental Mechanics, Inc. 2011

The key difference is that the varactor diodes are replaced by much smaller MEMS variable capacitors. A metamaterial design that incorporates a MEMS device is proposed by Hand and Cummer in [5]. However, their design uses a MEMS switch that, unlike our design, has only two states: open and close. The MEMS capacitor used in our design has six different states.

Theory

The structure presented in this paper attempts to adjust the resonant frequency of the metamaterial device by using an additional circuit element, a variable MEMS capacitor. If an equivalent circuit model of the metamaterial structure is used, the angular resonant frequency (ω_0) can be expressed as a function of the device's equivalent capacitance (C_{eq}) and inductance (L_{eq}) as [5]

$$\omega_0 = \frac{1}{\sqrt{L_{eq} C_{eq}}}. \tag{1}$$

A novel metamaterial structure that achieves frequency adaptability was proposed in a previous research effort [6]. The metamaterial structure is a basic DNG unit cell. The proposed structure incorporates a MEMS variable capacitor across the gaps of the SRR particles. Figure 1 shows the proposed design. The variable MEMS capacitor contains five cantilever beams. The capacitance of the device is adjusted by applying a voltage to the control lines of the cantilever beams. The control lines also serve as the wire lattice that gives the metamaterial its electric response.

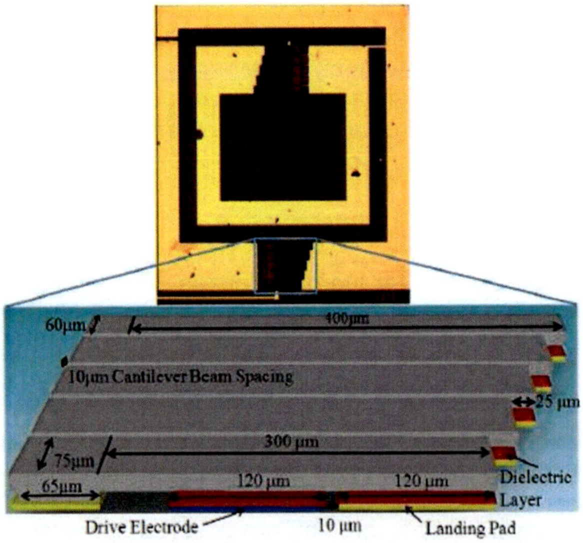

Figure 1: An adaptive DNG metamaterial design that uses cantilever beam variable capacitor [6, 52].

The array of cantilever beams is controlled by a single control line. When there is no voltage applied to the control line, all of the cantilever beams are in the up state. As the voltage on the control lines increase, the cantilever beams flex downwards. This has the effect of adding a capacitor to the circuit. Measurements of the MEMS device's capacitance are described in [6, 88] and are repeated in Table 1. Note that measurements did not include the 325 µm beam. For this paper, its capacitance is estimated from the capacitance of the other beams.

Table 1: Measured capacitances for MEMS device.

State	Activated Beams	Capacitance (pF)
1	None	0.12
2	300 μm	0.77
3	300 and 325 μm	1.36
4	300, 325, and 350 μm	1.90
5	300, 325, 350, and 375 μm	2.52
6	300, 325, 350, 375, and 400 μm	3.06

Computational Models

Computational models are presented to show the effect of changing the state of the MEMS capacitor. These models are run using CST Microwave Studio, a commercial full-wave electromagnetics solver. Since the metamaterial structures are resonant in frequency, the frequency-domain solver is used. The models are meant to simulate operation of the adaptive metamaterial structures in a stripline waveguide.

The adaptive metamaterial structure is modeled in a large stripline. The stripline is designed for single-mode operation below 4 GHz, so the resonant behavior of the original-size structure (expected to occur near 10 GHz) will not be visible. Figure 2 shows the model of the adaptive metamaterial structure in the stripline. All of the boundary conditions are set to open, and the *xz*-plane passing through the center of the center conductor is designated a magnetic symmetry plane. With the control wire traces arranged perpendicular to the stripline's center conductor as shown in Figure 2, a metamaterial slab that is 5 unit cells tall is the largest structure that can fit in the stripline. To ease some of the computational requirements, the slab is just four unit cells in the *z*-direction. The center conductor is scaled in length to be three times longer than the metamaterial slab.

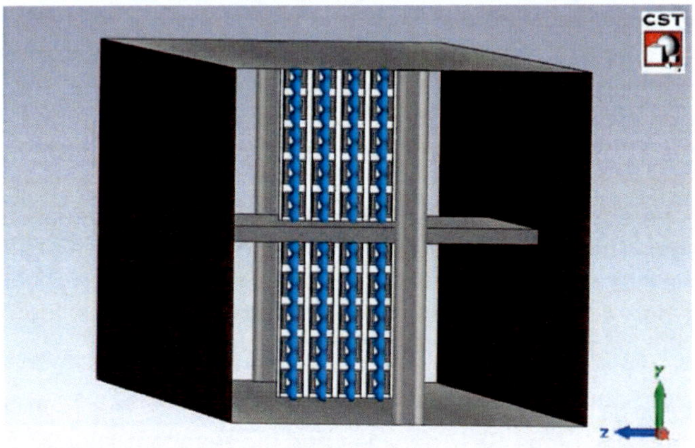

Figure 2: Model of the original-size adaptive metamaterial structure in a stripline.

The MEMS variable capacitor is represented by a lumped network element (the blue devices in Figure 2). Figure 3 shows the *S*-parameter results from this model for each of the capacitance values shown in Table 1. There is a lot of erratic behavior in the data, possibly indicating the presence of multiple modes. There appears to be a resonance point near 2.25 GHz, but there is too much erratic behavior to clearly determine the resonance points.

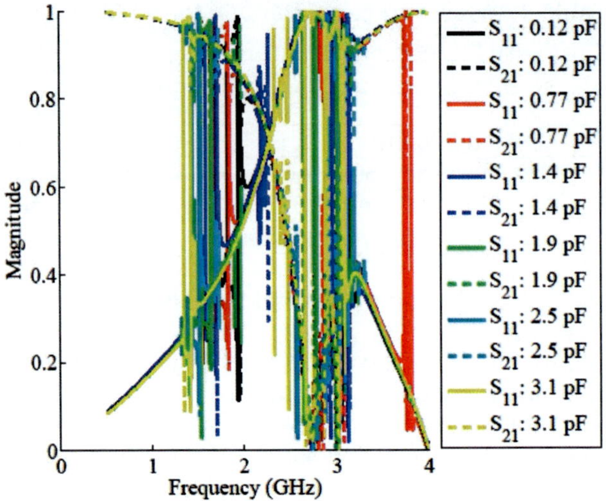

Figure 3: The magnitudes of S_{11} and S_{21} from the model of the original scale AFIT metamaterial structure.

The model was also modified to add two more metamaterial strips parallel to the two already in the model (two strips above the center conductor and two strips below the center conductor) to determine the effects of non-coplanar coupling. The strips are spaced such that they are separated by a distance equal to the width of one unit cell. The *S*-parameter results are very similar to the results from the previous model shown in Figure 3. The apparent resonance is shifted to about 1.8 GHz. Like the previous results, the results for this model display too much erratic behavior to accurately determine the resonance points, but introducing out-of-plane coupling by adding an additional strip typically results in the same resonant frequency. The fact that the four-strip model significantly changes the resonant frequency suggests that the crossing point of S_{11} and S_{21} seen in Figure 3 is not the resonant frequency of interest. This is expected since the original-size structure should not resonate at this low of a frequency. One thing is apparent in the results shown in Figure 3: the resonant frequency of the structure is not very responsive to changes in the MEMS capacitor. Simplified models of the adaptive design that provide accurate results above 10 GHz also confirm the small changes in resonant frequency.

Because the larger stripline operates in a lower frequency band, the metamaterial structure can be redesigned to match the new frequency band. This means increasing the size of the structure. Unlike a rectangular waveguide, the TEM mode of the stripline has no lower frequency cutoff. So, only the physical size of the stripline restricts how large the structure can be. Scaling the metamaterial structure between four to five times larger than the original design will nearly fill the height of the stripline with a single unit cell.

The relationship between the scale of the metamaterial structure and its resonant frequency can be seen in the relationship between the dimensions of the SRR particles and the resonant angular frequency given by [7]

$$\omega_0 = \sqrt{\frac{3dc^2}{\pi^2 r^3}}, \tag{2}$$

where *d* is the length of the gap between the inner and outer SRR, *c* is the speed of light in the substrate containing the SRR particles, and *r* is the radius of the inner SRR. It should be noted that this relationship is only approximately true for this structure since it assumes the SRR particles are round. Scaling both *d* and *r* by the scaling factor produces the approximate resonant frequency of the larger structure. This predicts that a structure that is five times larger than the current design will resonate near 2 GHz. Furthermore, it appears anything scaled at least 2.5 times larger than the current design will have a resonant frequency in the frequency band that is in the single-mode band for the stripline.

Models of the adaptive metamaterial design at the 4.0 and 4.9 scale factors in the stripline were created. The 4.0 scale factor model is shown in Figure 4. Like the previous models, lumped network elements are used to model the MEMS capacitors. The same capacitances from the original-size design are used in this model (see Table 1). The first model has two strips total – one above the center conductor and one below.

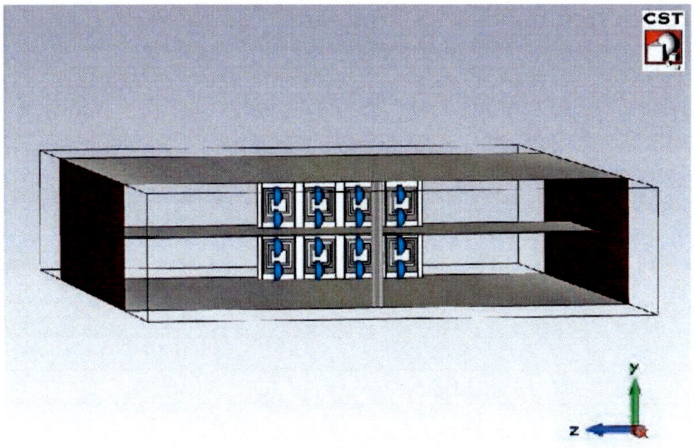

Figure 4: Model of the larger-scale adaptive metamaterial structure in a stripline.

The results of the simulation clearly show the change in resonant frequency at each capacitance value. Figure 5 shows the magnitude of S_{21} for the 4.0 and 4.9 scale factors. The dips in S_{21} show the resonant frequencies. The resonant frequencies for the 4.0 scale structure are approximately 2.17, 1.26, 0.995, 0.833, 0.752, and 0.680 GHz for additional capacitances of 0.12, 0.77, 1.36, 1.90, 2.52, and 3.06 pF, respectively. Likewise, the resonant frequencies for the 4.9 scale structure are approximately 1.93, 1.12, 0.883, 0.761, 0.667, and 0.608 GHz for additional capacitances of 0.12, 0.77, 1.36, 1.90, 2.52, and 3.06 pF, respectively.

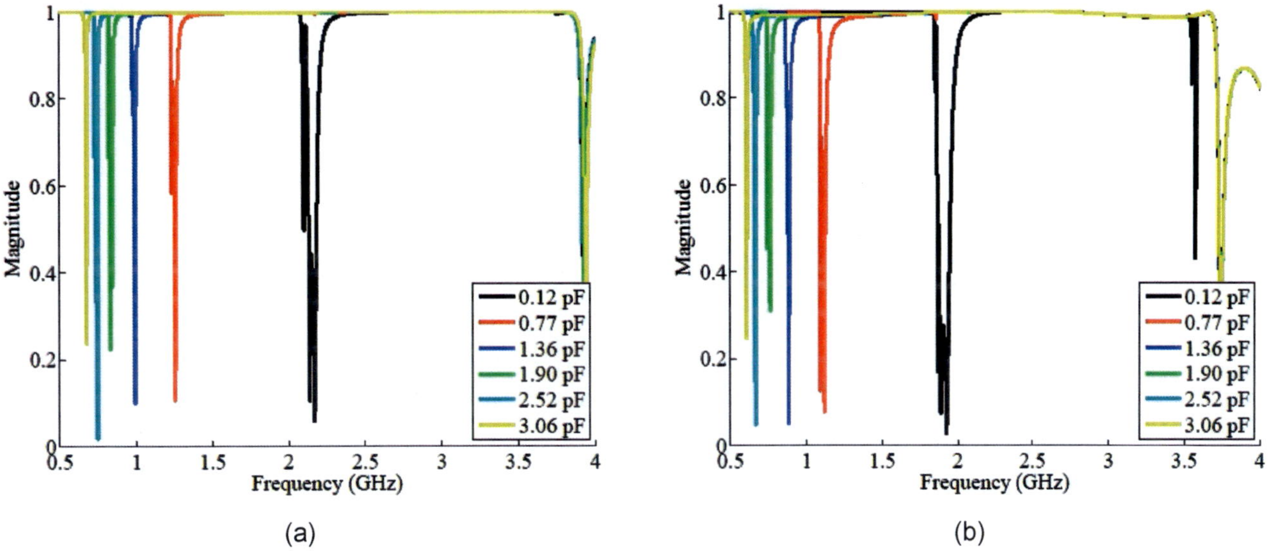

Figure 5: *S*-parameter results from the large-scale adaptive metamaterial model at the 4.0 (a) and 4.9 (b) scale factors.

Like before, two strips are added to each model to create a total of four strips (two above and two below the center conductor). The results of these models are very similar to the results of the two-strip models. The resonant frequencies are very close to the two-strip models. The resonant frequencies for the 4.0 scale structure are 2.15, 1.24, 0.977, 0.838, 0.739, and 0.671 GHz for additional capacitances of 0.12, 0.77, 1.36, 1.90, 2.52, and 3.06 pF, respectively. Similarly, the resonant frequencies for the 4.9 scale structure are 1.89, 1.11, 0.874, 0.752, 0.658, and 0.604 GHz for additional capacitances of 0.12, 0.77, 1.36, 1.90, 2.52, and 3.06 pF, respectively.

The results of the four-strip models match the two-strip model results closely. Table 2 shows the resonant frequencies and the percentage difference between the results of the four-strip and the two-strip models. The maximum difference is slightly over 2%.

Table 2: Resonant frequencies for large-scale adaptive metamaterial models.

Scale	Capacitance (pF)	Resonant Frequency (GHz)		Difference (%)
		Two-Strip Model	Four-Strip Model	
4.0	0.12	2.17	2.15	0.92
	0.77	1.26	1.24	1.59
	1.36	0.995	0.977	1.81
	1.90	0.833	0.838	0.60
	2.52	0.752	0.739	1.73
	3.06	0.680	0.671	1.32
4.9	0.12	1.93	1.89	2.07
	0.77	1.12	1.11	0.89
	1.36	0.883	0.874	1.02
	1.90	0.761	0.752	1.18
	2.52	0.667	0.658	1.35
	3.06	0.608	0.604	0.66

The resonant behavior of the structure is clearly evident below 4 GHz. Furthermore, the resonant frequency appears to be highly influenced by the parallel capacitor. Figure 6 shows the resonant frequency of the 4.0 and 4.9 scale models at the measured values of the MEMS capacitor. The resonant frequency is inversely proportional to the capacitance. The capacitance between the two SRR particles is in series with the MEMS capacitor, and their equivalent capacitance (C_{eq}) is given by

$$C_{eq} = \frac{C}{1 + C / C_{VC}}, \tag{3}$$

where C refers to the capacitance between the two SRR particles and C_{VC} is the capacitance of the variable capacitor.. As a consequence of (3), the equivalent capacitance increases with the MEMS capacitor. Since the resonant frequency of a metamaterial structure is given by (1), the trends in the resonant frequencies shown in Figure 6 are expected.

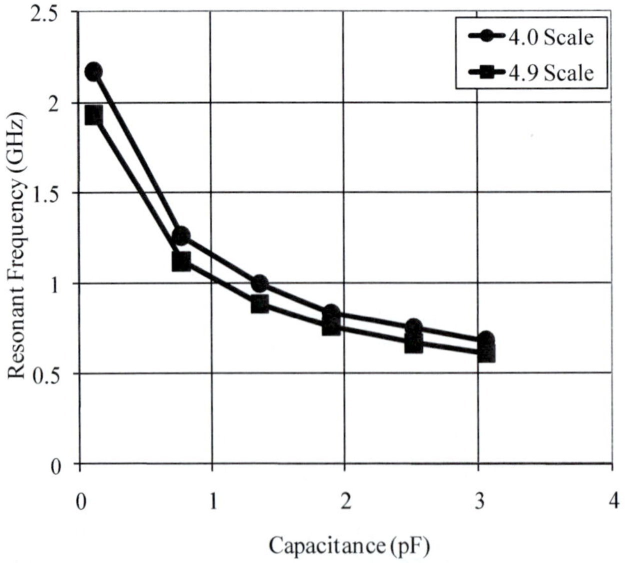

Figure 6: Resonant frequencies for the larger-scale designs as a function of MEMS capacitance.

Analyzing (3) reveals that if C is small, the denominator is close to one, resulting in $C_{eq} \approx C$. In that scenario, changes in the variable capacitor will have minimal effect on the equivalent capacitance. The capacitance per unit length between the two SRR particles (C_l) is given by [7]

$$C_l = \frac{\varepsilon_0}{\pi}\ln\frac{2w}{d}. \qquad (4)$$

Multiplying C_l by the length of the SRR particles (roughly $2\pi r$) yields $C = 2r\varepsilon_0 \ln(2c/d)$ where ε_0 is the permittivity of freespace. Therefore, multiplying the size of the structure by a scaling factor directly increases the capacitance between the SRR particles. This explains why the MEMS capacitances cause larger changes in the resonant frequency of the larger designs and not the smaller designs. The capacitance between the SRR particles in the smaller design is small and dominates (3). The capacitance of the larger design is bigger, allowing the variable capacitor to have more influence on the equivalent capacitance.

Conclusion

The models presented in this thesis provide an accurate and efficient way to characterize the behavior of electromagnetic fields in metamaterial devices. Despite their increased complexity, the simulation time is tolerable (the longest model took a little over 15 hours to compute – the computation included 12 individual runs), and they provide accurate results. Comparisons between the results from models and stripline measurements for passive metamaterial structures with no variable capacitors show the same resonant frequencies. Thus, the modeling approach used in this paper is valid. Samples of the adaptive metamaterial structure were not available for measurement before the publication of this paper.

The computational models of the original-scale metamaterial structure show that the variable capacitor has little effect on the resonant frequency of the device. They also show that the original-scale structures have no resonances below 4 GHz. Increasing the size of the metamaterial structures lowers the resonant frequency, and it is shown that scale factors between 4.0 and 4.9 are well suited for the large stripline. Structures larger than 4.9 times the original scale will not fit between the center and outer conductors of the stripline.

As mentioned above, the models show that the resonant frequency of the original-scale structure is not very responsive to changes in the variable capacitor. This is because the equivalent capacitance of the structure is dominated by the capacitance between the inner and outer SRR particles. At the larger capacitances, the capacitance between the inner and outer SRR particles is increased, so the variable capacitor has more influence on the resonant frequency. There is a strong relationship between frequency adaptability and the size of the metamaterial structure, and the model results show that the 4.0 and 4.9 scale structures' resonant frequency can be highly influenced by the MEMS capacitor design.

References

[1] Han, Jiaguang, Akhlesh Lakhtakia, and Cheng-Wei Qiu. "Terahertz metamaterials with semiconductor split-ring resonators for magnetostatic tenability". *Opt. Express*, 16(19):14390-14396, 2008.

[2] Gollub, Jonah N., David R. Smith, and Juan D. Baena. "Hybrid resonant phenomenon in a metamaterial structure with integrated resonant magnetic material". *Opt. Express*, 17(4):2122-2131, 2008.

[3] Gil, I., J. García-García, J. Bonache, F. Martín, M. Sorolla, and R. Marqués. "Varactor-loaded split ring resonators for tunable notch filters at microwave frequencies". *Electron. Lett.*, 40(21):1-2, Oct. 2004.

[4] Shadrivov, Ilya V., Steven K. Morrison, and Yuri S. Kivshar. "Tunable split-ring resonators for nonlinear negative-index metamaterials". *Opt. Express*, 14:9344, 2006.

[5] Hand, T. and S. Cummer. "Characterization of tunable metamaterial elements using MEMS switches". *IEEE Antennas Wireless Propag. Lett.*, 6:401-404, Jan. 2007.

[6] Rederus, Luke. *A MEMS Multi-Cantilever Variable Capacitor on Metamaterial.* Master's thesis, Air Force Institute of Technology, Air University, Wright Patterson AFB OH, 2009. (ADA497157).

[7] Pendry, J. B., A. J. Holden, D. J. Robbins, and W. J. Stewart. "Magnetism from conductors and enhanced nonlinear phenomena". *IEEE Trans. Microw. Theory Tech.*, 47(11):2075-2084, Nov. 1999.

Proceedings of the SEM Annual Conference
June 7-10, 2010 Indianapolis, Indiana USA

Design of a microfluidic pump, based on conducting polymers

Karthik Kannappan, Department of Chemistry, 23, Symonds St, University of Auckland, Auckland 1010. New Zealand. k.kannappan@auckland.ac.nz
Gib Bogle, Auckland Bioengineering Institute, University of Auckland
Jadranka Travas-Sejdic, Department of Chemistry, University of Auckland
David. E. Williams, MacDiarmid Institute for Advanced Materials and Nanotechnology, Department of Chemistry, University of Auckland

ABSTRACT

Electrochemically-active conducting polymers (ECP) swell or shrink in response to ion and solvent incorporation or ejection as a result of electrochemical reaction of the polymer. As a consequence, they are, in principle, attractive materials to consider for inducing fluid motion of electrolytes in microfluidic systems. When anodic potential is applied to an electrode attached to one end of ECP strip, the oxidation process starts from the electrode and proceeds along the polymer, propagating as a wave. This wave is driven as a consequence of the electrochemical reactions and would be coupled to a propagating front of compositional change. This property of the ECP can be used to design pumps and mixers for microfluidic systems. We in this paper set up a 2-D transport model to explain this wave phenomenon that includes both diffusion and electro-migration, that is coupled to the reaction at the polymer-solution interface, and that also includes the effects of change of polymer conductivity on the charge transport in the polymer layer. We explore the design for a microfluidic pump that uses this process and its efficiency to pump electrolytes.

1. Introduction:

When conducting polymers switch between oxidized and reduced state, the conductivity, color and volume undergo significant changes. There is a significant interest to use the volume change property of the ECP to build actuators and mixers for microfluidic devices. Smela et al [1] have summarized the estimates of the magnitude of the volume change that can be achieved in ECP actuators up to 120%, depending on the conditions. There have been attempts to construct actuators and valves utilizing the redox switching and volume change phenomenon of the conducting polymers [2, 3]. These actuators and valves involve complicated fabrication techniques and are somewhat difficult to control. Tezuka and others [4-7] reported an oxidation wave phenomenon in conducting polymers, where when anodic potential is applied to one end of a polymer strip, the oxidation process starts from that end and propagates as a wave. This oxidation wave would be coupled with a mechanical deformation due to the volume change property of the conducting polymer. We aim to utilize this property of the oxidation wave to design a simple pump for microfluidic devices.

We have constructed a numerical model to determine the conditions under which this surface wave can be made to propagate along the polymer surface. We have linked the effects of geometry of the system, conductivity of polymer/electrolyte and the applied potential to the redox switching and the conductive front wave propagation. We then have used the numerical modeling package COMSOL to design a pump utilising this property and have determined its pumping efficiency.

2. Modelling oxidation wave propagation in conducting polymers:

We consider a thin slab of electronically conducting polymer. The polymer slab is in contact with the electrolyte solution. When we apply an anodic potential at one end of the polymer slab it creates a potential gradient initially at one end of the solution. Due to this potential gradient, the polymer will be oxidized, releasing electrons to the

DOI 10.1007/978-1-4419-8825-6_19, © The Society for Experimental Mechanics, Inc. 2011

external circuit and exchanging ions with the solution in order to balance the charge. The oxidation of the polymer increases its electrical conductivity, so that this part acts as a contacting electrode to the rest of the polymer strip. The process continues until the entire polymer strip is oxidized. Qualitatively, one can observe that the way in which the polymer transformation propagates along the strip should depend on the conductivity of the polymer in the initial state, whether the polymer conductivity increases or decreases in response to the electrochemical change, and the ion transport in the solution. Under some conditions, the transformation should occur as a wave front propagating along the strip.

This process of the wave propagation has been explained and also been experimentally demonstrated by Tezuka and others [4-7]. The experimental set up [4] consisted of strip of reduced conducting polymer on a non conductive glass with one end clamped to a metal electrode. When an anodic potential was applied to the metal electrode an oxidation wave proceeded from the metal electrode end to the other end of the film.

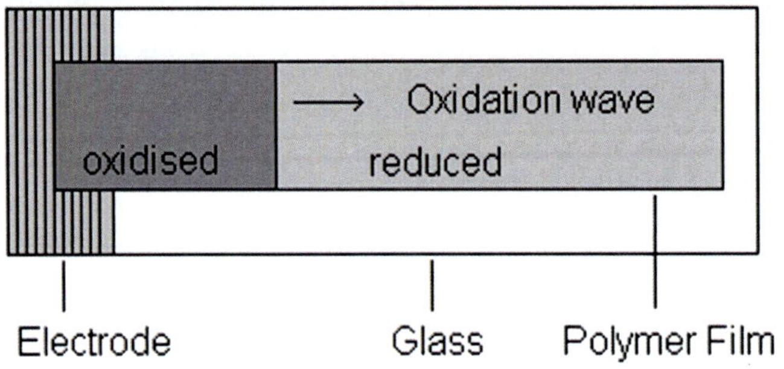

Fig 1 Oxidation wave (Tezuka et al. experimental setup)

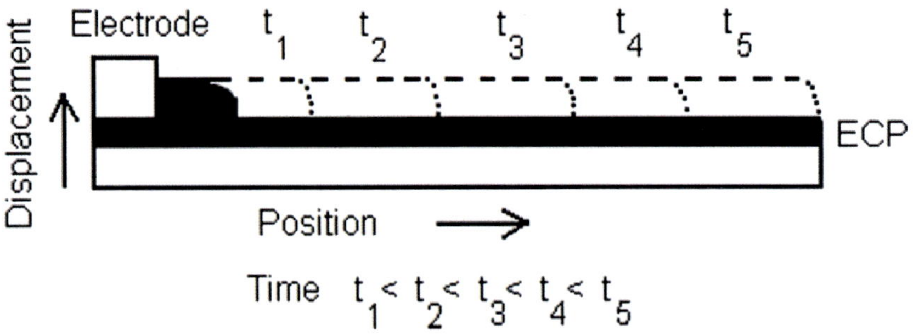

Fig 2 Oxidation wave with polymer swelling

In their experiment the oxidation wave was observed by measuring the film absorbance to a monochromatic light beam. Tezuka et al explained this phenomenon qualitatively. Madden and Warren [8] addressed the results of Tezuka et al and simulated the behavior using an empirical variable resistance transmission line model. Rather than adopt the empirical approach of Madden and Warren, we have developed and solved a complete 2-D model.

We have developed a mathematical model for the two dimensional geometry represented in fig 3. The following assumptions were made to derive the equations. The polymer layer is considered long in comparison to its thickness and is assumed to be uniform in composition in the *z* direction (right angles to its long axis). The ion exchange reaction at the polymer solution interface is assumed to be at equilibrium, so that the potential difference across the polymer solution interface at any position is given by the Nernst equation applied at that position. The electrolyte solution (represented as A^+B^-) has an initial concentration C_0 For simplicity; all the equations derived here assume that A^+ and B^- are the only ions in the solution. To keep system electrically neutral and balanced, we consider that the negative ion is injected from counter electrode to balance that absorbed into the polymer. The redox reaction of the polymer can be represented by equation (1)

$$\underset{(1-\varepsilon)}{Ppy + B^-} \underset{k^-}{\overset{k^+}{\rightleftharpoons}} \underset{(\varepsilon)}{(Ppy^+B^-) + e^-} \tag{1}$$

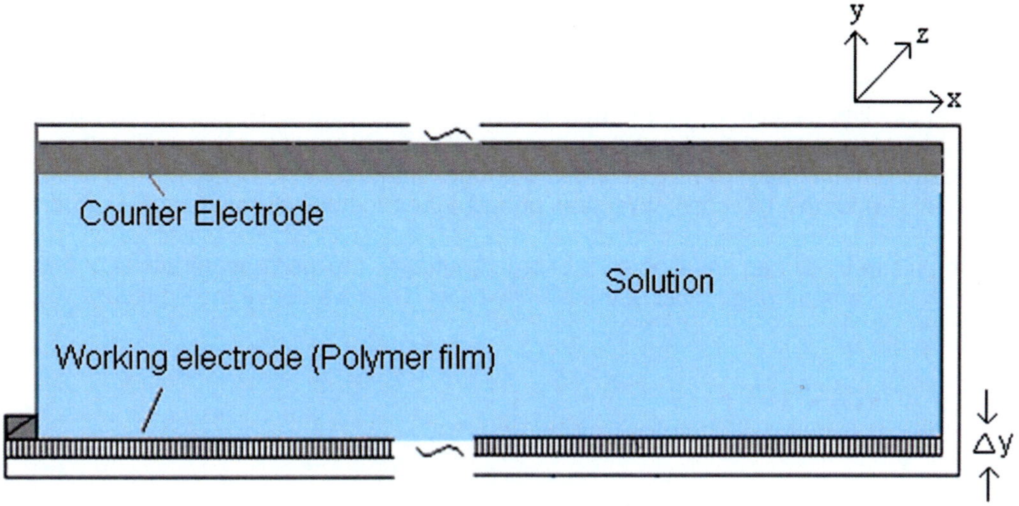

Fig 3. Geometry of the model

Where ε is the degree of oxidation. When a potential difference is created at the interface, the polymer is oxidized (or reduced), releasing (or consuming) electrons from the external circuit. Balancing this electron flux, there is a flux of negative ions into (or out of) the polymer. The local current density in the solution is related to the rate of oxidation of the polymer which in turn is directly related to the flux of the ions into the polymer. Hence the local current density in the solution is given by

$$i_{Sy} = MF\Delta y \frac{\partial \varepsilon}{\partial t} \tag{2}$$

Where F is the Faraday constant, M is the ion capacity of the polymer film expressed in Moles.m^{-3}, and i_{sy} is the current density in the y-direction (the surface normal) at the interface and Δy is the polymer layer thickness. Equation (2) expresses the assumption of uniformity of composition of the polymer in the y-direction. Note that the width in the z-direction, w, is assumed sufficiently large that the problem can be considered as a 2-D one, with variations only along the x- and y-axes. Similarly there will be a current flow I_P along the length of the polymer, removing the charge consumed by oxidation resulting from the ion flux into the polymer.

$$\frac{\partial I_P}{\partial x} = -i_{Sy} w \tag{3}$$

The potential gradient in the polymer film is determined by this current I_P:

$$\frac{\partial \phi_P}{\partial x} = -\left(\frac{\rho}{w\Delta y}\right) I_P \tag{4}$$

Where ρ is the resistivity of the polymer, which varies with the state of oxidation of the film. (Note that i_{sy}, I_P, ϕ_P and ρ are all functions of x.) Feldman [9] and others [10] suggest that the resistivity switches from a very high (insulating) to a very low (conducting) value when the state of oxidation increases. Warren and Madden [8] propose an empirical sigmoidal function model from which we can determine the conductivity σ_p of the polymer from its fractional oxidation state.

$$\sigma_P = \frac{\sigma_{\max} - \sigma_{\min}}{1 + e^{(\varepsilon - \varepsilon_T)/W_T}} + \sigma_{\min} \tag{5}$$

Where ε_r is the oxidation state at which the transition occurs and W_T is the transition width. The oxidation state and hence the conductivity varies along the length of the polymer.

The Nernst-Planck equation describes the dynamics of ion transport under the influence of an applied electric field and diffusion. Now to solve for the potential distribution in the solution one has to solve Poisson's equation along with this equation. In practice the high degree of non-linearity of these equations, and the very different distance and time scales of variations in the potential and variations in the concentration make it difficult to solve the resulting set of equations numerically. So we here use an approximate method to handle this situation, developed by Cohen and Cooley [16] for the 1-D case. We have adapted their method to solve our 2-D problem.

The key assumption made by Cohen and Cooley is charge neutrality, i.e. the concentrations of the positive and negative ions remain the same through most of the bulk solution. Thus we derive the equations (6) and (7) and (8)

$$\frac{\partial C}{\partial t} = D_A \nabla^2 C + U_A \nabla \cdot (C \nabla \phi_S) \tag{6}$$

$$i_{S\,y} = (J_A - J_B)F \tag{7}$$

$$(U_A - U_B)\frac{RT}{F}\nabla^2 C + (U_A + U_B)\nabla \cdot (C \nabla \phi_S) = 0 \tag{8}$$

The polymer potential ϕ_p and the polymer current I_P together satisfy Ohm's law (4), where the resistivity ρ is a function of ε (ρ is the inverse of σ_p which is given by (5)). The reaction at the polymer solution interface is assumed to be at equilibrium so the potential difference across the interface, ϕ_p - ϕ_s is assumed given by the Nernst equation (9)

$$\phi_p - \phi_s = \Delta\phi^{01} - \frac{RT}{F}\ln\left(\frac{C_B(1-\varepsilon)}{C_0 \varepsilon}\right) \tag{9}$$

For a given set of boundary conditions, the coupled equations (2), (3), (4), (5), (6), (7), (8) and (9), completely describe the time evolution of the system. The non-linear time-dependent partial differential equations linking the spatial distributions of potential, ionic concentrations, and polymer oxidation are solved in Matlab using the Scharfetter-Gummel finite difference method. These methods are based on approaches developed in semiconductor analysis [12] and bio-physics [13]. The results show how the concentrations and polymer oxidation evolve with time over the simulation.

3. Results and discussion determining oxidation wave properties:

The ratio of the electrolyte conductivity to the minimum polymer conductivity determines the shape and velocity of the wave propagation (Fig. 4, 5). Solution concentration determines the electrolyte conductivity and minimum conductivity of the polymer depends on its properties. Here length of the polymer is 1cm and the distance between electrodes is 1mm. When the applied voltage was increased the average velocity of the wave increased linearly (Fig.6)

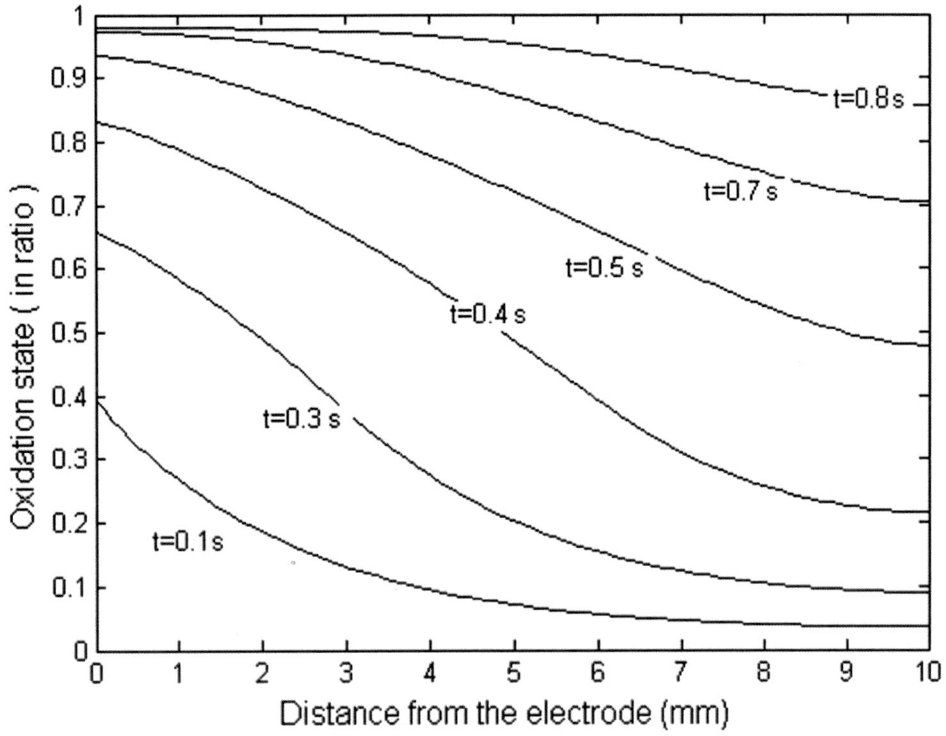

Fig 4 Variation of oxidation wave shape when electrolyte to minimum polymer conductivity is 0.001

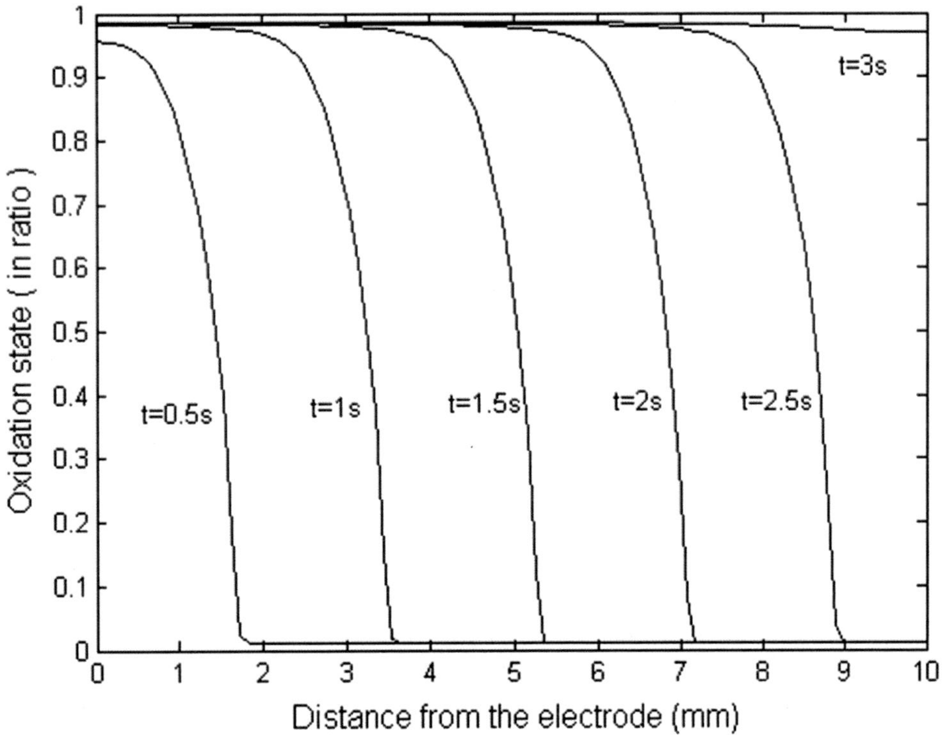

Fig 5 Variation of oxidation wave shape when electrolyte to minimum polymer conductivity is 1

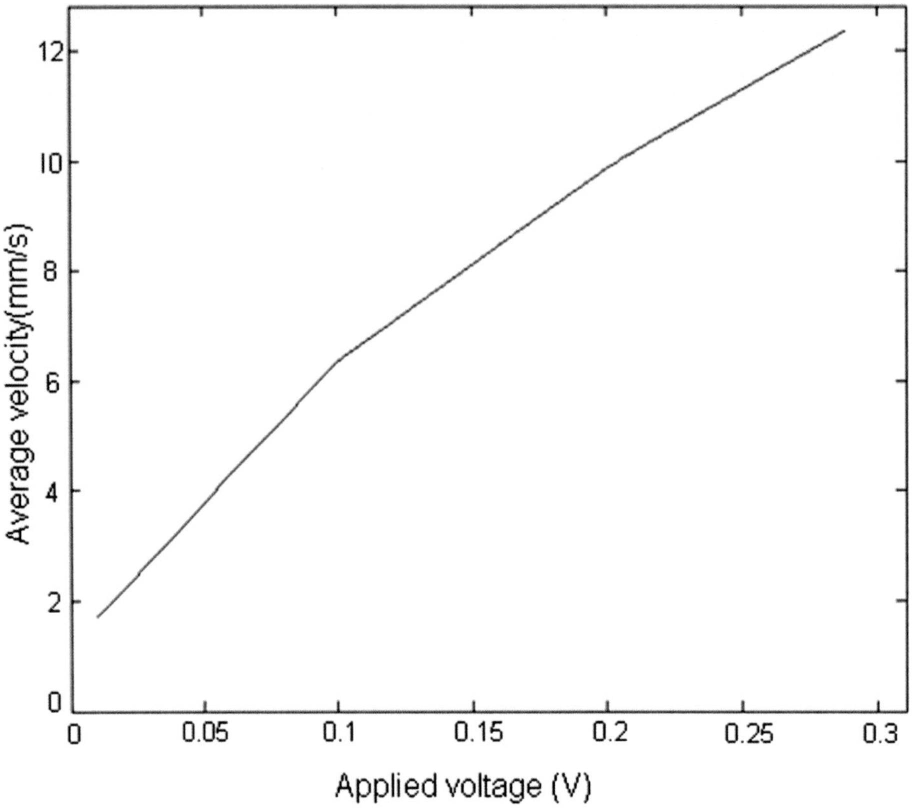

Fig 6.Variation of oxidation wave velocity with applied voltage

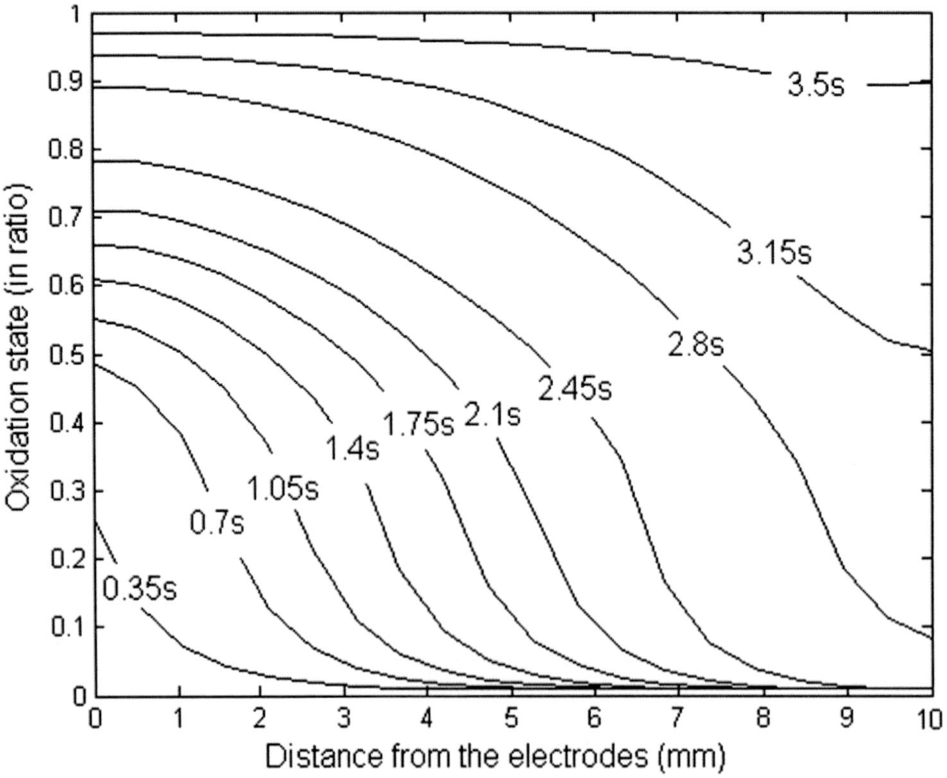

Fig 7.Variation of oxidation wave shape when distance between electrodes is 5mm

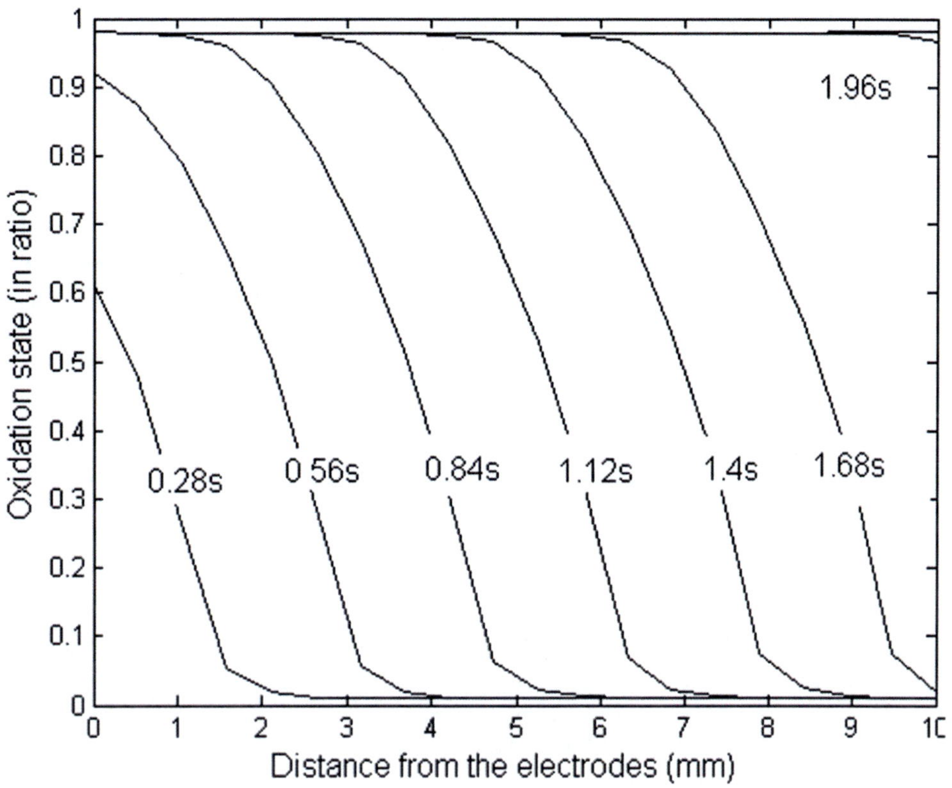

Fig 8.Variation of oxidation wave shape when distance between electrodes is 1mm

When distance between polymer and counter electrode is increased, the rate of the oxidation is reduced and the wave propagates slowly (Fig. 8). Here, the length of the polymer is 1cm and the solution concentration is 0.1M. The reverse process results in a uniform contraction of the polymer. The reason is that the switch is from a high conductivity state to a low conductivity state, so the current is initially distributed uniformly across the polymer surface [8].

The two dimensional numerical model helps us to understand the oxidation wave process in conducting polymers. It can be used to describe the effects of various parameters such as the concentration of the solution, conducting polymer properties, geometry of the cell and the applied voltage and to determine the conditions at which oxidation wave can be made to propagate along the conducting polymer surface. Thus by varying the properties of electrolyte, polymer, applied voltage and geometry we can control the shape and velocity of the propagating oxidation wave and that implies we can control the shape and properties of the polymer swelling.

4. Design and simulation of micro-pump:

In this section we have designed a micro-pump utilising the oxidation wave property discussed in the previous sections. We then find out the pumping rates by varying two different parameters of the pump namely, thickness of polymer swelling and slope of the propagating wave front. Rather than use the exact wave shape derived from the previous analysis, we have simplified the problem by approximating the wave front as a sigmoid function. We have used the finite element method to solve the Navier-stokes equation to predict the fluid velocity in the microchannel. This method gives us a fair idea of the fluid movement in the channel and ways to optimize the channel to obtain the desired pumping volume. Here we have designed a 30 micrometer thick channel with conducting polymer strips in the base of the channels. The model was created using Comsol multiphysics finite element solver.

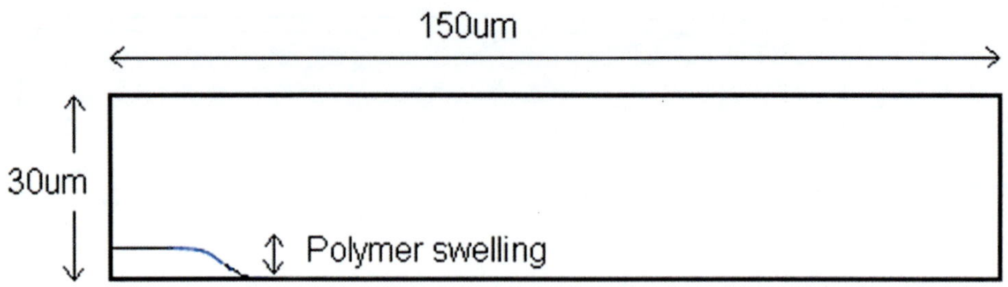

Fig 9. Geometry of the microchannel modelled

We have solved the incompressible Navier-Stokes equation (10) for the 2D domain illustrated in fig.9.

$$\rho \frac{\partial v}{\partial t} + \rho v \cdot \nabla v = \nabla(-p + \eta(\nabla v)) \tag{10}$$

Where v is the velocity vector, p is the pressure, ρ is the density, and η is the viscosity. We also solved the mass continuity equation (11) along with the Navier-Stokes equation.

$$\frac{\partial \rho}{\partial t} + \nabla \cdot (\rho v) = 0 \tag{11}$$

All the boundaries were considered as no slip boundaries except for the inlet and outlet. The inlet and outlet were set as open boundaries with zero stress. The boundaries that define the polymer swelling were made into a moving mesh boundary. Here we have set the polymer swelling to move with a defined velocity of 150µm/s. We assume the channel width to be same as the height of the channel, which is 30µm: that is, a chamber volume of 135pL, which is swept by the polymer wave in 1s.

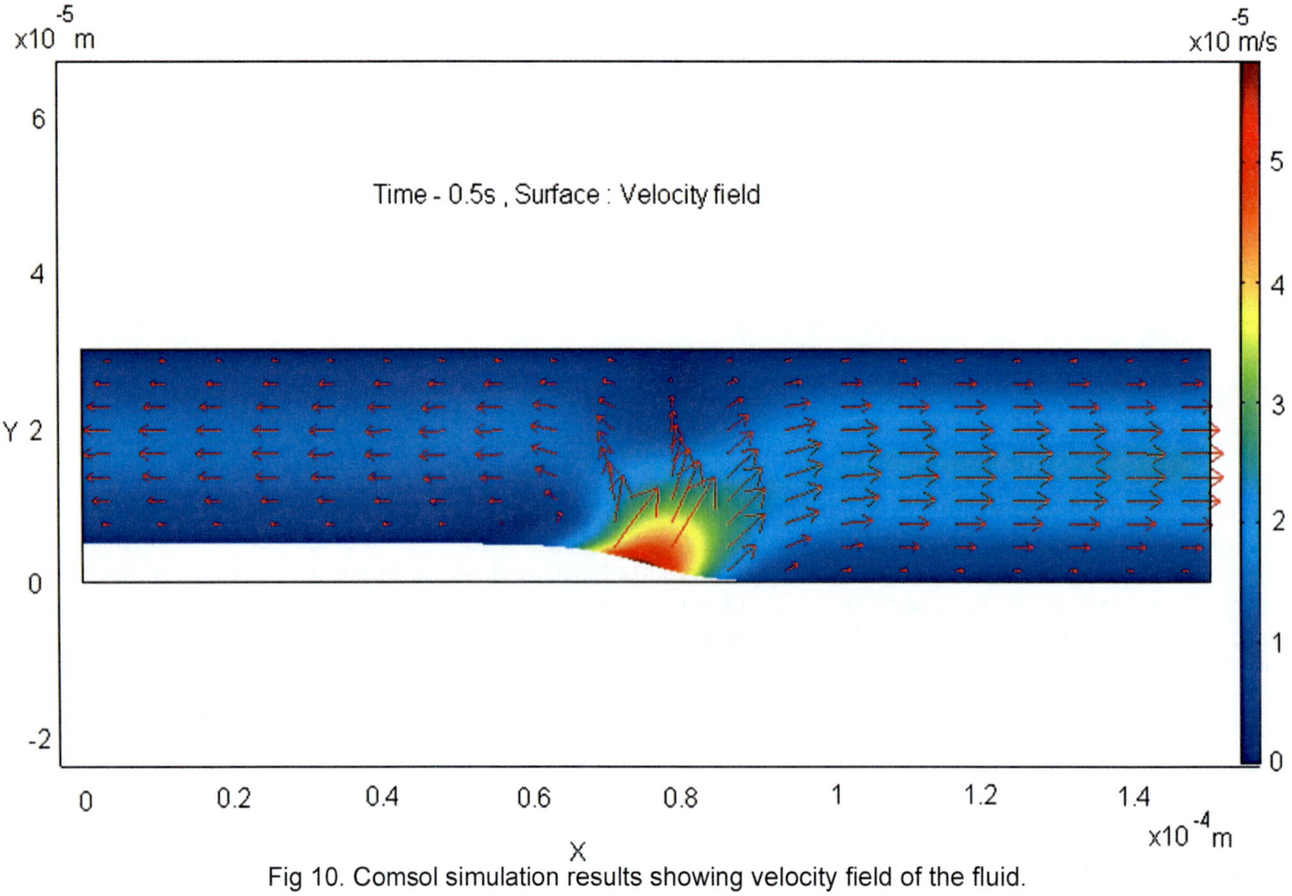

Fig 10. Comsol simulation results showing velocity field of the fluid.

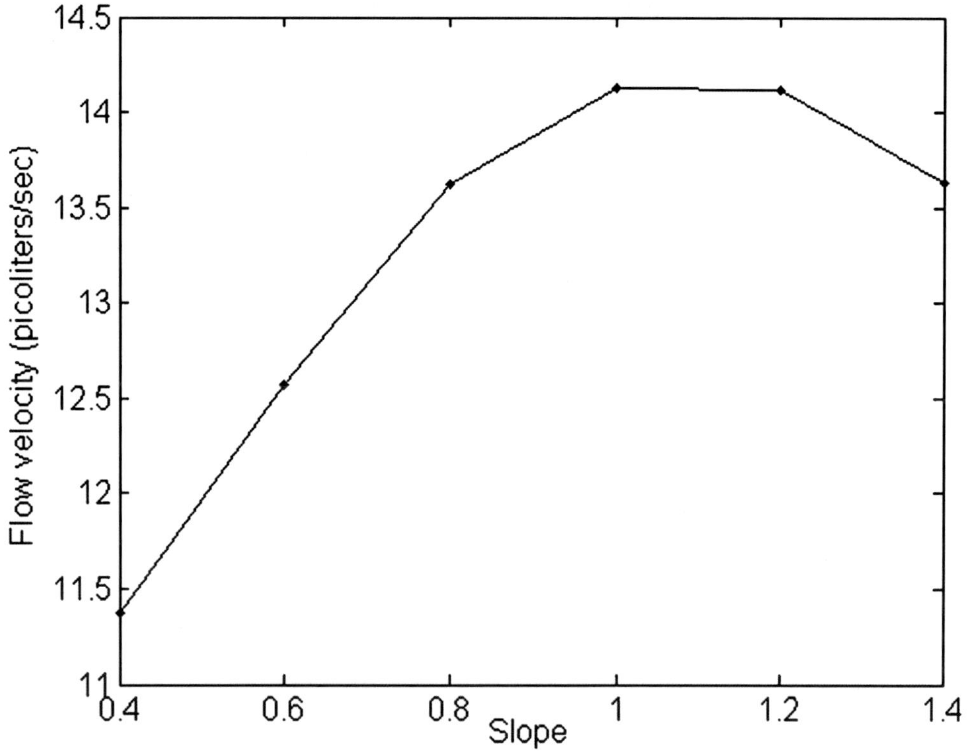

Fig 11. Variation in flow rate with slope of the polymer wave.

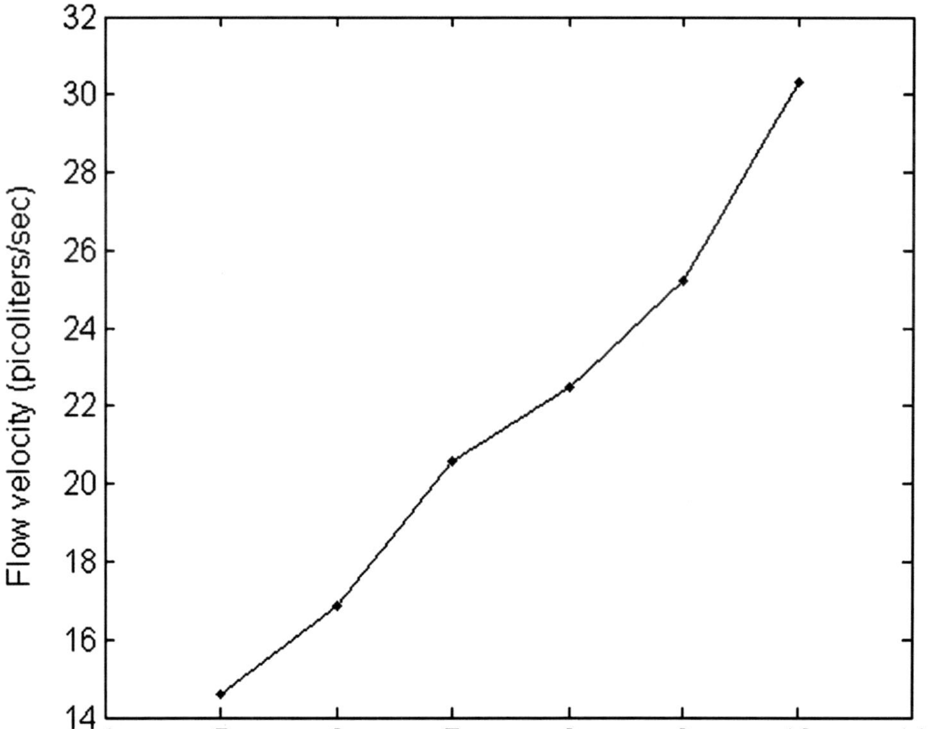

Fig 12. Variation in flow rate with amount of polymer swelling.

From the simulations we were able to predict the maximum flow rate that can be achieved using this design, which was around 30 picoliters per sec, or approximately 22% of the chamber volume in each pass of the wave. The pumping rate peaked when the slope of the wave was around 45 degrees (Fig. 11) and it increased linearly with the polymer swelling (Fig. 12).

5. Conclusion:

A propagating wave of deformation of an electrochemically-active conducting polymer can be launched along a polymer strip as a consequence of the electrochemical reaction. The shape of the wavefront and its velocity depend upon the relative conductivities of the electrolyte and of the polymer in its oxidised and reduced states. A key element determining the wave propagation is the proportion of the ionic flux in the solution near the interface that is carried by electromigration rather than diffusion. A micropump can be designed that utilises this phenomenon.

6. References:

1. Smela E, Gadegaard N *Volume Change in Polypyrrole Studied by Atomic Force Microscopy,* Journal of physical chemistry B, 2001. 105: p. 9395-9405.
2. Wu Y, Zhou D, Spinks G M, Innis P C, Megill W M, Wallace G G, *TITAN: a conducting polymer based microfluidic pump.* Smart materials and structures, 2005. 14: p. 1511-1516.
3. Low L, Seetharaman S, He K, Madou M J, *Microactuators toward microvalves for responsive controlled drug delivery.* Sensors and Actuators B, 2000. 67.
4. Tezuka Y, Ohyama S, *Observation of propogation speed of conductive front in electrochemical doping process of polypyrrole films.* The Chemical Society of Japan, 1991(64): p. 2045-2051.
5. Tezuka Y, Aoki K, *Direct demonstration of the propagation theory of a conductive zone in a polypyrrole film by observing temporal and spatial variations of potentials at addressable microband array electrodes.* Journal of Electroanalytical Chemistry, 1989(273): p. 161-168.
6. Aoki K, Tezuka Y. *Kinetics of oxidation of polypyyrole-coated transparent electrodes by in situ linear sweep voltammetry and spectroscopy.* Journal of Electroanalytical Chemistry, 1989.
7. Wang X, Smela E., *Experimental Studies of Ion Transport in PPy(DBS).* Journal of Physical Chemistry, 2009. 113.
8. Warren M R, Madden J D, *Electrochemical switching of conducting polymers: A variable resistance transmission line model.* Journal of Electroanalytical Chemistry, 2006(590): p. 76-81.
9. Feldman B J, Burgmayer P, Murray R W, *The potential dependence of electrical conductivity and chemical storage of Polypyrrole films on electrodes.* Journal of American Chemical Society, 1984(107).
10. Audebert P, Miomandre F, *Handbook of conducting polymers.* Electrochemistry of conducting polymers. 2006.
11. Newman J S, Thomas-Alyea K E, *Electrochemical Systems.* Electronuetrality and Laplace equation. 2004. 286-289.
12. Scharfetter D L, Gummel H K, *Large-Signal Analysis of a Silicon Read Diode Oscillator.* IEEE Transactions on electron devices, 1969. ED-16(1): p. 64-77.
13. Cooley J W, Cohen H, *The numerical solution of the time-dependent Nernst -Planck Equations.* Bio-Physics Journal, 1965. Volume 5.

Proceedings of the SEM Annual Conference
June 7-10, 2010 Indianapolis, Indiana USA

Micromotor Fabrication by Surface Micromachining Technique.

Author
Mr.Dhananjay Barbade,
SE (Mechanical Engg),
NKOCET, Solapur,
Maharashtra, India.

Co-author
Mr.Rohit Soni,
SE (Mechanical Engg),
NKOCET, Solapur,
Maharashtra, India

Guide
Prof Shriniwas Metan
Asst.prof.Mech.engg.
NKOCET, Solapur
Maharashtra, India.

Abstract:

From the last decade MEMS has emerged as one of the major area of interest, not only in India but also in the globe. This is basically due to its wide application in various field including Bio-Engineering, Automotive system (transducers, accelerometers), manufacturing and fabrication etc. MEMS is an approach that conveys the advantages of miniaturization, multiple components and microelectronics, to design and to construct integrated electromechanical systems.

This paper is a high level discussion of surface micromachining and its applications amongst many such as sensors, actuators and micro motor etc. Micro fabrication can be done by the various techniques such as Lithography, Bulk micromachining, LIGA and surface micromachining. Surface micromachining enables the fabrication of complex multicomponent integrated micromechanical structures that would not be possible with other methods. This technique encases specific structural parts of a device in layers of a sacrificial material during fabrication process. In other methods even small misalignment between mask and structure can pose a problem. Further all other processes are subtractive micromachining processes but surface micromachining is an additive micro structuring technology.

In the present work, fabrication of micro motor by using surface micromachining technique is explained. One of the major advantage of surface micromachining is to fabricate quite small structural heights typically 2-5 µm and seldom exceeds 20 µm .The small structural height also allows for much lateral dimensions compared to other micromachining techniques. Though micromachining of micro motor is a complex process with different micro fabrication techniques, surface micromachining excels through this.

Keywords: MEMS, surface micromachining, Micromotor, additive technique, etching, sacrificial material.

Introduction:

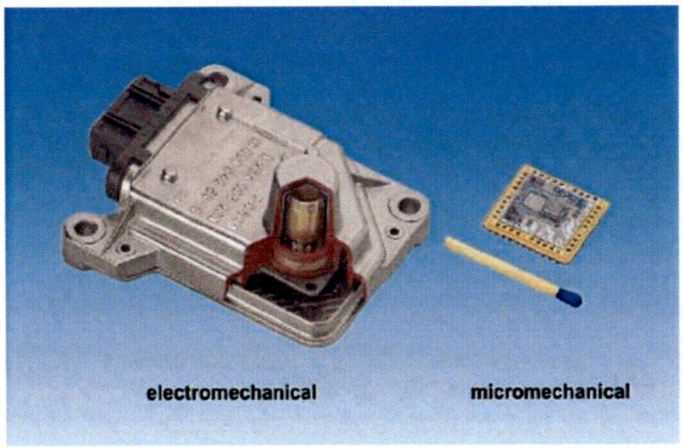

Fig 01: Comparison of MEMS with Conventional method

Attention to Micro Electro Mechanical Systems (MEMS) continues to increase worldwide and MEMS has a phenomenal growth in the last decade .The inspiration for MEMS came from a CALTECH University

Professor Richard P.Feynman's in 1959. Visionary talk at CALTECH"There's Plenty of Room at the Bottom". Great strides have been made in understanding, visualizing and controlling matter at the micro scale. In particular, substantial progress has been made towards the construction of micro devices. [
MEMS promises to revolutionize nearly every product category by bringing together silicon based microelectronics with micromachining technology, making possible the realization of complete *systems on a chip* .MEMS is an enabling technology allowing the development of smart products, augmenting the computational ability of microelectronics with the perception and control capabilities of micro sensors and micro actuators and expanding the space of possible designs and applications. The successful miniaturization and multiplicity of traditional electronics systems would not be possible without micro fabrication. Furthermore micro fabrication provides an opportunity for integration of mechanical systems with electronics to develop high –performance closed- loop-controlled MEMS. The various micro fabrication technologies have two distinctly different origins. Different technologies, such as lithography and thin film techniques, have been derived from microelectronics fabrication technologies. Those techniques are mainly used for the production of Microsystems based on silicon. Another group of micro structuring technologies are advancements of conventional precision technologies which allow the micro structuring of a large range of materials, such as metals, polymers, ceramics and glasses.

In the present work, fabrication of micro motor by using surface micromachining technique is explained. One of the major advantages of surface micromachining is to fabricate quite small structural heights typically 2-5 μm and seldom exceeds 20 μm. One of the important issues in the design of micro motor structures is how to reduce friction and surface sticking between rotor and bearing.(or hub)it is found that the contact of rotor and bearing in prior designs is a surface contact or plane contact resulting in a relatively large frictional torque for micro motor. Though micromachining of micro motor is a complex process with different micro fabrication techniques, surface micromachining excels through this.

Bibliography:

Around 1982, the term micromachining came into use to designate the fabrication of micromechanical parts, such as pressure- sensor diaphragms and accelerometer suspension beam etc, for Si micro sensor. The micromechanical parts were fabricated by selectively etching areas of Si substrate away in order to leave behind the desired geometries. Isotropic etching of Si was developed in the early 1960's for fabrication of transistor. Anisotropic etching of Si then came in existence around 1967. Various etch–stop techniques were subsequently developed to provide further process flexibility. This technique also forms the basis of the bulk micromachining techniques. However, the need for flexibility in device design and performance improvement has motivated the development of new concepts and techniques for micromachining.
Prior to 1987, these micromechanical structures were limited in motion. During 1987 to 1988, a turning point was reached in micromachining, when for the first time techniques for integrated fabrication of mechanisms (i.e., rigid bodies connected by joints for transmitting, controlling, or constraining relative movement) on Si were demonstrated. Some publications, for example W.S.Trimmer and K.J.Gabriel "Design Considerations for a Practical Electrostatic Micro motor", Sensors and Actuators, and S.F.Bart et al.,"Design Considerations for Micro fabricated Electric Actuators" , Sensors and Actuators have discussed possible designs for micro motors based on electrostatic-drive principles. In recent years, there has been a continuous development of silicon micromachining technology. Some micromechanical devices, such as micro-turbines and microelctro-static motors, which are not accessible by conventional machining technologies have been successfully made out by using technology derived from IC(integrated-circuit) manufacturing processes.

What is MEMS?

MEMS can be defined in following ways:
"The MEMS is the batch-fabricated integrated micro scale system that:
- Converts physical stimuli, events and parameters to electrical, mechanical and optical signals and vice versa;
- Performs actuation, sensing and other functions;
-Comprise control (intelligence, decision-making, evolutionary learning, adaptation, self-organization,etc.), diagnostics, signal processing, and data acquisition features."
"MEMS are small integrated devices or systems that combine electrical and mechanical components. They range in size from sub micrometer level to the millimeter level, and there can be any number, from a few to millions, in a particular system."

Micro fabrication Techniques:

The schematic representation of basic micro fabrication steps are as shown below.

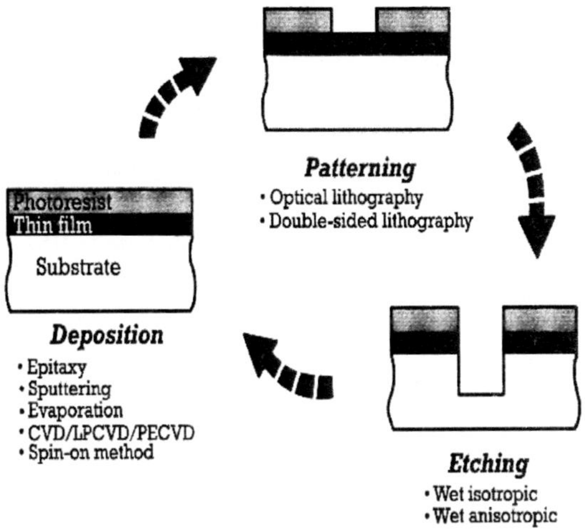

Fig 2: Basic steps in micro fabrication process

Micro fabrication can be done by the various techniques such as photolithography, Bulk micromachining, LIGA and Surface micromachining.

Photolithography:

Lithography is one of the micro structuring technology and part of many MEMS processes. It consists of three essential steps. First the substrate to be structured needs to be covered with a light sensitive film (photo resist). The second step consists of irradiating the resist with light through a mask, i.e. printing an image of the mask into the resist. Finally this latent image is developed by application of a suitable solvent, which e.g. dissolves the exposed parts of the resist film.

Bulk micromachining:

Bulk micromachining is an extension of IC technology for the fabrication of 3D structures. Bulk micromachining of Si uses wet- and dry-etching techniques in conjunction with etch masks and etch stops to sculpt micromechanical devices from the Si substrate. The two key capabilities that make bulk micromachining a viable technology are: 1) Anisotropic etchants of Si, such as ethylene-diamine and pyrocatechol and 2) Etch masks and etch-stop techniques that can be used with Si anisotropic etchants to selectively prevent regions of Si from being etched.

LIGA technology:

The LIGA process has originally been developed in the research center Karlsruhe during the 1980's. In order to generate microstructures with very high aspect ratios, i.e. the ratio between the height and the width of a structure, the first step of the original process consists of fabricating the primary microstructures by deep x-ray lithography. Deep x-ray lithography can generate structures with extreme structural heights up to some mm and extreme lateral precision in the sub-µm range in a polymeric resist, such as PMMA. These PMMA structures are then transformed into metallic microstructures by filling the PMMA relief by micro electroplating. After removing the remaining PMMA the metal structure can be used as a high precision tool in further replication processes, such as micro mold injection of thermoplastic polymers

Surface Micromachining:

Introduction-

Surface micromachining is characterized by the fabrication of micromechanical structures from deposited thin films. Originally employed for integrated circuits, films composed of materials such as low-pressure chemical-vapor-deposition, polycrystalline silicon, silicon nitride, and silicon dioxides can be sequentially deposited and selectively removed to build or "machine" three-dimensional structures whose functionality typically requires that they be freed from the planar substrate. Although the process to accomplish this fabrication dates from the 1960's, its rapid extension over the past few years and its application to batch fabrication of micromechanics and of monolithic microelectromechanical systems (MEMS)make a thorough review of surface micromachining appropriate at this time. Four central issues of consequence to the MEMS technologist are: i) the understanding and control of the material properties of micro structural films, such as polycrystalline silicon ,ii) the release of the microstructure, for example, by wet etching silicon dioxide sacrificial films, followed by its drying and surface passivation, iii) the constraints defined by the combination of micro machining and integrated-circuit technologies when fabricating monolithic sensor devices, and iv) the methods, materials, and practices used when packaging the completed device. Last, recent developments of hinged structures for post release assembly, high aspect-ratio fabrication of molded parts from deposited thin films, and the advent of deep anisotropic silicon etching hold promise to extend markedly the capabilities of surface-micromachining.

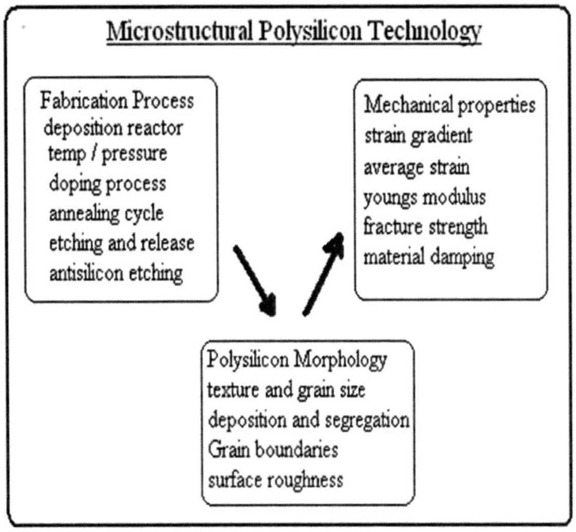

Fig 3: Micro structural polysilicon mechanical property dependency on film morphology and fabrication process parameters

Basic Surface micromachining steps:

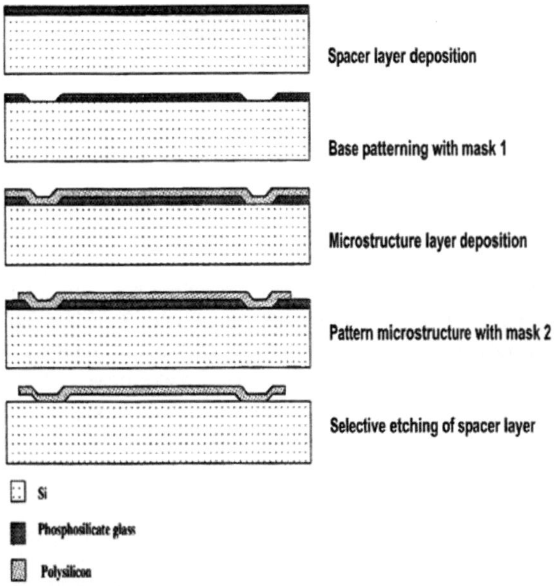

Fig 4; Surface micromachining: basic process steps

To illustrate the fabrication process by surface micromachining, process is divided into four modules:

1) ***Substrate Passivation and Interconnect*****:** After a blanket diffusion, to define the substrate ground plane, a silicon wafer is passivated with a layer of LPCVD silicon nitride deposited over a layer of 0.5m thick thermal SiO. Contact windows to the substrate ground plane are then opened to yield the cross section shown in Fig 4. Deposition, definition, and patterning of an *in situ* phosphorus-doped polysilicon interconnection layer follows Fig.4 to form a second electrode plane and an interconnection to the n diffusion and the microstructure to be built above.

2) Sacrificial Layer Deposition and Patterning: A 2 m thick LPCVD sacrificial phosphosilicate glass (PSG) layer is deposited and patterned in two separate masking steps. The first is a timed etch to create "dimples" (mechanical standoffs), as shown in Fig. 4. The second masking step etches through the PSG layer in windows, which allow the formation of the anchors of the polysilicon structure, as shown in Fig 4.

3) Structural Polysilicon Deposition, Doping, and Stress:
Anneal: The thick polysilicon structural layer is then deposited by LPCVD (undoped) at 610 C in Fig. 4).This layer is doped by depositing a 0.3m thick PSG film Fig.4 and then annealing at 1050 degree Celsius in N for one hour. This step causes the polysilicon to be doped symmetrically by simultaneous diffusions from both the top and the bottom PSG layers in order to achieve a uniform grain texture and avoid gradients in residual strain. If a strain gradient is present, it generates bending in the microstructure upon release. The top PSG layer is then stripped, and the structural polysilicon is then patterned by reactive-ion etching (RIE) in order to achieve the nearly vertical sidewalls illustrated in Fig 4.. An alternative is to implement an *in situ* phosphorus-doped polysilicon layer or an ion-implanted and annealed film to achieve a more conductive film.

4) Microstructure Release, Rinse, and Dry: In the last steps, the wafer is immersed in aqueous hydrofluoric acid (typically) 10 : 1 diluted hydrofluoric acid (HF) or buffered HF) to dissolve the sacrificial PSG layer. The wafer is rinsed in de ionized water and dried under an infrared lamp, or rinsed in de ionized water, and then dried in away to avoid its collapse and adhesion to the substrate—a phenomenon known as "stiction." The final cross section is shown in fig 4; By repeating these basic sacrificial oxide and structural polysilicon fabrication steps while including provisions for anchor points and other structural features, one can build extremely complex structures.

Surface Micromachining Applications-

Micro motor:

Multiple structural layers can for instance be used for the fabrication of freely moving components, such as the rotor in the electrostatic micro motor by using surface micromachining process as shown in Fig5.

The process used to fabricate this motor is a three layer polysilicon process, where the first polysilicon layer is only used as a base plate. Stator and rotor are fabricated from the second polysilicon layer, while the bearing is made from the third polysilicon layer. Such electrostatic micro motors have served as excellent demonstrators for the technological potentials of surface micromachining. However, due to the extremely small torque generated and the short lifetime of the bearings no commercial applications have yet been found. In the future, optical applications needing only very limited torques, like turning a micro mirror, might be feasible.

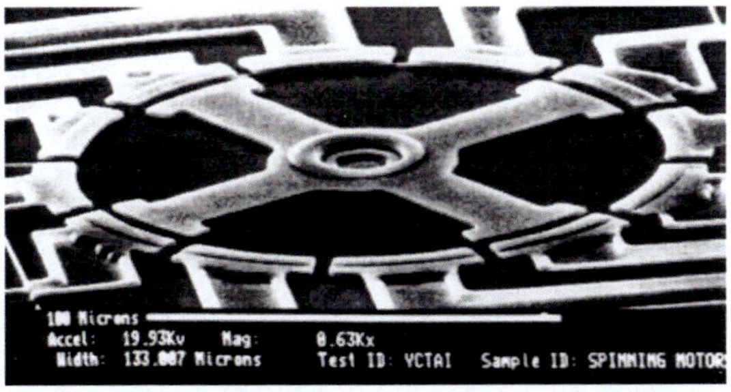

Fig 5; Surface micro machined micromotor

Fabrication of Micro motor:

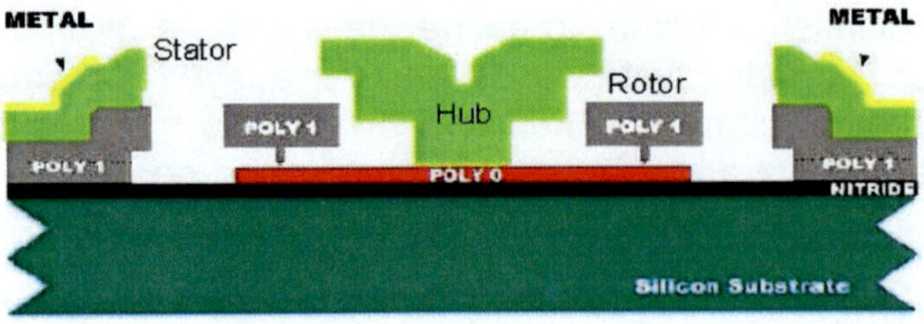

Stepwise diagrammatic representation of fabrication of micro motor is given below:

STEP 1-

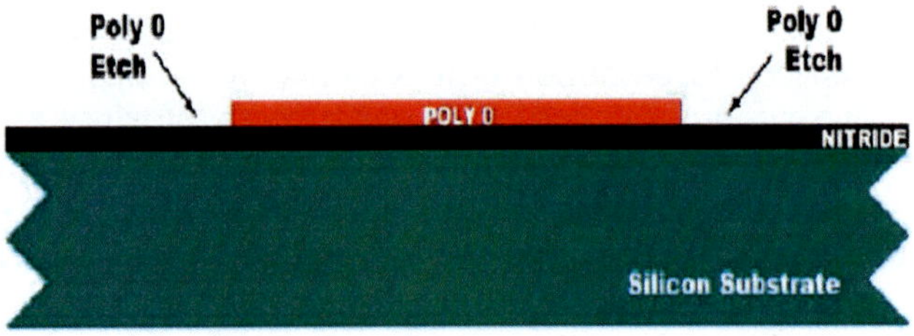

a) Providing a silicon substrate.
b) Forming an isolation layer (silicon nitride) over the silicon substrate.
c) Forming a base layer of polysilicon (poly0) on a predetermined surface region of the isolation layer by etching techniques

STEP 2-

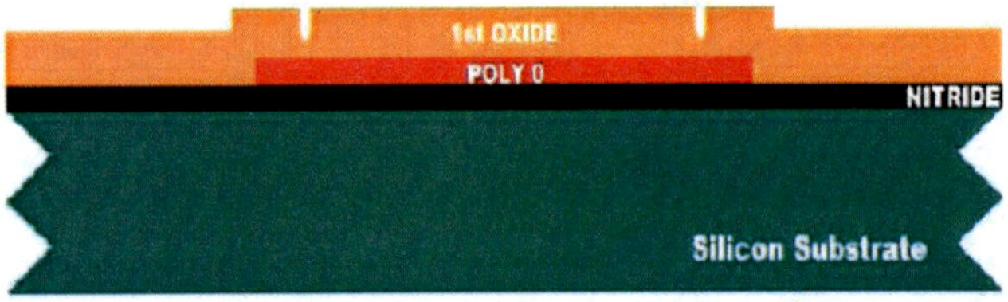

a) Depositing a layer of phosphosilicate glass [PSG oxide-1) over the structural layer (poly0) and the exposed surface of the isolation layer.

STEP 3-

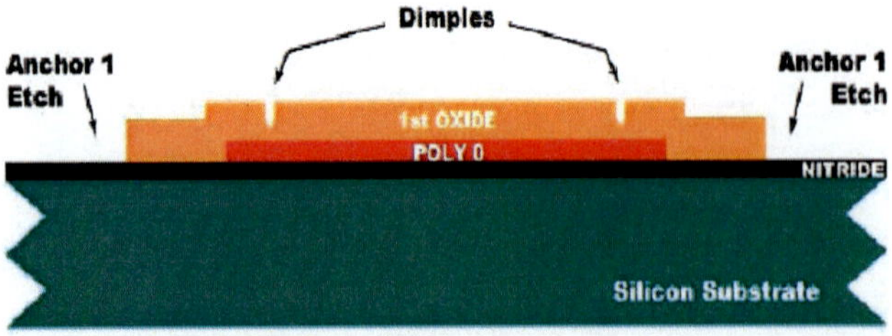

a) Patterning the layers of the PSG (oxide-1) to form rotor dimples and stator anchors by lithography and etching techniques.

STEP 4-

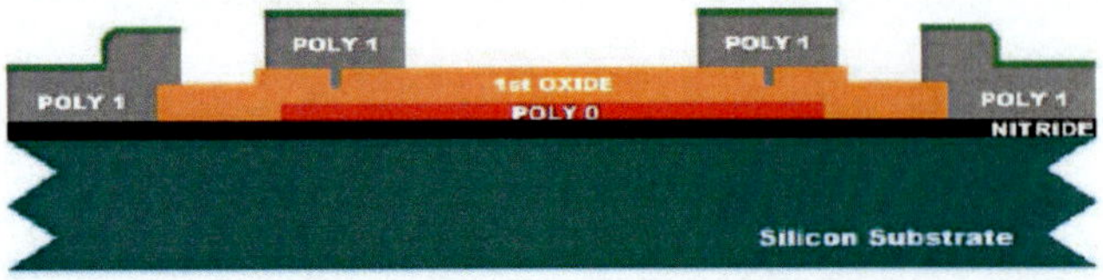

a) Depositing a structure layer (poly1) over the PSG (oxide-1) and the exposed surface of the isolation layer.
b) Patterning the first structure layer (poly1) by lithography and etching techniques to form a rotor and a stat of the micro electrostatic motor.

STEP5-

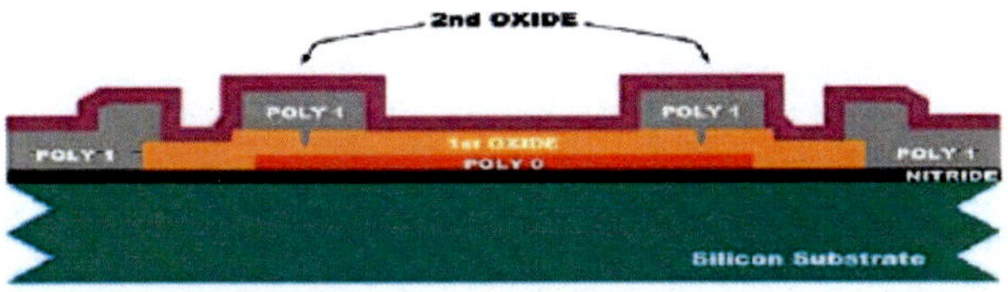

a) Forming a 2nd oxide layer over the rotor and stator regions and the no planned bearing region

STEP 6-

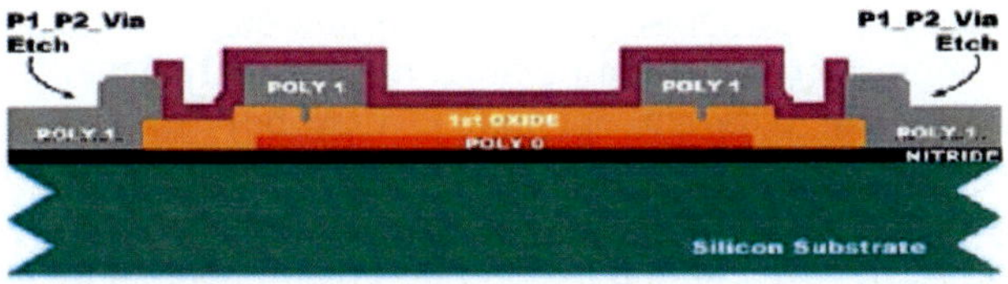

a) Patterning the exposed oxide layer (oxide-2) for stator design.

STEP 7-

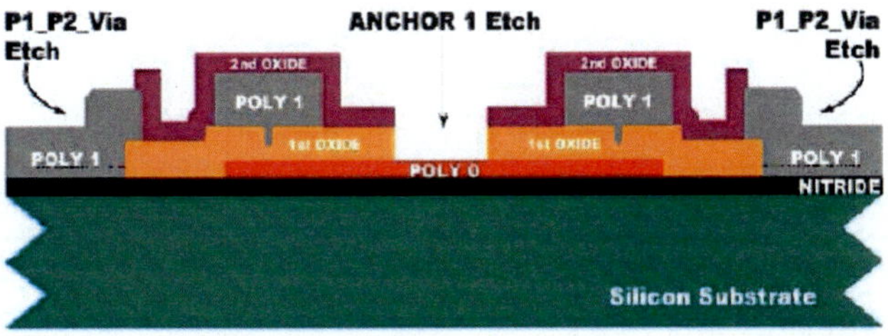

a) Etching the exposed PSG (oxide-2) sacrificial layer to form a bearing opening.

STEP 8-

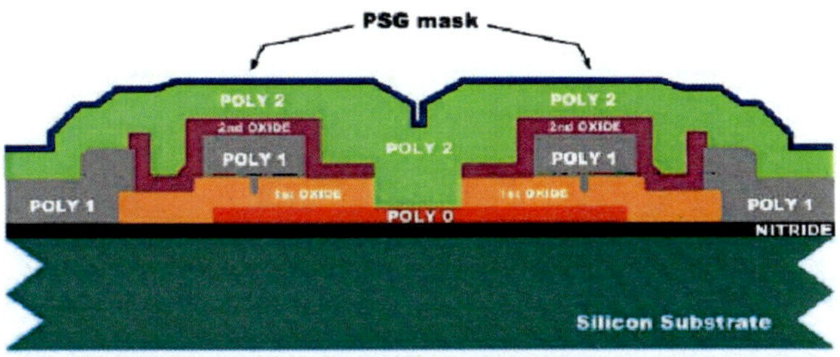

a) Removing the first mask& depositing a second structure layer (poly-2) and defining it to form a bearing residing in the bearing anchor opening.

STEP 9-

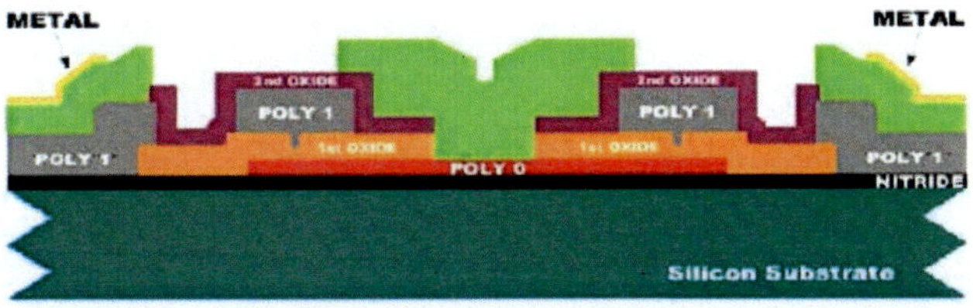

a) Micro structuring the deposited second structure layer.

STEP 10-

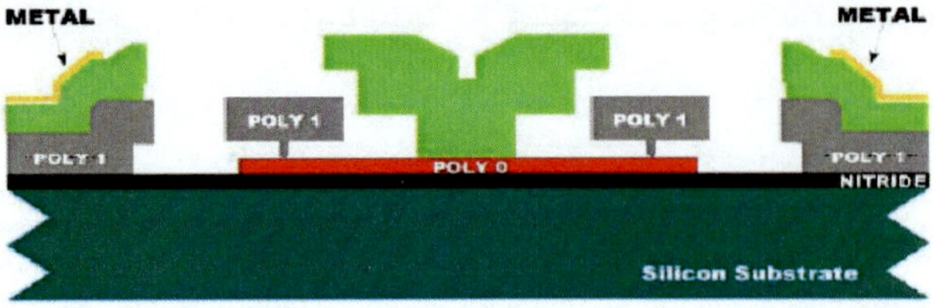

a)Removing all of the PSG(oxide) sacrificial layer and the second sacrificial layer by etching to release the rotor, so that the rotor is rotatbly supported by the bearing.

Conclusion:

Surface micromachining has developed since the early1980's from academic "proof-of-concept" demonstrations into a commercially important technology. Material properties are now understood sufficiently that polysilicon based surface-micro machined MEMS are a commercial success. MICROFABRICATION of polysilicon microstructures has made monolithic structures feasible. In these sophisticated interface and control circuits, one can fully exploit simple mechanical elements. These integrated technologies continue to evolve, with different options on how best to merge microstructure and electronic processes

From the above work it is observed that the rotor is held off the base layer of polysilicon by the bearing which leads to a line contact of rotor bottom with the bearing instead of a plane contact as in the prior designs.Therefore, the bearing structure of the present work can greatly reduce the frictional torque during micro motor motion avoiding the quick wear of the bearing and the wobbling of the rotors.

References:

[1] Fan,L.-S.Tai,Y.-C.,&Muller,R.S.,"IC-Processes for Electrostatic Micro-motors",IEEE International Electronics Devices Meeting, Dec1988, pp.666-669.

[2] Trimmer, W.S.N., &Gabriel, K.J.,"Design considerations for Practical Electrostatic Micro motors", Sensors&Actuators, vol-11, 1987, pp.189-206

[3] Mehragny, M.Gabriel, K.J., &Trimmer, W.S.N.,"Integrated Fabrication of Polysilicon Mechanisms", IEEE Transactions on electronic devices,vol-ED-35, no-6,1988,pp.719-723

[4] "MEMS" Wikipedia online encyclopedia.
http://www.wikipedia.org/wiki/surfacemicromachining

[5] Livemore Carol,"Micro fabrication of MEMS: PART 2"
http://ocw.mit.edu/

Proceedings of the SEM Annual Conference
June 7-10, 2010 Indianapolis, Indiana USA

Improvement of Piezoresistive Microcantilever Beams for Gas Detection and Sensing

Ni Wang[1]
Bruce W. Alphenaar[2]
Robert S. Keyton[1,3]
Roger D. Bradshaw[1]

[1]Department of Mechanical Engineering
[2]Department of Electrical and Computer Engineering
[3]Department of Bioengineering
J.B. Speed School of Engineering
University of Louisville, Louisville, KY 40292

ABSTRACT
Past work associated with this project has focused on both symmetric and asymmetric base arrangement piezoresistive microcantilever beams used for gas detection. Symmetric piezoresistive microcantilever beams have been shown to detect the presence of surrounding gas through changes in the resonance behavior of the beam; however, the device sensitivity was relatively low, leading to challenges due to noise and clear identification of resonance changes. An asymmetric base arrangement has been shown to improve the sensitivity of the device dramatically by changing the nature of the stress state in the piezoresistive base. The current work is focused on both improved fabrication methods for the asymmetric piezoresistive microcantilever beams and their resonance behavior in vacuum. The new fabrication method employs new equipment and photoresistors to speed the process. Furthermore, the critical step in beam writing is upgraded to whole wafer writing rather than each individual diced component. This eliminates at least two time-consuming steps and has shown promising results. The new process is being applied to several different types of SOI wafers to assess the importance of resistivity and device/insulation layer thickness on device outcomes. Future studies using these new devices will focus on detection and sensing with gases with suitable modeling and validation to demonstrate their efficacy for general use.

INTRODUCTION
In recent years, rapid development in micro/nano actuation-based sensor technology has become widely used for gas detection and biochemical analysis. [1-2] Devices based on resonating cantilever beams represent one popular type of sensor due to its simple geometry, fast response times and high sensitivity.[3] Cantilever beams operating in a gas or fluid environment experience a resonance frequency shift and change in vibration amplitude around resonant frequency due to the damping effect.[4] Similarly, coated cantilever beams that selectively attach to various compounds will experience a resonance frequency shift due to the added mass.[5] In order to assess resonance, a method for monitoring the vibration amplitude is required. One common approach uses laser-based optical methods for this measurement.[4, 6-9] Another approach is to use piezoresistive methods to detect the beam vibration behavior.[10-12] This method avoids complicated optical equipment but introduces other challenges as it relies on the stress state in the monitored structure, requires knowledge of piezoresistive material properties, and can be sensitive to temperature variations.

For resonance changes caused by a surrounding fluid or gas, suitable models must be used to relate the observed changes to appropriate media properties such as viscosity and density. Blom et al in 1992 utilized a model based on a string of spheres moving sinusoidally in an infinite media to simulate damping effects for a cantilever

T. Proulx (ed.), *MEMS and Nanotechnology, Volume 2*, Conference Proceedings of the Society for Experimental Mechanics Series 2,
DOI 10.1007/978-1-4419-8825-6_21, © The Society for Experimental Mechanics, Inc. 2011

beam; this model indicated that the shift in resonance frequency of a microcantilever was a function of the molar mass, viscosity and pressure of the gas.[13] This finding provided an avenue to assess the type(s) of gas in which the microcantilever was operating by measuring the resonance frequency shift.[12, 14] Of course, a cantilever beam of rectangular (or other) cross-section does not really resemble a string of spheres. As such, various researchers have created a variety of other fluid-structure models to assess the damping effect; these have included both inviscid and viscous fluids, cylindrical and thin blade geometries, and media flow rate effects.[15-24] Other researchers have used CFD models to assess the behavior.[20]

Device fabrication represents another challenge. For piezoresistive microcantilever beam devices, common fabrication approaches include photolithography and combinations of dry and wet etching.[2] As with many MEMS-based devices, there is a desire to reduce the size of that cantilevers towards the nanoscale to increase sensitivity of the beam.[25-26] This paper presents current work in this area, focused on alternative approaches for simplified fabrication.

PREVIOUS RESEARCH AT UNIVERSITY OF LOUISVILLE

Previous research work at the University of Louisville demonstrated methods to create narrow piezoresistive microcantilever beams capable of detecting the presence of various gases via changes in the resonance characteristics of the beam. These structures were all symmetric about the beam midplane.[12, 14] One style of microcantilever considered was the T-shaped resonator; top and tilted views of one such structure are shown in Figure 1. The beam (128 μm long) is attached to a shorter base (9 μm wide) that connects two electrodes. Near the end of the beam, another electrode is used to actuate the beam via electrostatic forces; by applying an AC voltage at the driving electrode, the beam will enter a state of resonance when the AC frequency is equal to certain specific values that depend upon the geometry and material properties of the beam.The resistance across the base of the beam was monitored using a DC circuit; changes in this value, caused by piezoresistivity and stresses in the beam base during oscillation, indicate the deflection of the beam tip. When the AC voltage at the driving electrode has frequency f, the beam base resistance changes occur at frequency 2f. The magnitude of this resistance change can be detected by a lock-in amplifier by seeking a frequency that is twice the reference AC frequency. This value describes the behavior of the beam during resonance, which is then used to ascertain the nature of the surrounding gas medium.

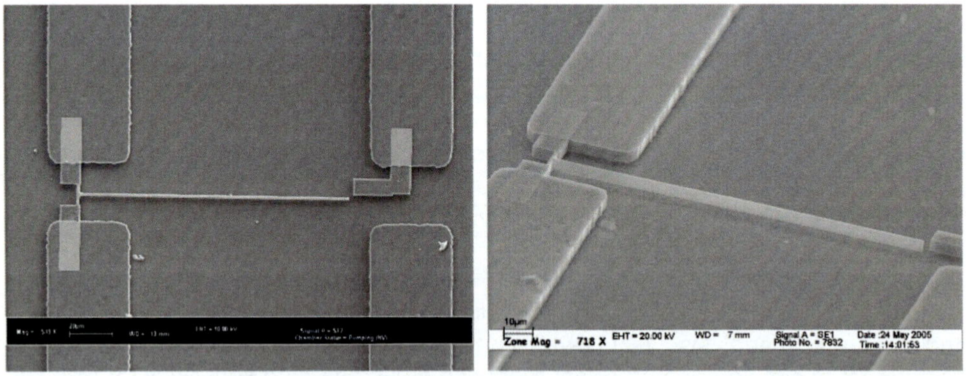

Figure 1. SEM images of the T-shaped resonators: (a) top view; (b) tilted view [12].

While the results above detected gases as intended, the piezoresistive changes the beam base were relatively low; beam resonance lead to relative resistance changes on the order of 1-10 parts per million (not far above the system noise level). In the T-shaped design, the beam base is in a state of bending and many of the resistance changes cancel out due to the tension/compression nature of the stress state across the base legs. The alternate configuration uses an asymmetric base leg arrangement. This supports loads due to beam deflection primarily via a force couple in each leg; this leads to both legs having a similar state of stress and therefore a much larger resistance change compared to the original device. This effect was studied using numerical methods to validate the concept; an example of a 3D finite element model and its associated current density plot used to determine base resistance is shown Figure 2.[27] These numerical studies indicated relative resistance changes on the order of 1%, a 3-4 magnitude order increase compared to the original design. Limited experimental studies using this configuration have shown base resistance changes that are 15-200 times greater than the original design.[28-29] An example of a asymmetric resonator in resonance and the associated experimental setup are shown in Figure 3.

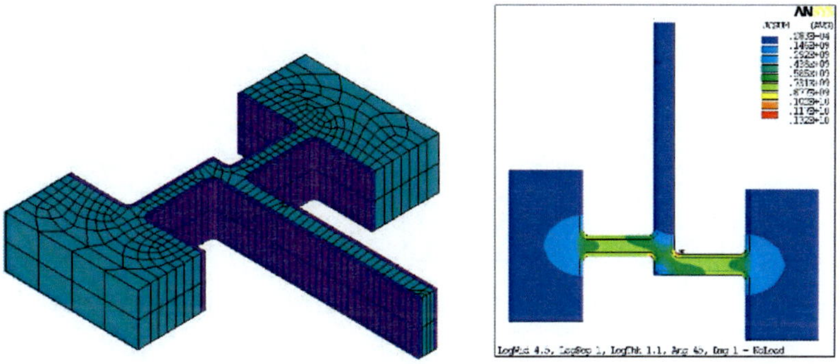

Figure 2. Finite element modeling of asymmetric cantilever:
(left) coarse 3D mesh (part of beam shown); (right) current density plot in undeflected state.[27]

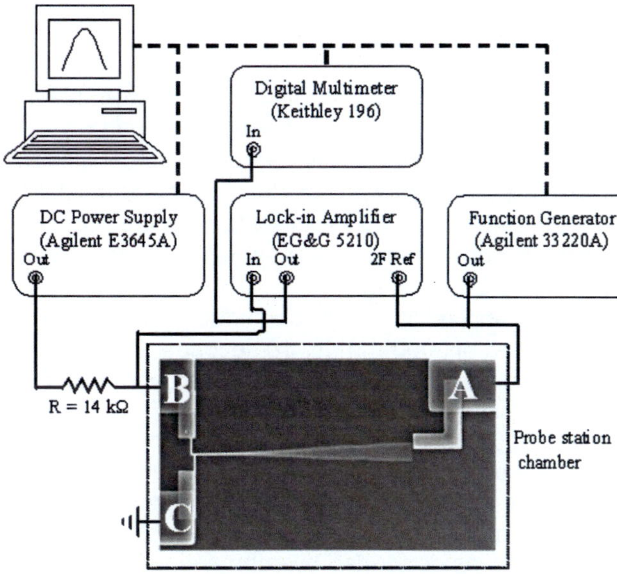

Figure 3. Schematic of general instrumentation setup.
(——) BNC cable, (- - -) GPIB cable.[28]

In the original approach, a piezoresistive boron-doped silicon wafer was cut into squares using a dicing saw after the electrical circuit features were created using standard photolithography methods (electrical pads, large connecting paths). Each square (referred to here as a "dice") consists of 10 asymmetric piezoresistive beams as shown in Figure 4. The pattern for each beam was written using (e-beam) lithography; because of the very low magnifications needed in the systems used, each microcantilever beam was aligned and exposed separately. An iron layer was then sputtered onto the dice; once liftoff was completed, the remaining iron defined the pattern of the beams. Deep reactive ion etching (DRIE) was then remove exposed silicon down to the silicon oxide layer. Buffered oxide etching to remove the silicon oxide layer followed by critical point drying leads to the final free-standing beam configuration. Detailed fabrication steps are described elsewhere.[12, 14] While this process was successful in fabricating both symmetric and asymmetric beam configurations, it was quite time-consuming.

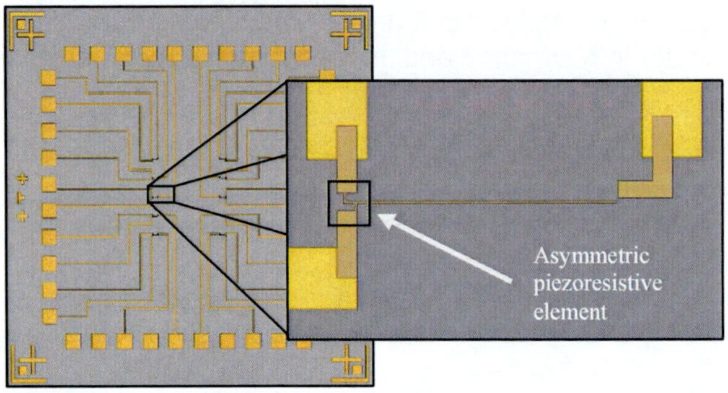

Figure 4. Device layout with 10 microcantilever beams in an array.[28]

IMPROVEMENT OF DEVICE FABRICATION

Based on the promising results of previous work, there is huge motivation to improve the device fabrication methodology for more general use. The new approach simplifies fabrication by eliminating several steps, especially the e-beam writing step that required manual intervention for each beam to be created. The new approach is described in detail in the next several sections.

METHODOLOGY

A commercial, 4" silicon-on-insulator (SOI) wafer is utilized for device fabrication. This wafer was comprised of three layers. The top layer is a 2 um±0.5 μm thick layer of boron-doped crystal silicon to be acting as pizoresistive layer. The middle layer is buried silicon dioxide, acting as an electrical insulator. The base one is a boron-doped crystal silicon too. The overall fabrication process includes gold bond pads and leads patterned, beam patterned, and beam released. Three lithography steps are required though the requirement for each layer is different. The three masks used in the optical lithography process are shown in Figure 5(d, e, and f).

First, Shipley 1827 photoresist was applied to the silicon surface using a wafer spinner at a spread speed of 500 rpm for 0.2 seconds and a spin speed of 3000 rpm for 10 seconds. Then SOI wafer was soft baked on a hotplate for 1 minute at 115°C to remove excess solvent. Next the substrate was exposed to UV light for 15 seconds using the lead mask (Figure 5(d)) in a mask aligner (SUSS as shown in Figure 6(a)). The pattern image was reversed in the YES oven for 45 minutes at 90 °C. After image reverse, the floor exposure is performed in the SUSS for 22 seconds. The resulting pattern was developed in MF319 for 1 minute with lateral agitation and rinsed in DI water. After drying, an adhesion layer of 10 nm thick chromium was RF sputtered onto the substrate at 350W for 27 seconds and a 35 nm thick layer of gold was DC sputtered onto the substrate at 120W for 30 seconds. After sputtering, a gold/chrome lift-off process was performed by submerging the wafer in a recirculating acetone bath for 20 minutes to remove the excess gold and chromium; this leaves thedesired electrode lead pattern on the substrate. The same procedure and parameters were utilized for bonding pads except that the bond pad mask is used (Figure 5(e) and the DC sputtering time is increased to 103 seconds; this leaves the gold bond pads thicker than the gold layer desired for the leads. These steps are quite similar to the original process.

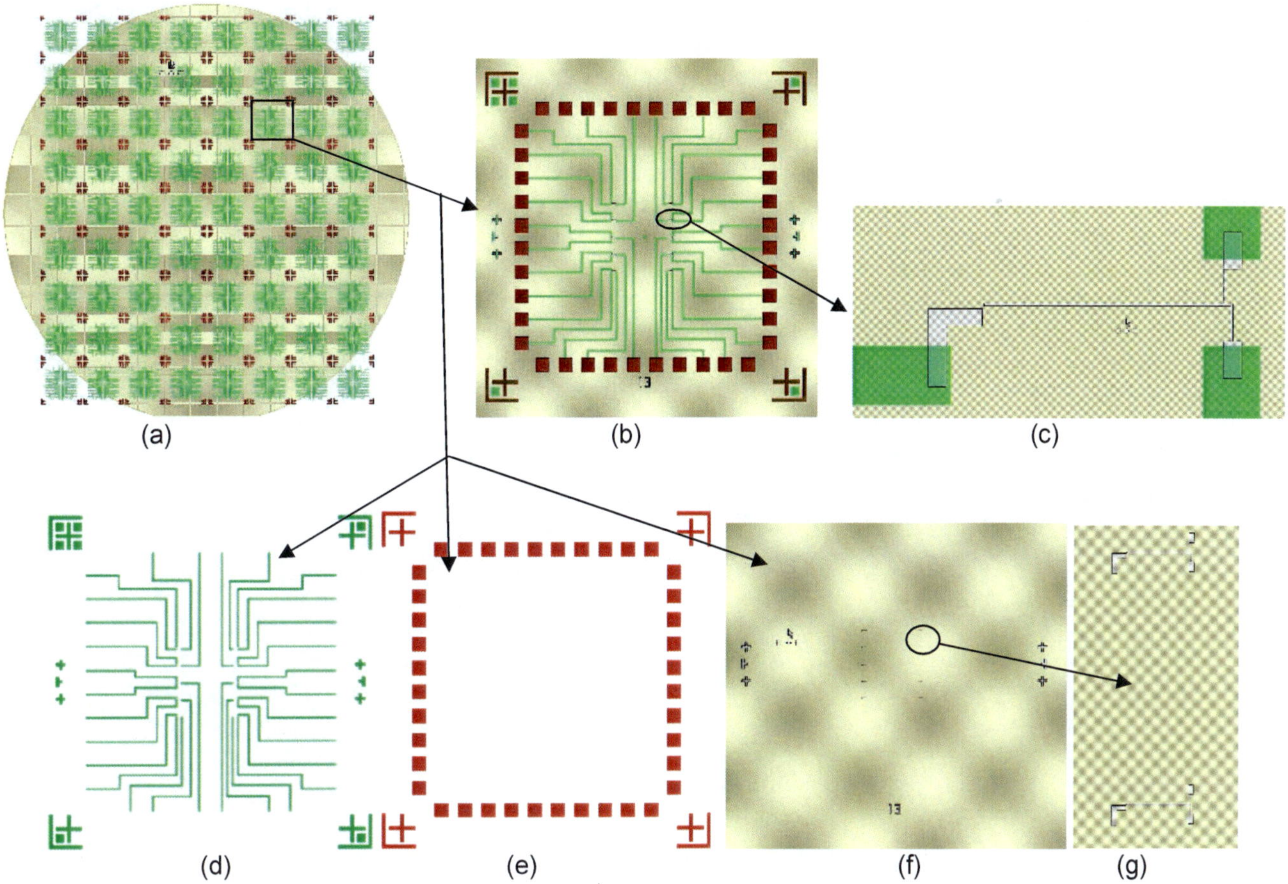

Figure 5. Masks for device fabrication. (a) Whole wafer with three masks together; (b) One device in the wafer with three masks together; (c) Pattern demonstrated on device; (d) Metal pad mask for only one device in the wafer; (e) Bond pad mask for only one device in the wafer; (f) Beam pattern mask for only one device in the wafer; (g) Pattern demonstrated on beam pattern mask.

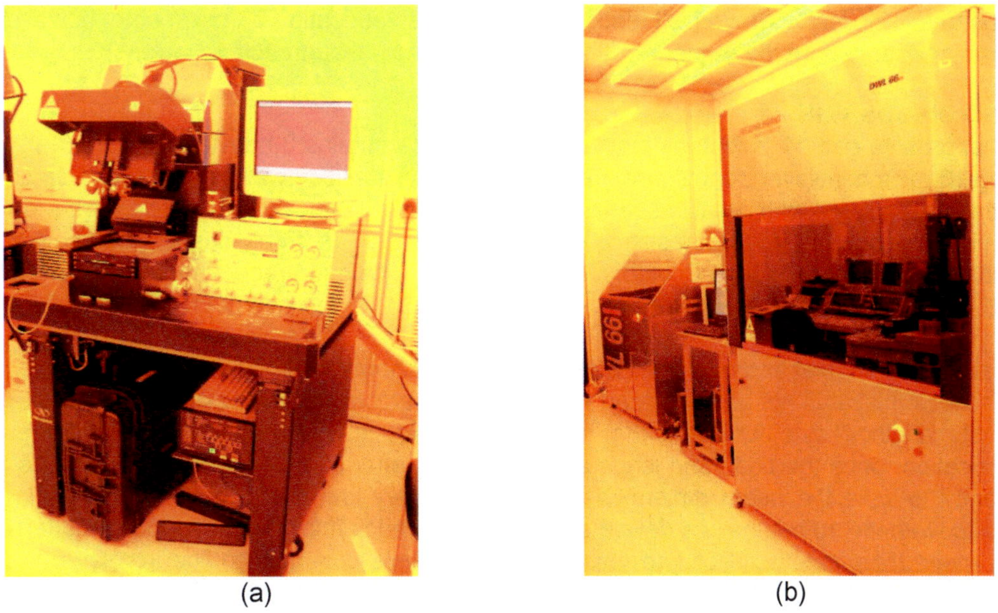

Figure 6. Equipment for writing beam pattern: (a) SUSS; (b) HEIDELBERG DWL 66FS

The remaining steps are quite different from the original fabrication method. To begin the photolithography process for beam pattern, the cleaned SOI wafer with gold pads and leads was "baked" for 5 minutes at 115 ºC on a hotplate to remove excess moisture and promote potoresist adhesion. Then 1805 photoresist was applied to the silicon surface using a wafer spinner at a spread speed of 500 rpm for 0.2 seconds and a spin speed of 3000 rpm for 10 seconds to achieve a thickness of 200 nm. Subsequently, the SOI wafer was soft baked at 115 ºC for 1 minute to remove excess solvent from the positive resist. Next, the substrate was exposed to UV light for 5 seconds using beam pattern mask (Figure 5(f)) in SUSS. The resulting pattern was developed in MF 319 for 30 seconds stably and rinsed softly in DI water. After developing, the beam pattern must be checked under microscope to make certain that the pattern is clear and neat. Figure 7 shows the pattern under microscope after beam written and development. This step eliminates the need for e-beam writing the pattern beam-by-beam; the photoresist used also eliminates the need for the iron sputtered layer as it is sufficient to act as a mask for the next step (DRIE). It should also be noted that this step is done for the entire wafer at once; the end result after these steps is shown in Figure 5(a).

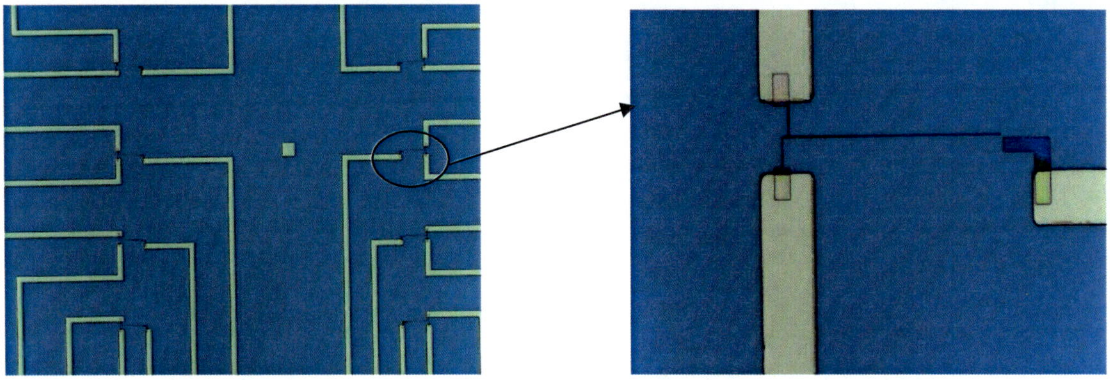

Figure 7. Patterns after beam written and developing

Once the pattern is verified, an anisotropic DRIE process was performed to etch the silicon device layer; this is capable of forming high aspect ratio vertical sidewalls in silicon without etching the silicon-dioxide layer as described in previous work.[12, 14]. The wafer is then sawed into the desired devices. Afterwards, the cantilever beams are released from the substrate through an buffered isothropic wet etching of the sacrificial SiO_2 under the large pads and electrode leads; this is followed by critical point drying to prevent cantilever beams from sticking to the silicon substrate. This last operation is identical to the original fabrication process; note, however, that the wafer is sawed into devices before the e-beam writing step in the original approach.

RESULTS AND DISCUSSION

In the current approach, each 4" silicon-on-insulator (SOI) wafer contains 60 devices (Figure 5(a)), with each device containing 10 microcantilever beams arranged in an array on a silicon wafer (Figure 4). Each microcantilever beam has 3 pads associated with it – two spanning the beam base (DC circuit) and one for the driving electrode close to the end of the beam. The piezoresistive base legs and beam are freely-suspended above the silicon substrate; the gap is the thickness of the silicon oxide layer (2 mm). The end product should be similar between both the new and original fabrication processes.

Figure 8 compares the original and new fabrication processes schematically. The chief difference when viewed in this manner is the iron layer is not used; instead a resist layer is employed to mask for the DRIE process. As mentioned previously, another key difference is that the original approach first dices the whole wafer (cut into squares for device fabrication) once the pads and leads are written (i.e. after step (3) in Figure 8). This is followed by e-beam lithography to write the beam pattern; because of the very low magnifications needed in the SEM and NPGS system to write the 10 beams in a device simultaneously, each microcantilever beam pattern was manually aligned and exposed separately. Of course, this procedure is time consuming. The iron layer is then sputtered and after liftoff it asks as a mash for the DRIE step.

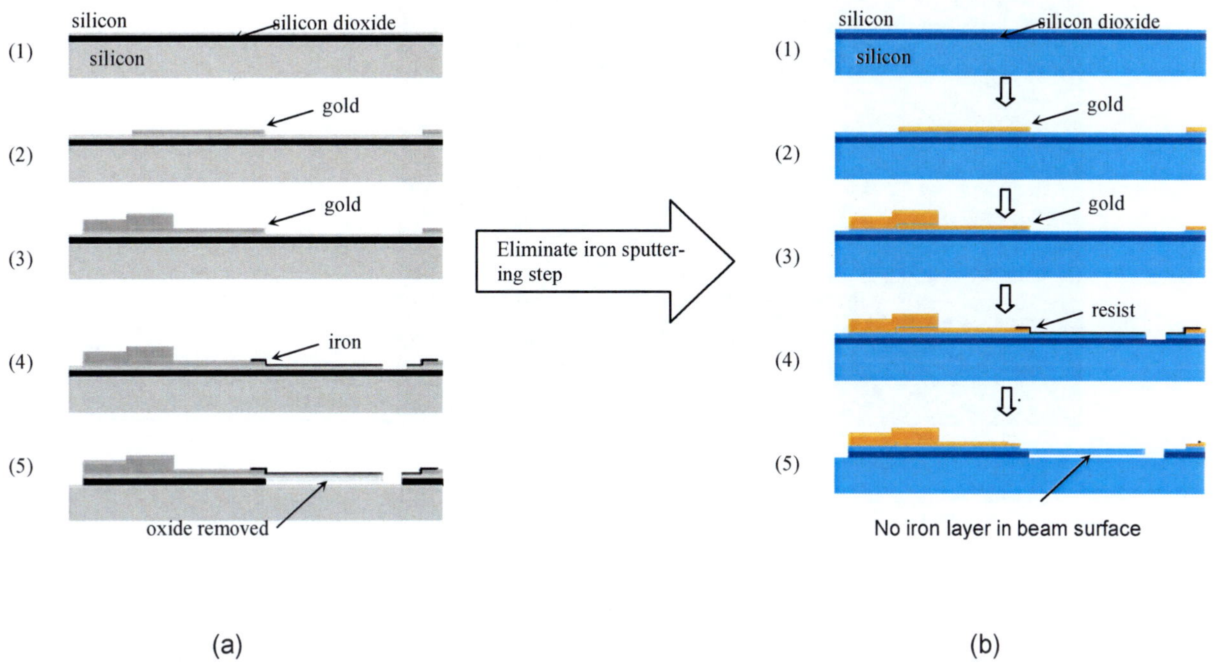

Figure 8. Fabrication steps: (a) the original approach [28]; (b) the new approach – (1) initial wafer; (2) application of gold leads; (3) application of bonding pads; (4) structure after DRIE; (5) final configuration after removal of silicon oxide layer

The new approach made changes from step 3. After gold pond pads and leads are done, a third mask (Figure 5(f)) is used for whole wafer lithography. The reason why the original approach did not adopt this method is that traditional lithography techniques (AM-B) available at that time were insufficient to create the small, high resolution feature sizes. Recently a new SUSS mask aligner system has been obtained by the UofL micro/nano technology center (MNTC). This system permits smaller features to be created via traditional photolithography methods (Figure 6(a)). More importantly, the MNTC also acquired an HEIDELBERG DWL66FS as shown in Figure 6(b). This machine is used to write the beam mask and its minimal resolution size is 500 nm; this makes fabrication much easier by permitting traditional lithography methods to be used to create structures with larger feature sizes (the beams shown in this work have minimum widths on the order of 1100 nm). By adopting traditional lithography for beam writing, not only the whole wafer does not need to be diced to pieces, but also beam pattern can be written in several seconds for a whole wafer (60 devices in total). Compared with the E-beam writing, the speed of fabrication in this step gets increased by at least 3 orders of magnitude! Furthermore, since the photoresist 1805 could be used as mask for DRIE, the iron evaporation could be eliminated (this step usually takes more than 1 hour per device). From above discussion, the new beam pattern writing step is very critical for improving the speed of fabrication.

The dry etch was performed with a base pressure of 0.2 mTorr and a process pressure of 10 mTorr at room temperature. The etching process was performed in multiple short-time steps to prevent over-etching and damaging the cantilever beams. In order to ascertain the proper DRIE time, an SOI die substrate was initially etched for 6 minutes and checked to see if the purple silicon dioxide was visible. Subsequent tests on a pure silicon test wafer were performed using the identified durations. The beams on this wafer were measured after DRIE and the thickness of the beam (from base to surface) is 20.24KA, which equates 2.024 μm (

Figure **9**). These results match the thickness of the top layer on the SOI. Therefore, it is anticipated that the DRIE duration will be sufficient to remove the top (silicon) layer down to the silicon oxide layer.

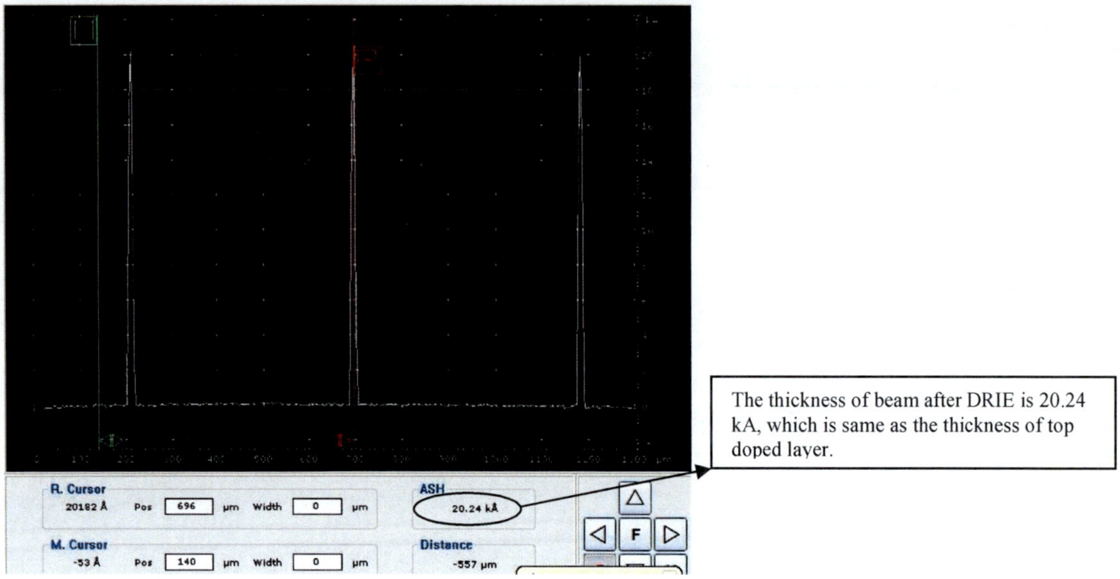

Figure 9. The thickness measurement of beam after DRIE

At present, all fabrication has been performed on low-cost pure silicon wafers; as such, free standing beams cannot be created but the process can be validated before moving on to SOI wafers. An example of a beam fabricated using this approach is shown in Figure 10. Beam widths on the order of 1 – 2 microns have been obtained successfully. More devices need to be fabricated to clarify what minimum feature size and tolerance can be attained with this method. From the current data, a 1.382 µm beam width has been obtained for a designed width of 1.1 µm. Further experience will provide an approach that relates mask sizes to the final structure dimensions (such as beam width). Thus far, some minor defects have also been observed such as waviness in the beam edges along the length of the beam; this also needs to be further investigated to understand whether it is an issue with the mask or another part of the process.

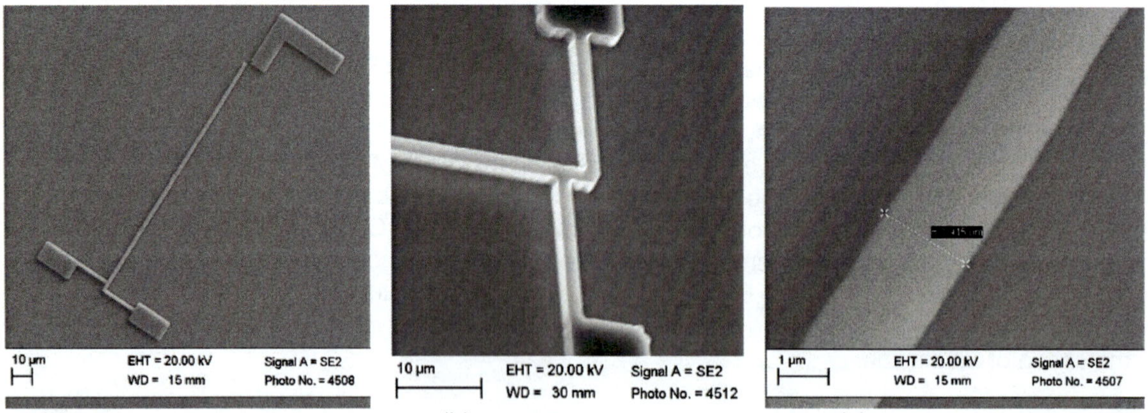

(a) (b) (c)

Figure 10.SEM images of beam fabricated using new approach:
(a) full beam and electrode; (b) close-up of beam base; (c) measurement of beam width

Currently, fabrication of devices on SOI wafers is underway. Once these devices are complete, the beams will be operated in both full vacuum and several gas environments at various pressures to understand how resonance behavior is affected. It is anticipated that correlations between the frequency shift and gas pressure will be observed.

CONCLUSIONS

Resonating microcantilever beams have been shown in previous work to be capable of detecting gases via changes in the resonance behavior. Previous fabrication methods were extremely time-consuming, requiring beam mask patterns to be written one-by-one. A new approach has been developed which uses traditional photo-lithography approaches to create the beam mask pattern much more quickly and efficiently. Initial results on pure silicon wafers are positive; fabrication is currently underway on SOI wafers to verify that the new method leads to results that are comparable with the old approach.

REFERENCES

1. Gfeller, K.Y., N. Nugaeva, and M. Hegner, *Rapid Biosensor for Detection of Antibiotic-Selective Growth of Escherichia coli.* Applied & Environmental Microbiology, 2005. **71** (5): p. 2626-2631.
2. Lavrik, N.V., M.J. Sepaniak, and P.G. Datsko, *Cantilever transducers as a platform for chemical and biological sensors.* Review of Scientific Instruments, 2004. **75**(7): p. 2229-2253.
3. Lang, H.P., M. Hegner, and C. Gerber, *Cantilever array sensors.* Materials Today, 2005. **8**(4): p. 30-36.
4. Chen, G.Y., et al., *Resonance response of scanning force microscopy cantilevers.* Review of Scientific Instruments, 1994. **65**(8): p. 2532.
5. Gupta, A., D. Akin, and R. Bashir, *Single virus particle mass detection using microresonators with nanoscale thickness.* Applied Physics Letters, 2004. **84**(11): p. 1976-1978.
6. Chen, G.Y., et al., *Adsorption-induced surface stress and its effects on resonance frequency of microcantilevers.* Journal of Applied Physics, 1995. **77**(8): p. 3618.
7. Lang, H.P., et al., *A chemical sensor based on a micromechanical cantilever array for the identification of gases and vapors.* Applied Physics A: Materials Science & Processing, 1998. **66**(7): p. S61.
8. Chen, G.Y., et al., *Harmonic response of near-contact scanning force microscopy.* Journal of Applied Physics, 1995. **78**(3): p. 1465.
9. Shin, S., et al., *Gas Sensor Application of Piezoelectric Cantilever Nanobalance; Electrical Signal Read-Out.* Ferroelectrics, 2005. **328** (1): p. 59-65.
10. Bargatin, I., et al., *Sensitive detection of nanomechanical motion using piezoresistive signal downmixing.* Applied Physics Letters, 2005. **86**(13): p. 133109.
11. Tortonese, M., R.C. Barrett, and C.F. Quate, *Atomic resolution with an atomic force microscope using piezoresistive detection.* Applied Physics Letters, 1993. **62**(8): p. 834.
12. Xu, Y., *Fabrication, Development and Characterization of Micro/Nano Actuation Based Sensors and Applications for Gas Composition Analysis*, in *Department of Mechanical Engineering.* 2006, University of Louisville: Louisville, KY.
13. Blom, F.R., et al., *Dependence of the quality factor of micromachined silicon beam resonators on pressure and geometry.* Journal of Vacuum Science & Technology B: Microelectronics and Nanometer Structures, 1992. **10**(1): p. 19-26.
14. Xu, Y., et al., *Viscous damping of microresonators for gas composition analysis.* Applied Physics Letters, 2006. **88**(14): p. 143513.
15. Chon, J.W.M. and P. Mulvaney, *Experimental validation of theoretical models for the frequency response of atomic force microscope cantilever beams immersed in fluids.* Journal of Applied Physics, 2000. **87** (8): p. 3978.
16. Dufour, I. and L. Fadel, *Resonant microcantilever type chemical sensors: analytical modeling in view of optimization.* Sensors & Actuators B: Chemical, 2003. **91**(1-3): p. 353.
17. Dufour, I., S.M. Heinrich, and F. Josse, *Theoretical Analysis of Strong-Axis Bending Mode Vibrations for Resonant Microcantilever (Bio)Chemical Sensors in Gas or Liquid Phase.* Journal of Microelectromechanical Systems, 2007. **16** (1): p. 44-49.
18. Gorman, D.G., et al., *Analytical modelling and extraction of the modal behaviour of a cantilever beam in fluid interaction.* Journal of Sound & Vibration, 2007. **308**(1/2): p. 231-245.
19. Green, C.P. and J.E. Sader, *Torsional frequency response of cantilever beams immersed in viscous fluids with applications to the atomic force microscope.* Journal of Applied Physics, 2002. **92**(10): p. 6262-6274.
20. Jana, A., et al., *Microcantilever mechanics in flowing viscous fluids.* Applied Physics Letters, 2007. **90** (11): p. 114110-1.
21. Kaabi, L., et al., *Modelling and analysis of MEMS sensor based on piezoresistive effects.* Materials Science & Engineering: C, 2007. **27**(4): p. 691-694.
22. Ma, R.-H., et al., *A microcantilever-based gas flow sensor for flow rate and direction detection.* Microsystem Technologies, 2009. **15**(8): p. 1201-1205.
23. Weigert, S., M. Dreier, and M. Hegner, *Frequency shifts of cantilevers vibrating in various media.* Applied Physics Letters, 1996. **69** (19): p. 2834.
24. Van Eysden, C.A. and J.E. Sader, *Frequency response of cantilever beams immersed in viscous fluids with applications to the atomic force microscope: Arbitrary mode order.* Journal of Applied Physics, 2007. **101** (4): p. 044908.
25. Huang, X.M.H., et al., *Nanomechanical hydrogen sensing.* Applied Physics Letters, 2005. **86**(14): p. 143104.
26. Kleimann, P., J. Linnros, and R. Juhasz, *Formation of three-dimensional microstructures by electrochemical etching of silicon.* Applied Physics Letters, 2001. **79**(11).
27. Bradshaw, R.D., et al., *Model Optimization of a Piezoresistive Microactuation-Based Sensor for Gas Composition Analysis.* 2007.
28. Fletcher, P.C., et al. *Piezoresistive Geometry for Maximizing Microcantilever Array Sensitivity.* in *IEEE Sensors.* 2008. Lecce, Italy.
29. Fletcher, P.C., *Alternative Piezoresistor Designs for Maximizing Cantilever Sensitivity*, in *Department of Mechanical Engineering.* 2008, University of Louisville: Louisville, KY.

Proceedings of the SEM Annual Conference
June 7-10, 2010 Indianapolis, Indiana USA

Investigation of the Young's Modulus of Fibers in an Electrospun PCL Scaffold Using AFM and its Correlation to cell Attachment

Nandula Wanasekara[1], Ming Chen[2], Vijay a Chalivendra[3] and Sankha Bhowmick[2,3]
[1]Materials & Textiles Department
[2]Bioengineering & Biotechnology Program
[3]Department of Mechanical Engineering
University of Massachusetts Dartmouth, MA 02747
*Corresponding author, vchalivendra@umassd.edu, 508-910-6572

ABSTRACT

Seeding a layer of cells at specific depths within scaffolds is an important optimization parameter for bi-layer skin models. Experimental investigation has been performed to investigate the effect of fiber diameter and its mechanical property on the depth of cell seeding of for electrospun fiber scaffold. Polycaprolactone (PCL) is used to generate scaffolds that are submicron (400nm) to micron (1100nm) using electro-spinning. 3T3 fibroblasts were seeded on the electro-spun fiber scaffold mat of 50-70 microns thickness in this study. In order to investigate the effect of fiber diameter on cell migration, first, the electrospun fiber scaffold was studied for variation of mechanical properties as a function of fiber diameters. Atomic force microscopy (AFM) was used to investigate the Young's modulus (E) values as a function of fiber diameter. It was identified that as the fiber diameter increases, the Young's modulus values decreases considerably from 1.9GPa to 600MPa. The variation in E is correlated with cell seeding depth as a function of vacuum pressure. A higher E value led to a lower depth of cell seeding (closer to the surface) indicating that nano-fibrous scaffolds offer larger resistance to cell movement compared to microfibrous scaffolds

1. INTRODUCTION

Electro-spinning is one of the approaches that allow the fabrication of natural synthetic materials into fibrous structures in the submicron nanometer scale [1-2]. These scaffolds provide excellent framework for cell growth as widely reported in the literature for a variety of material-cell combinations including skin, bone, cartilage, tendons, blood vessels and heart valves [3-6]. Nano-fibrous scaffolds are suitable for replicating the physical structure of extra cellular matrix (ECM). However, the exact role played by the fiber diameter in modulating cellular attachment, proliferation, differentiation and eventual tissue architecture is not clearly understood.

Great advances have been made over the past few years in the development of techniques for probing the mechanical properties of materials on the submicron scale. The two mechanical properties measured most frequently measured using load and depth sensing indentation techniques are the elastic modulus, *E,* and the hardness, *H* [7]. R. Lopez reported the nanoindentation studies on metal fibers of YAG, YAG: Nd and YAG: Eu fibers with different compositions [8]. There are a few literatures on indentation on sub-micron fibers. Mao Wang et. al. [9] have electrospun *B. mori* silk/poly (ethylene oxide) (PEO) fibers with diameters less than 1 μm. The mechanical properties of single fibers were characterized by AFM nanoindentation. The results have been consistent with uniaxial tensile tests and with the morphological analysis. E. P. S. Tan and C. T. Lim [10] have done nanoindentation studies of a single poly (L-lactic acid) nanofiber. Issues concerning the use of AFM for nanoindentation of polymer nanofibers have been discussed. The Hertz theory of contact mechanics was used to analyze the indentation results. It was found that the elastic modulus was comparable to that obtained from the nanoscale three-point bend test done in their previous study, after roughness correction was made.

A typical starting point for *in-vitro* cell growth on 3-D scaffolds is seeding them either on or within the scaffold. Cell seeding represents a fundamental step in any tissue engineering application that incorporates cells in or on scaffolds prior to implantation [11]. Desirable features for any successful cell seeding include minimized cell injury, uniform cell distribution, high seeding efficiency, reduced seeding time [12]. For example bilayer skin model-fibroblasts would be residing at an intermediate layer and the kerotinocytes would be on the top. Therefore, carefully seeding fibroblasts at a given depth within the scaffold is important. A similar requirement exists for other multi cellular tissues and organs. Precise placement of cell layers in a matrix is important and they

T. Proulx (ed.), *MEMS and Nanotechnology, Volume 2*, Conference Proceedings of the Society for Experimental Mechanics Series 2,
DOI 10.1007/978-1-4419-8825-6_22, © The Society for Experimental Mechanics, Inc. 2011

are a strong function of the fiber architecture. For electrospun scaffolds, we have shown that cell seeding is a function of fiber diameter [13]. In the current study we extend our work to measure Young's modulus of the fibers as a function their diameter and correlate them to the cell seeding depth.

There are two methods for seeding cells in a 3-dimensional matrix with the aim of achieving uniform distribution: passive and active [14, 15]. Typically in the passive methods, cells are laid on top of the scaffold and allowed to infiltrate the scaffold over time. This is a simple process in which cells are not subjected to potentially damaging mechanical forces like high shear stresses [16, 17] resulting in a greater viability. In active seeding methodologies, an external force is applied to enable the cells to infiltrate the scaffold at a faster rate. This method allows good control on quantitative loading and a homogeneous distribution of cells.

The goal of the current study was to correlate the depth of cell seeding and the corresponding attachment and proliferation to the mechanical property of the fiber. PCL was the material that was electrospun. For the study; we have used a simple, rapid and controllable vacuum seeding protocol for uniform seeding of NIH 3T3 fibroblasts. We have compared the depths of seeding for different vacuum pressures to the Young's modulus of the fibers.

2. EXPERIMENTAL DETAILS

2.1. Materials

Chemical reagents from Sigma (St. Louis, MO) were used. Dulbecco' modified eagle medium (DMEM) and penicillin-streptomycin were from Invitrogen (Carlsbad, CA). Bovine calf serum (BCS) was from Hyclone (Logan, UT)

2.2. Scaffold Preparation and Cell Culture

PCL with an average molecular weight of 80000g/mol was acquired from Sigma-Aldrich (St. Louis, MO) and dissolved in acetone (Fisher Scientific, Pittsburg, PA) under 30oC heat and mechanical agitation to obtain homogenous clear, viscous solution (10-20 wt %) suitable for spinning. A 10 ml plastic syringe fitted with a needle with a tip diameter of 0.8 mm was used for electrospinning. PCL nanofibers were fabricated by electrospinning at different applied voltages between 10 to 25 kV using a high voltage power supply (Gamma High Voltage Research, Ormand Beach, FL). The ground collection plate was located 7.5 to 25 cm from the charged metal needle tip. A syringe pump (Braintree Scientific Co, Braintree, MA) was used to feed a polymer solution into the needle at a feeding rate of 3 ml/hr. It was found that cell attachment and proliferation behaviors were significantly different for scaffold fiber diameters over 1 µm and therefore 400 nm and 1100 nm scaffolds were used as model scaffolds for this study.

The electrospun scaffolds (50 µm) thickness was cut with a razor blade into 1x1 cm^2 squares. To prepare the scaffold for seeding they were each placed in a well of 24 well plate (Becton Dickinson Co. Franklin Lakes, NJ), sprayed with alcohol, rinsed twice with PBS, and sterilized under ultraviolet light overnight before cell seeding to facilitate cell attachment onto the other fiber surfaces.

NIH 3T3 mouse fibroblast cells (obtained from Shriners Hospital, Boston, MA) were used in this study. A monolayer of the cells were deposited in 75 cm^2 tissue culture flasks (BD Falcon, Bedford, MA) and cultured to confluence in Dulbecco's modified Eagle Medium (DMEM) (Gibco-BRL, Gaithersburg, MD) containing 10 % Bovine serum(Hyclone, Logan, UT) and 1 % penicillin-streptomycin (Gibco-BRL, Gaithersburg, MD). The medium was replaced every 3 days and cultures were maintained in an incubator at 37oC with 10 % CO_2.

2.3. Vacuum Chamber set-up and Cell Seeding

A CFM capacity Robinair vacuum pump (Robinair division, Montepelier OH) was connected by flexible tubing secured to the outside of the valve going into the hood. A surfactant free cellulose acetate Nalgene filter unit (Fisher Scientific, Pittsburg, PA) was used. The seeding density was 3 x 10^5 cells/cm^2. The control scaffolds were placed in the seeding chamber, the solution was pipetted on to the scaffold surface, and gravity was allowed to drain the solution through the scaffold. Cells were then subjected to vacuum pressure of 2.5, 8 and 20 Hg. Initial cell loss was 8.9 % during regular seeding and 29. 2% for 20"Hg pressure.

2.4. AFM Characterization

Atomic force microscope (XE100, Park systems Inc, CA, USA) was used in the topographic imaging and nano-indentation of scaffold fibers. A contact mode scan was made prior to the indentation of the sample and typical scanned images are shown in figure 1. This scanned image was used in locating the fibers with different diameters for indentation. The micro scale in the AFM software was used to measure the width of fibers. A methodology provided by Reynaud et al., [18] was employed to calculate the Young's Modulus values. First indentation was performed on a steel sample that provided linear force-deflection diagram with pure cantilever deflection. The indentation depth was determined by subtracting cantilever deflection of steel from that of fibers for a given load. The initial slope of the unloading curve in force vs. indentation depth diagram was used to obtain the tip-to-sample stiffness (S) which is then used to calculate Young's modulus value using the equation (1). The h_{max} and F_{max} correspond to the maximums indentation depth and maximum force. The requirement to calculate the projected area was eliminated. The constant C was calculated using indentation on a sample with known Young's modulus. The number of fibers with diameter distribution of scaffolds was calculated. These data were used in calculating the average Young's modulus values. The scaffold with average fiber distribution of 400 nm had an average E value of 1.8959 GPa while the scaffold with 1100nm average diameter distribution had an average E value of 0.6567 GPa.

$$E = \frac{0.84\sqrt{\pi}.S}{2\sqrt{A}} = \frac{0.84\sqrt{\pi}}{2\sqrt{C}}\left[\frac{S^2}{(Sh_{max}-\varepsilon F_{max})}\right] \qquad (1)$$

Electropsun scaffolds have free fibers standing out and this makes the scanning a challenging task using AFM. An initial effort to scan the mat as it is resulted with undue vibrations of the cantilever. The trials were made to stick the fibers to a double scotch tape; however this method was not successful. Then we utilized the method of gluing the fibers with a room temperature curing epoxy. The control of layer thickness presented a challenge and after with few trials we were successful in controlling the epoxy layer with few epoxy droplets to hold the fibers together in immobilized position. The surface roughness of the scaffold was maintained to be less than 6 μm.

3. RESULTS AND DISCUSSION

3.1. AFM Analysis

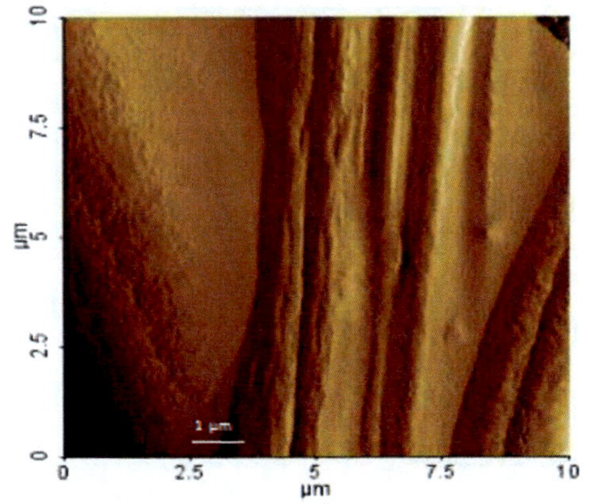

Figure. 1: AFM non-contact images of PCL electrospun fibers

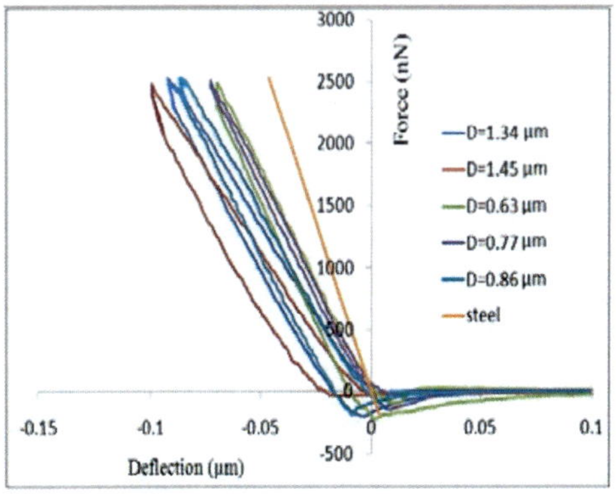

Figure. 2: Typical Force Deflection diagrams for fibers with different diameters

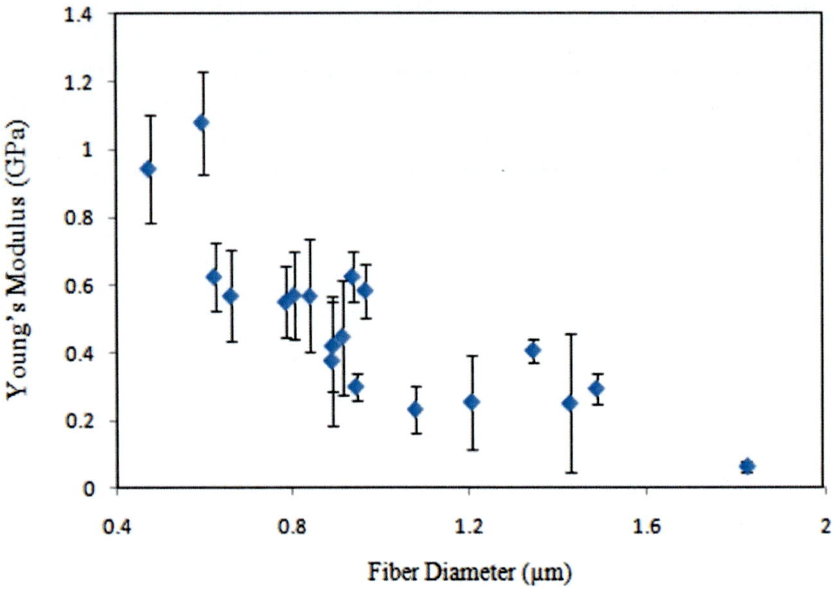

Figure. 3: Variation of young's modulus with fiber diameter for nanoindentation of PCL fibers.

Typical force deflection diagrams obtained for fibers with range of diameters is shown in figure 2. Different slopes of the initial portion of unloading curves verify the calculated difference of Young's modulus values. The minimum diameter of the fibers that could be indented was around 400 nm since the fibers with smaller diameters were immersed and covered with the epoxy. The variation of Young's Modulus as a function of diameter was plotted in figure 3. Young' modulus values for higher fiber diameters were lower than that of smaller fiber diameters.

Cell spatial distribution analysis was made at 3 and 5 days of cell culture. The cells were rinsed with PBS and stained with SYTOX green fluorescent dye. The specimens were observed under a laser scanning confocal microscope (Leica TCS SP2 AOBS, Wetzlar, Germany). The average intensity of images was analyzed by ImageJ software. The average intensity profile of each scaffold was plotted to represent the cell spatial distribution as shown in figure 3.

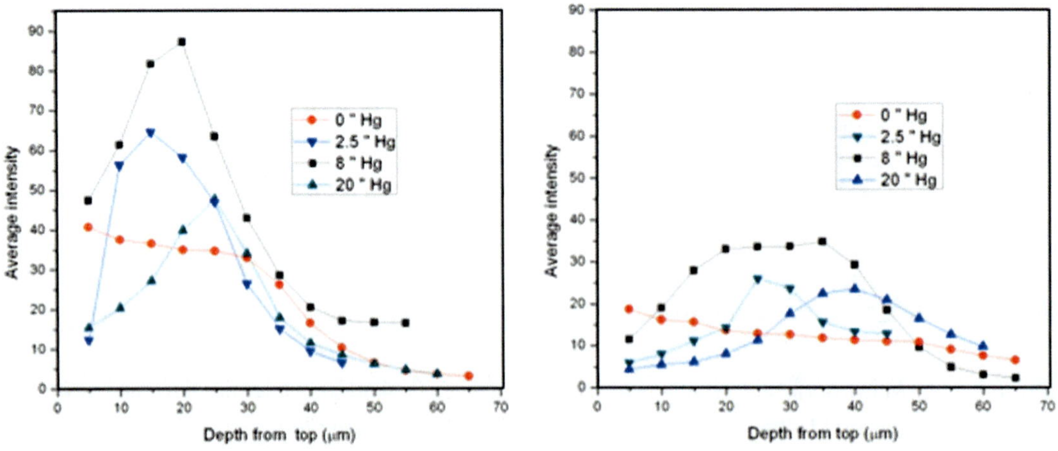

(a) (b)

Figure. 4: (a) cell intensity of scaffolds having an average E of 1.9 GPa at day 3; (b) cell intensity of scaffolds having an average E of 0.6 GPa at day 3.

Figure 4 shows cell fluorescence intensity profile distribution in the scaffolds. It is observed that cell intensity profile shifts from the top layer towards the interior of the scaffold with increasing vacuum pressure. A comparison between 400 nm and 1100 nm scaffolds shows that overall intensity is always higher for 400 nm scaffolds indicating higher cell density at day 3 for all the seeding conditions. We also observe that for a given vacuum pressure, the peak intensity corresponding to the plane where maximum cells reside is shallower for the 400 nm scaffold compared to the 1100 nm scaffolds. These results indicate that smaller diameter fibers which have higher Young's modulus or stiffness offer more resistance for filtration for a given vacuum pressure. The larger diameter fibers which have less stiffness or lower Young's modulus offer less resistance for cell filtration during vacuum seeding.

4. CONCLUSIONS

The electrospun fiber scaffold was studied for variation of mechanical properties as a function of fiber diameters using Atomic Force Microscope. It was identified that as the fiber diameter increases, the Young's modulus values decreases considerably from 1.9GPa to 600MPa. The 3T3 fibroblasts were seeded on the electro-spun fiber scaffold mat of 50-70 microns thickness in this study. The cell spatial distribution analysis of scaffolds was made to investigate the cell movement into middle and bottom layers. The cells were seeded deeper within the 1100 nm scaffold for a given vacuum pressure. It was observed that a fiber with higher stiffness led to a lower depth of cell seeding (closer to the surface) indicating that nanofibrous scaffolds offer larger resistance to cell movement compared to microfibrous scaffolds.

ACKNOWLEDGMENTS

Financial support of National Science Foundation (NSF) through the grant CMS0618119 for acquiring atomic force microscope. In addition, this work was supported through NTC funds F06-MD04.

REFERENCES

1. Nair LS, Bhattacharyya S, Laurencin CT, "Development of novel tissue engineering scaffolds via electrospinning" *Expert Opinion on Biology Therapy*. 4, 659, 2004.
2. Fong H, Chun I, Reneker D, "Beaded nanofibers formed during electrospinning" *Polymer* 40, 4585, 1999.
3. Matthews J, Wnek G, Simpson D, Bowlin G, "Electrospinning of collagen nanofibers" *Biomacromolecules*. 3, 232, 2002.
4. Jin H, Chen J, Karageorgiou V, Altman G, Kaplan D, "Human bone marrow stromal cell responses on electrospun silk fibroin mats" *Biomaterials*. 25, 1039, 2004.
5. Yoshimoto H, YShin, Terai H, Vacanti J, "A biodegradable nanofiber scaffold by electrospinning and its potential for bone tissue engineering" *Biomaterials*, 24, 2077, 2003.
6. Shin M, Ishii O, Sueda T, Vacanti J, "Contractile cardiac grafts using a novel nanofibrous mesh." *Biomaterials*. 25, 3717, 2004.
7. Oliver WC and Pharr GM, "A new improved technique for determining hardness and elastic modulus using load and sensing indentation experiments" *Journal of Material Research*., 7, 1564, 1992
8. Nascimento EM and Lepiensk CM, *Journal of non-crystalline Solids*, 352, 3556, 2006
9. Wang M, Jin H, Kaplan DL, and Rutledge GC, *Macromolecules*. 37, 6856, 2004.
10. Tan EPS and Lim CT "Nanoindentation study of nanofibers", *Applied Physics. Letters*. 87, 123106, 2005
11. Langer R, "Tissue engineering." *Molecular Therapy*. 1, 5, 2000
12. Soletti L, Nieponice A, Guan J, Stankus JJ, Wagner WR, Worp DA,"A seeding device for tissue engineered tubular structures." *Biomaterials*, 27, 4863, 2006.
13. Chen M, Michaud H and Bhowmick S, *Journal of Biomechanical engineering- ASME* 131, 1, 2009
14. Dar A, Shachar M, Leor J, Cohen S, "Optimization of cardiac cell seeding and distribution in 3D porous alginate scaffolds." *Biotechnology and Bioengineering*. 80, 305, 2002.
15. Saini S, Wick TM, "Concentric cylinder bioreactor for production of tissue engineered cartilage: effect of seeding density and hydrodynamic loading on construct development." *Biotechnology Progress*. 19, 510, 2003.
16. Carrier F, Owens RA, Nebert DW, Puga A, "Dioxin-dependent activation of murine Cyp1a-1 gene transcription requires protein kinase C-dependent phosphorylation." *Molecular Cell Biology*. 12, 1856, 1992.

17. Merchuk JC, "Shear effects on suspended cells." *Advances in Biochemical Engineering Biotechnology*. 44. 65, 1991
18. Reynaud C, Sommer F, Quet C, Bounia NI and Duc TM, "Quantitative determination of Young's modulus on a biphase polymer system using Atomic force microscopy", *Surface and Interface Analysis*.30, 185, 2000.

Proceedings of the SEM Annual Conference
June 7-10, 2010 Indianapolis, Indiana USA

Recent Progress in E-Beam Lithography for SEM Patterning

Ning Li[1], Siming Guo and Michael A. Sutton

Department of Mechanical Engineering, University of South Carolina
Columbia, SC 29208
[1]Presenting Author, lining@sc.edu

ABSTRACT

An improved e-beam lithography technique has been developed to generate a high quality micro/nano-scale random speckle pattern on various specimens for metrology and characterization of specimens using Scanning Electron Microscope (SEM) images with Digital Image Correlation (DIC) for image analysis. In this application, a mathematical algorithm has been integrated into the e-beam control system to greatly reduce the time for etching a dual-layer photo-resist coating to obtain a random distribution with optimal size distribution in the resulting pattern. It was determined that the thickness of photo-resist must be carefully controlled to obtain the desired pattern spot size after completing the development process. The resulting beam lithography technique has been employed by the authors to obtain high quality Au random pattern ranging from 150nm to 500nm on Al and Si specimens.

In addition to the local application of high quality random pattern, the authors developed a marker-placement methodology so that the local pattern area could be readily located through specimen translations. Extension of the approach to the production of high quality 50nm pattern on silicon wafer using various metals (e.g., Al, Ti, Ni, Cu, Zn, W, Ag, Pt) is in progress.

INTRODUCTION AND OBJECTIVE

The DIC technique[1] has been widely used in mechanical engineering area for non-destructive, accurate measurement of full-field surface deformations. In its generic form, the methodology employs a high contrast speckle pattern on the surface of specimen for a digital camera to acquire images during the deformation process.

The 2D version of this methodology has been extended and used effectively with SEM images[2~5]. In these applications, the micro/nano-scale random pattern on the specimen was generated using a chemical method[6]. Heating of a specimen in a specific chemical environment resulted in islandization of the coated metal film. Since it was relatively difficult to control the size of spot for a proper SEM imaging in this approach, the authors developed the methodology outlined in the following sections to generate a more controlled pattern.

PATTERN DEVELOPMENT

E-beam lithography technique is mostly used for manufacturing of IC (Integrated Circuits) chips and micro-scale devices. Since e-beam imaging is performed in a rastering mode, it is clear that the e-beam position can be controlled such that the e-beam impinges on a surface at pre-defined locations. To do so, the authors implement a LabView program cooperated with the original e-beam etching program to control the e-beam motion and "write" a random pattern with specific size and spatial distribution. Because the e-beam spot size is on the order of a few nanometers, the specimen should be well polished. For the aluminum specimen, the authors polish the surface using fine grit sand paper with water to achieve surface roughness of 5μm. Then, fine Al_2O_3 powder is used with water to reduce surface roughness to <1μm.

To obtain high quality micro/nano-scale pattern, the authors use e-beam etching of two layers of PMMA (PolyMethyl MechAcrylate) that are applied as the positive photo-resist by spin coating of the specimen. The top layer is 950K PMMA with a typical thickness of 300nm. The bottom layer is 496K PMMA with a typical thickness of 50nm. Spinning rates and times for deposition of the top and bottom layers are 3,000RPM for 30s and

T. Proulx (ed.), *MEMS and Nanotechnology, Volume 2*, Conference Proceedings of the Society for Experimental Mechanics Series 2,
DOI 10.1007/978-1-4419-8825-6_23, © The Society for Experimental Mechanics, Inc. 2011

4,000RPM for 40s, respectively. After coating each layer onto the specimen, the specimen should be baked for 2 minutes at 180°C to bond the layers firmly to each other and to the substrate. It is important to note that the optimal combination of spin coater rotation speed and total thickness of PMMA are a function of the required pattern size; smaller pattern size requires thinner PMMA layers and higher rotation speed. It is noted that the e-beam dot size limits the smallest spot size in the random pattern. For the older SEM system used in this work, the smallest random spot achievable is ≈50nm.

To locate the area on the specimen that is to be studied, a series of large markers are created also by e-beam etching. The markers extend from one edge of specimen to the pattern area, and are placed on the specimen prior to pattern etching. The etching process for pattern typically takes 1~2 hours depending on the size of pattern area.

Once the pattern is etched on the PMMA layers, the specimen is immersed in solvent of MIBK (Methyl IsoButyl Ketone) : IPA (IsoPropyl Alcohol) = 1:3 (in volume) for 30 seconds at 21°C then dried by N_2 thoroughly. With a light filter in place, the specimen can be imaged by an optical microscope to identify defects and the correctness of the final position of the etched area.

The next step in the pattern development process is to deposit a thin metallic layer over the etched specimen. Since the SEM generates image contrast according to the difference in atomic weight of the elements in the pattern and the substrate, Au is commonly used for most substrates. In the current process, 2nm of Ti is evaporated onto the Si specimens prior to deposition of Au to improve adherence. The deposition time should be strictly monitored to ensure that the appropriate Au layer thickness is obtained.

To remove the residual photo-resist and evaporated metal from the specimen, it is immersed in acetone overnight. To keep the pattern usable for a longer period of time, the authors have found that immersion for extended periods of time will result in reduced quality of the pattern and increased loss of contrast with time.

RESULTS

Fig. 1 shows an SEM image at 20,000X for the center part of a 60μm×60μm pattern area of Au on an Al substrate. The spot size is 150nm and the contrast is adequate for further tensile or thermal loading experiments.

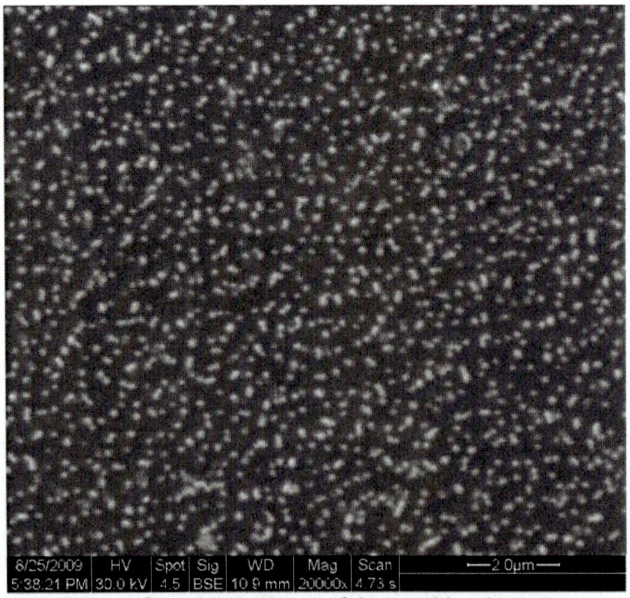

Fig. 1: Speckle pattern of Au on Al substrate.

Fig. 2 shows a calibration grid created on the same specimen by the same lithography technique, where the LabView program is modified for this purpose.

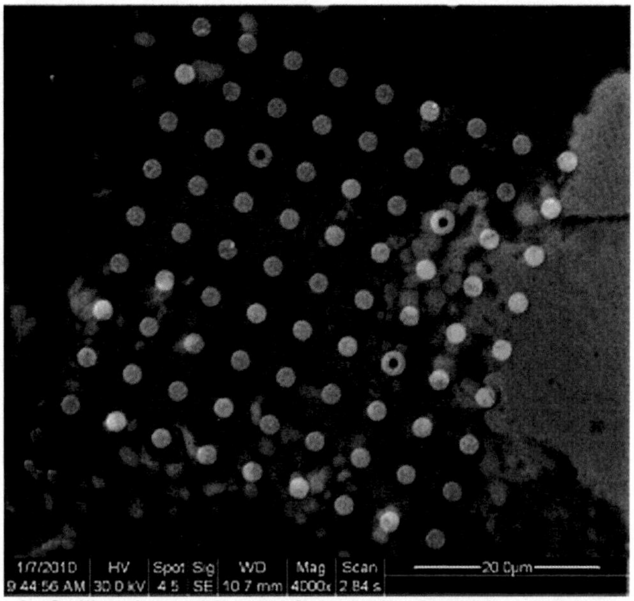

Fig. 2: Calibration grid manufactured using the same process

Fig 3 shows a ring pattern which is to determine orientation and spacing of calibration grid by the post-processing software. The authors SEM originally can not write this kind of ring pattern. The LabView control program of e-beam etching has been modified again to adapt this special need.

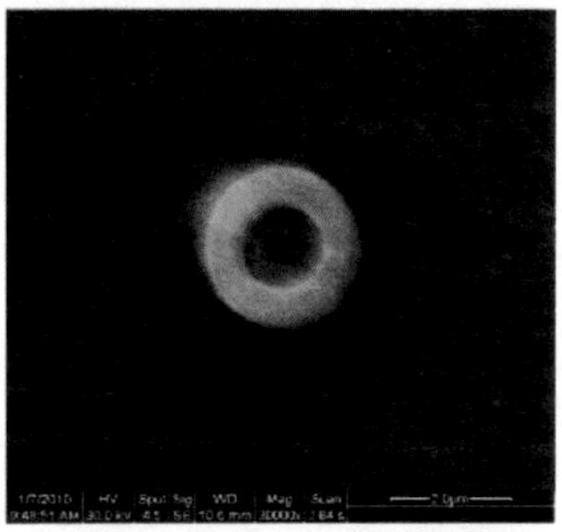

Fig. 3: Ring pattern dot in the grid.

CONCLUSIONS

LabView software has been written to control the e-beam motion for use in e-beam lithographic development processes to apply micro/nano-scale pattern, including high quality random pattern as well as structured dot pattern for use in calibration studies. The process can control both the spot size and the spot distribution in the pattern area. The resulting patterns on aluminum and silicon specimen surfaces have been shown to have good adherence and good contrast for several months after application.

REFERENCES

[1] Sutton MA, Orteu JJ and Schreier HW (2008) Image Correlation for Shape, Motion and Deformation Measurements. Springer Science+Business Media, LLC, ISBN: 978-0-387-78746-6.

[2] Sutton MA, Li N, Garcia D, Cornille N, Orteu JJ, McNeill SR, Schreier HW and Li XD (2006) Metrology in SEM: Theoretical Developments and Experimental Validation. Measurement Science and Technology, 17(10)2613~2622.

[3] Li N, Sutton MA and Li XD (2008) Full-Field Thermal Deformation Measurements in a Scanning Electron Microscope by 2D Digital Image Correlation. Experimental Mechanics 48(5)635~646.

[4] Sutton MA, Li N, Joy DC, Reynolds AP and Li XD (2007) Scanning Electron Microscopy for Quantitative Small and Large Deformation Measurements, Part I: SEM Imaging at Magnification from 200 to 10,000. Experimental Mechanics 47(6)775~787.

[5] Sutton MA, Li N, Joy DC, Reynolds AP and Li XD (2007) Scanning Electron Microscopy for Quantitative Small and Large Deformation Measurements, Part II: Experimental Validation for Magnifications from 200 to 10,000, Experimental Mechanics 47(6)789~804.

[6] Collette SA, Sutton MA, Miney P, Reynolds AP, Li XD, Colavita PE, Scrivens WA, Luo Y, Sudarshan T, Muzykov P and Myrick ML (2004) Development of Patterns for Nanoscale Strain Measurements: I. Fabrication of Imprinted Au Webs for Polymeric Materials. Nanotechnology 15(12)1812~1817.

Proceedings of the SEM Annual Conference
June 7-10, 2010 Indianapolis, Indiana USA

Analysis of Scattering-type Scanning Near-field Optical Microscopy for Residual-strain Measurements

Chia-Chi Liao and *Yu-Lung Lo
Department of Mechanical Engineering
National Cheng Kung University
No.1, Ta-Hsueh Road, Tainan 701, Taiwan ROC
*Corresponding author: loyl@mail.ncku.edu.tw

ABSTRACT

An analytical model for residual-strain measurement based on the Scattering-type scanning near-field optical microscopy (s-SNOM) has been developed in this study. A-SNOM has a capability for inspection properties of materials in nanometer-scale and with resolution up to 10 nm. However, the scattering signals in s-SNOM are highly complex and contaminated by the background noise critically. To overcome the problem, we have proposed a mathematical model to improve the near-field signals by eliminating the background noise in heterodyne detection. According to the mathematical model, the study will discuss the signal in s-SNOM in detail, analyze the spectrum of measurements, and explore more methods to get better signal. Then, the mathematical model will be combined with other modified near-field ones to construct a novel near-field analytical model to fit the experimental data on phonon-polariton as possible. Based on the new analytical model, the dielectric constants of materials can be obtained more precisely, and the residual stress and strain relative to the variation of dielectric constants of SiC which most often utilized in micro- and nano-electromechanical system (MEMS and NEMS) can be determined more distinctly.

1. Introduction

In order to overcome the restriction in resolution for optical microscopy, the aperture scanning near-field optical microscopy (SNOM) which has been developed [1]. In SNOM, the illumination light is confined by the metallic aperture to enhance near-field effect and achieve a nano-measurement in chemical, structural, and conduction properties. However, the resolution of SNOM is still restricted by the size of the tapered metal-coated optical fiber aperture and the waveguide cut-off effect [2]. The very small aperture of SNOM also severely restricts the light throughput, and the light intensity can not be simply increased to compensate the loss because of the risk of damaging due to heating effect [3]. Accordingly, an alternative SNOM was proposed. Scattering-type scanning near-field optical microscopy (s-SNOM, also called apertureless-SNOM) replaces the optical fiber with a sharp vibrating tip. In this configuration, the incident light illuminates the tip to get a small scatter and induce a local enhancement of the electric field between the tip and the specimen in near-field. The enhancement due to the near-field interaction depends on the dipole effect and is similar as the sphere with a nanometer-scale and makes an optical resolution at the sub-10 nm scale [4-6].

Comparing to the conventional SNOM, s-SNOM can obtain the measurement with better resolution but has some problems to get over. For the scattering detection, the signal of s-SNOM is the form as the interference between the near-field electric field and the background ones. Hence, the desired signal in s-SNOM will be affected by the background noise seriously. As a result, it is necessary to develop techniques for suppressing the background–scattering noise from the detected signal and improving the precision and reliability of the measured results. In s-SNOM, the signal often be obtained by lock-in amplifier and acquired by signal modulated process, such as heterodyne detection [7-8] or homodyne one [8-9]. Various formulas have been proposed to describe the characteristics of both techniques [10-12]. By analyzing the tip enhancement and modulated background components, the signal in s-SNOM can be improved. However, these studies not only neglect the near-field enhancement, but also utilized models overly complex. Recently, the current group proposed an analytical investigation into the modulation of s-SNOM homodyne and heterodyne signals [13-14]. This analytical model can

T. Proulx (ed.), *MEMS and Nanotechnology, Volume 2*, Conference Proceedings of the Society for Experimental Mechanics Series 2,
DOI 10.1007/978-1-4419-8825-6_24,

provide new methods to improve signal and demonstrate that heterodyne detection will magnify the near-field signal and obtain the better recognition than homodyne one.

Accordingly, this study combines the models proposed in ref. [13-14] and [15] and makes some adjustment to develop a new analytical model. The new model considers the modified theory for near-field interaction electric field, and makes it to interfere with all considerable background fields. This model can be demonstrated by experiments proposed in ref. [16]. Finally, since the dielectric constants of materials can be obtained more precisely by utilizing the new mathematical model, the further application such as measuring the residual stress and strain relative to the variation of dielectric constants of SiC which most often utilized in micro- and nano-electromechanical system (MEMS and NEMS) can be determined more distinctly.

2. Analytical Model

Fig. 1(a) presents a schematic illustration of a Mach-Zehnder interferometer-type s-SNOM. As shown, the illuminating light from light source is split into two beams. One served the incident electric field, E_i, will pass through the objective lens and focus on the probe tip of atomic force microscopy (AFM) to produce the near-field enhancement. The enhancement field and other scattering ones will be detected as the s-SNOM signal itself, I_{hom}. Another beam in system will be modulated by the frequency modulator such as the acousto-optic modulator (AOM), and this beam will shift a radian frequency $\Delta\omega$ and serve as a reference signal. The s-SNOM signal and the reference signal will interfere as heterodyne signal, I_{het}, and can be acquired by lock-in amplifier. Beside the reference signal, other electric fields are all shown in Fig. 1(b), the schematic illustration describing all considerable scattering electric fields in near-field region in AFM. For analysis on the near-field signal, all the desired lights are assumed that they will pass through the objective lens.

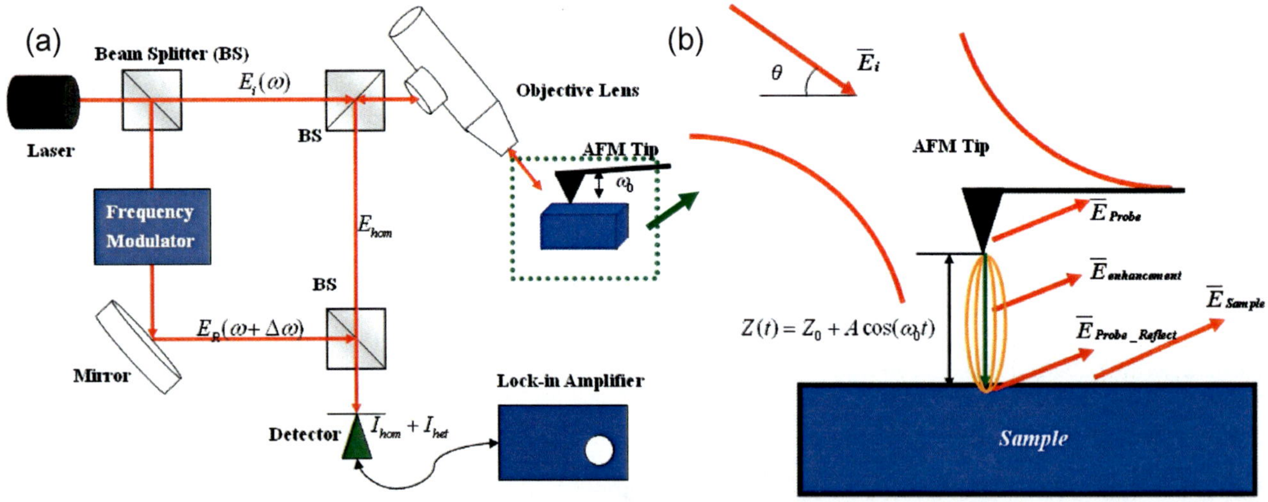

Fig. 1(a) Schematic illustration of heterodyne s-SNOM; 1(b) Electric field distributions in near-field region

In all fields, the field represents the near-field interaction (or enhancement) one is the most important and should be inserted in different situations. The interaction field can be expressed as:

$$\vec{E}_{enhancement} = \alpha_{eff}\vec{E}_{in} = \left|\alpha_{eff}\right| E_{in} e^{i(\omega t+\phi_{\alpha_{eff}}+\phi_{in})} = E_{enhance} e^{i(\omega t+\phi_{enhance})} \tag{1}$$

where E_{in} is the amplitude, ϕ_{in} is the phase, and ω is the radian frequency; α_{eff} is the effective polarizability; $E_{enhance}$ is the total amplitude of the interaction field; $\phi_{enhance}$ is the total phase of the interaction field; α_{eff} is a critically important factor since it contains the relative contrasts observable in s-SNOM. According to the general model of quasi-electrostatic theory [7-9], it can be expressed as

$$\alpha_{eff} = \frac{\alpha(1+\beta)}{1-\dfrac{\alpha\beta}{16\pi r^3}} \quad \text{with} \quad \alpha = 4\pi a^3 \frac{\varepsilon_p - 1}{\varepsilon_p + 2}, \quad \beta = \frac{\varepsilon_s - 1}{\varepsilon_s + 1} \tag{2}$$

where a is the radius of enhanced dipole; r is the distance between the dipole and the sample surface; ε_p and ε_s are complex dielectric numbers of probe and sample, respectively. This model depending on quasi-electrostatic theory can explain the phenomenon in near-field interaction but still is different than the exactly resonance spectrum and dielectric function. Therefore, recent study proposed a modified model which considers the size of probe, the distance between probe and sample surface, and wavelength [15]. The modified model presumes that the length of probe, $2L$ should be much longer than the wavelength due to electrostatic approximation. Hence, the modified model can be formulated as [15]:

$$\alpha_{eff} = a^2 L \frac{\frac{2L}{a} + \ln\frac{a}{4eL}}{\ln\frac{4L}{e^2}} \left[2 + \frac{\beta\left(g - \frac{a+H}{L}\right)\ln\frac{4L}{4H+3a}}{\ln\frac{4L}{a} - \beta\left(g - \frac{3a+4H}{4L}\right)\ln\frac{2L}{2H+a}} \right] \left(1 + R_p^2\right) \quad (3)$$

where H is the distance between the dipole and sample surfaces; L is the half of the probe tip length, g is the factor related to the proportion of the total induced charges. By inspection in actual experiments, the best values can be found in different situations. At later section, the background signal will be taken into account in both conventional theory and modified one to make a comparison and discussion.

The total electric field coupled into the objective lens is equivalent to the sum of the reference signal and the four electric fields shown as Fig. 1(a) in near-field region. And then, the signal is modulated by the frequency modulator and called heterodyne modulated signal [14].

3. Demonstration in the analytical model by previous experiments

The analytical formulation for explanation and prediction the s-SNOM polariton-resonance spectrum has been provided. The s-SNOM signal detected in lock-in amplifier from $\Delta\omega t+\omega_0 t$ to $\Delta\omega t+4\omega_0 t$ by heterodyne method can been obtained from the analytical model. For demonstration of the model, this section of paper will use the analytical model to simulate heterodyne detection signal and compare the simulation with the experimental result present in [16].The experiment presented in [16] employed s-SNOM to measure the near-field optical phonon-polariton resonance spectrum of 6H-SiC crystal.

Fig. 2 displays the comparison between the simulation and the experimental results. The black solid and the blue dash lines represent the simulation and experimental results, respectively. From the figure, it shows that the analytical model developed in this study has a good agreement with the experiment in [16]. Furthermore, Fig. 2 also illuminates the pure interaction signal without considering background signals by red line. It shows that the simulation which only considers the dipole-interaction effect but neglects background noise is different to the modulated signal detected by actual experiments. Therefore, the analytical model containing the interference with background fields presented in this study can provide the more precise description of the real detected signal.

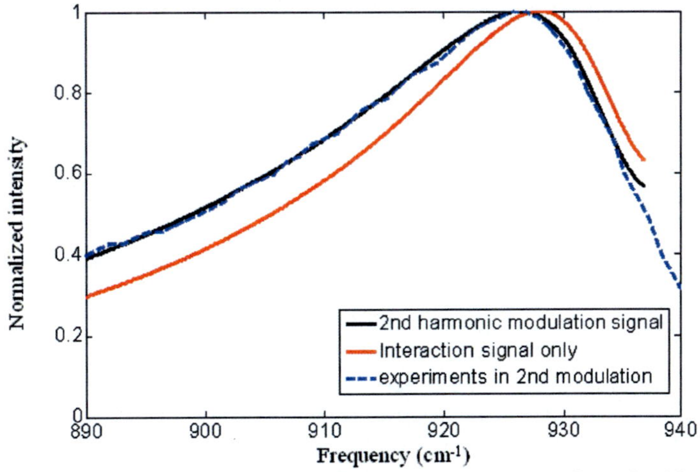

Fig. 2 Simulation results compared with experimental ones (the detection of the 2nd harmonic modulated signal)

4. Advanced application in measuring the residual strain and stress

In this section, an advanced application by the demonstrated analytical model for s-SNOM is introduced. A silicon carbide (SiC) crystal is one of the most robust semiconductor materials. In manufactured process, the materials will be suffered from the stress by fabricating and get a residual strain related to the stress. Therefore, measurements in local strain are necessary for monitoring and optimization of the fabrication processes. Since s-SNOM is a non-invasive measurement in nano-scale and requires the sample made only by minimum preparation, the strain mapping by s-SNOM is a worthy of technique. In the section 3, the s-SNOM signal of SiC has been simulated. It can provide the prediction of the nanoscale mapping in the polariton-resonance spectrum curve in SiC. For advanced measurement, this analytical model also can be employed to describe the shift of the detected spectrum due to the residual strain.

S-SNOM signals are very relative to the response of the dielectric function of the sample. The frequency response of the unstrained SiC is shown as the black line in Fig. 3. To provide an estimation of the shift related to the strain, the spectral shift $\Delta\omega$ compared to unstrained SiC is determined. The determined $\Delta\omega$ can be assigned to the shift of $\Delta\omega_{LO}$ of the longitudinal optical phonon frequency by $\Delta\omega \approx \Delta\omega_{LO}$ [17]. Therefore, the strain can be calculated by using the distinct relation between LO shifts and strain in SiC[18].

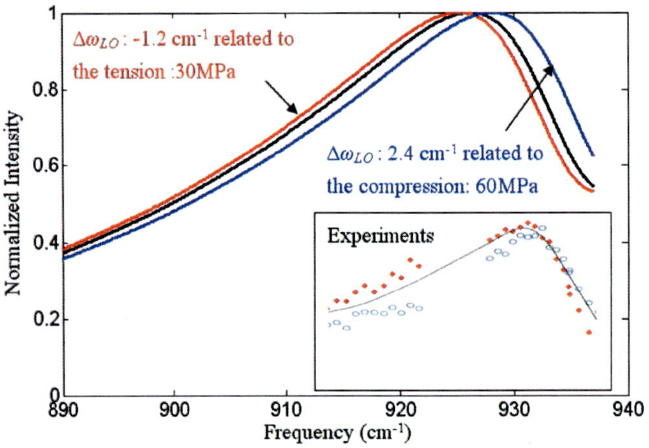

Fig. 3 Simulation of the measurement in the shift of the s-SNOM spectra induced by residual stress and strain in SiC; Black line represents the initial spectrum of SiC; Red line represents SiC suffered from the tension and around 30MPa [16]; Blue line represents SiC suffered from the compression around 60MPa [16].

From the experiments achieved by the study in [16], two points that represent sample suffered from the tension at 30 MPa and compression at 60 MPa are selected to figure out the frequency response. According to the estimation of the frequency shift related to the residual stress and strain, the shift of the longitudinal optical phonon frequency, $\Delta\omega_{LO}$ in dielectric function of SiC can be found. In Fig. 3, the frequency response of the part suffered from the tensile strain is simulated as the red line, while that suffered from the compressive strain is simulated as the blue line. The red line which represents the tensile strain has a peak at ~926 cm^{-1}, while the blue line which represents the compressive strain has a peak at ~928 cm^{-1}. The simulated result is accepted by compared to the experimental one presented in [16].

5. Conclusions and Discussions

S-SNOM signals are usually acquired by interference and modulation techniques and can not be exactly described by only considering the dipole-interaction model without other interference background. Hence, this study has developed a comprehensive analytical model established by the dipole-interaction theory interfering with background signal. The model can figure out the exactly different harmonic modulated signals in the s-SNOM and has a good agreement with the experiments presented in [16]. Therefore, the model can provide the method to eliminate the undesired noise and obtain the pure interaction signal in different conditions. Since this model has been confirmed that it can fit in with the exactly polarition-resonance response spectrum, it also can provide the rough value of the design in factors how to get the better signal in experiments.

Finally, a local nano-scale measurement in residual stress and strain related to the shift of the frequency response spectrum in s-SNOM is simulated according to the demonstrated analytical model. The simulation has an acceptable result compared to the study in [16] and will provide more identifying and accurate measurement in monitoring the residual strain effect.

Acknowledgements

The authors gratefully acknowledge the financial support provided to this study by the National Science Council of Taiwan under Grant No. 98-2221-E-006-053-MY3, 2009.

Reference

[1] D. W. Pohl, S. Denk, and M. Lanz, "Optical stethoscopy: image recording with resolution," J. Appl. Phys., Vol. **44**, pp. 651-653, 1984.
[2] J. D. Jackson, *Classical electrodynamics*, Wiley, 1999.
[3] S. Patane, G. G. Gucciardi, M. Labardi, and M. Allegrini, "Apertureless near-field optical microscopy," Rivita Del Nuovo Cimento, Vol. **27**, pp.1-46, 2004.
[4] J. Wessel, "Surface-enhanced optical microscopy," J. Opt. Soc. Am., Vol. **2**, pp. 1538-1540, 1985.
[5] H. K. Wickramasinghe and C. C. Williams, "Apertureless near field optical microscope," US Patent 4 947 034, 1990.
[6] Y. Inouye, and S. Kawata, "Near-field scanning optical microscope with a metallic probe tip," Opt. Lett., Vol. **19**, p.p. 159-161, 1994
[7] R. Hillenbrand and F. Keilmann, "Complex optical constants on a subwavelength scale, " Phys. Rev. Lett., Vol. **85**, pp. 3029-3032, 2000.
[8] R. Hillenbrand, B. Knoll, and F. Keilmann, "Pure optical contrast in scattering-type scanning near-field microscopy," J. Microsc., Vol. **202**, pp.77-83, 2000
[9] B. Knoll, and F. Keilmann, "Enhanced dielectric contrast in scattering-type scanning near-field optical microscopy," Opt. Commun., Vol. **182**, pp.321-328, 2000
[10] S. Hudlet, S. Aubert, A. Bruyant, R. Bachelot, P. M. Adam, J. L. Bijeon, G. Lerondel, P. Royer, and A. A. Stashkevich, "Apertureless near field optical microscopy: a contribution to the understanding of the signal detected in the presence of background field," Opt. Commun., Vol. **230**, pp.245-251, 2004.
[11] F. Formanek, Y. D. Wilde, and L. Aigouy, "Analysis of the measured signals in apertureless near-field optical microscopy," Ultramicroscopy, Vol. **103**, pp.133-139, 2005
[12] P. G. Gucciardi, G. Bachelier, and M. Allegrini, "Far-field background suppression in tip-modulated apertureless near-field optical microscopy," J. Appl. Phys., Vol. **99**, Art. No. 124309, 2006
[13] C. H. Chuang, and Y. L. Lo, "Analytical analysis of modulated signal in apertureless scanning near-field optical microscopy," Opt. Express, Vol. **15**, pp. 15782-15796, 2007.
[14] C. H. Chuang, and Y. L. Lo, "An analysis of heterodyne signals in apertureless scanning near-field optical microscopy," Opt. Express, Vol. **16**, pp. 17982–18003, 2008.
[15] A. Cvitkovic, N. Ocelic, and R. Hillenbrand, "Analytical model for quantitative prediction of material contrasts in scattering-type near-field optical microscopy," Opt. Express, Vol. **15**, pp. 8550-8565, 2007
[16] A. J. Huber, A. Ziegler, T. Köck, and R. Hillenbrand, "Infrared nanoscopy of strained semiconductors," Nature Nanotechnology, Vol. **4**, pp. 153-157, 2008.
[17] A. Huber, N. Ocelic, T. Taubner, and R. Hillenbrand, "Nanoscale resolved infrared probing of crystal structure and of plasmon–phonon coupling," Nano Lett., Vol. **6**, pp. 774–778, 2006
[18] J. Liu and Y, K. Vohra, "Raman modes of 6H polytype of silicon-carbide to ultrahigh pressures—a comparison with silicon and diamond," Phys. Rev. Lett., Vol. **72**, pp. 4105–4108, 1994

Proceedings of the SEM Annual Conference
June 7-10, 2010 Indianapolis, Indiana USA

Experimental Methods for Tensile Testing of Metallic Thin Films at High Temperatures

N.J. Karanjgaokar[a], C.S. Oh[b], I. Chasiotis[c]

a- Mechanical Science and Engineering, University of Illinois, Urbana, IL
email: nkaranj2@illinois.edu
b- School of Mechanical Eng., Kumoh National Institute of Technology, Gumi, South Korea
c- Aerospace Engineering, University of Illinois, Urbana, IL

ABSTRACT

Two methods for microscale tension experiments with microscale freestanding thin films at elevated temperatures were evaluated by means of optical microscopy/Digital Image Correlation (DIC) and Infrared (IR) imaging. The two methods employed uniform and resistive specimen heating. Optical images processed by DIC were used to calculate the strain along the specimen gauge section of specimens subjected to both experimental methods, while IR imaging was used to measure the temperature distribution along the specimens' gauge sections. The axial strain and temperature distributions were compared qualitatively to evaluate the efficacy of each method. Uniform specimen heating provided uniform temperature and axial strain distributions that were not affected in any measurable way by the use of the "cold" external probe employed to pull on the specimens. However, the resistively heated specimens had highly non-uniform temperature and axial strain distributions along their gauge sections. The associated high temperature gradients resulted in strain localization and significant reduction in yield and ultimate strength measured during resistive heating experiments compared to uniformly heated samples. The experiments revealed that the latter method provides high fidelity measurements at elevated temperatures.

INTRODUCTION

Metallic thin film materials such as Ni, Au, Ag, Cu and Pt have been used as interconnect and structural elements in a variety of applications in the microelectronics and Microelectromechanical Systems (MEMS). Their reliability is affected by the mechanical behavior of the particular thin film materials, which, in turn, is subject to factors such as grain size [1-2], specimen size [3], strain rate [1,4] and substrate effects. The aforementioned considerations are further accentuated at temperatures higher than room temperature. Common operating temperatures for thin film applications such as RF-MEMS switches are in the range -20 to 120 ^{0}C. In the past thermal expansion coefficients were measured from thin microbeams and films by using both resistive and uniform heating methods [5-8]. Emery et al [9] investigated the strain rate dependence of gold thin films of 2.1 μm thickness by placing their setup in a custom built furnace at temperatures between 200 ^{0}C and 400 ^{0}C. The experimental methods used either thin film materials attached to a large substrate or freestanding structures with thicknesses varying from the submicron scale to hundreds of microns. The sample size and the presence of a substrate provided different degrees of ability to determine the sample temperature and its strain distribution but not both.

Gruber et al [10] performed high temperature tension tests with a custom heating assembly from 123 to 473 K on metallic thin films deposited on thick polyimide and calculated full-field strains by using digital image correlation (DIC). In terms of freestanding structures, Oh et al [11-12] developed a technique for freestanding thin

T. Proulx (ed.), *MEMS and Nanotechnology, Volume 2*, Conference Proceedings of the Society for Experimental Mechanics Series 2,
DOI 10.1007/978-1-4419-8825-6_25, © The Society for Experimental Mechanics, Inc. 2011

polysilicon specimens at temperatures up to 700 °C by using resistive heating. Temperature scans along the gauge section by an optical pyrometer showed significant variations, but the temperature in the 250 μm long center region was uniform within ±10 °C. Zupan et al [13,14] employed the same approach to test Nickel super-alloy samples that were hundreds of microns in thickness which allowed for high specimen temperatures. Evidently, the use of pyrometers is appropriate for sub-millimeter wide and thick samples but it suffers from background emission in surface micromachined specimens. Haque et al [15] performed uniaxial tension tests on freestanding ultra thin aluminum films at temperatures up to 160 °C by gluing a bulk micromachined chip onto a hot plate. The authors measured the chip temperature using a thermocouple because direct temperature measurements were not possible due to the small specimen size. They also used the specimen resistance to verify that the temperature on the hot plate and in an oven were similar. Kalkman et al [16] developed a bulge testing setup to perform tensile tests and creep tests on thin film samples at temperatures up to 300 ^{0}C. In [16], the sample was heated using 8 resistances in parallel and temperature was measured using a thermocouple placed closed to the sample. Researchers have also performed creep [17,18], stress relaxation [19] and fatigue experiments on thin films [20] at elevated temperatures using one of the aforementioned techniques.

The techniques addressed the high temperature properties of thin film in ways that are not appropriate for surface micromachined specimens because of non-uniform or not directly measurable temperature profiles. Furthermore, in most high temperature measurements, cross-head displacements instead of full-field strains were measured. In the recent years, a number of researchers have established the usefulness of non-contact full-field strain measurements via DIC for micro and nanoscale experimentation [21-23]. This paper presents a method for uniform heating of surface micromachined specimens in order to conduct experiments at elevated temperatures with direct knowledge of the specimen temperature along the specimen gauge sections. The method is compared to experiments conducted with resistive heating applied to the same surface micromachined Au thin film specimens. The stress vs. strain curves obtained by the two methods are used as a metric to compare their utility qualitatively and quantitatively.

EXPERIMENTAL METHODOLOGY

The Au films fabricated by the procedures described in [24] had an average grain size of 55 nm and a preferential <111> texture in the grain growth direction as revealed by X-ray diffraction (XRD). The test specimens had dog bone geometry with 1 mm long gauge section that was 100 μm wide and 830 nm thick. The films were annealed at 150 °C for about 12 hours before the high temperature experiments. For the present experiments, the authors modified the setup used by Jonnalagadda et al [4] to include the different heating methods and conducted tension experiments under optical and infrared (IR) microscope at a nominal strain rate of 9 x10^{-5} /s. A thorough calibration of the infrared camera was carried out at relevant temperatures by using thermocouples attached to the sample surface. The IR images were subsequently used to obtain full-field temperature measurements on the thin film samples and determine the temperature uniformity along the gauge section.

A. Uniform Specimen Heating

A home-built microscale heater capable of achieving sample temperatures up to 160 ^{0}C was developed. A thermocouple element was attached inside the microscale heater and a second onto its surface to monitor the temperature as a function of voltage. A 1×1 cm^2 die carrying the freestanding specimens was attached to the top surface of the microscale heater. Special care was needed to thermally isolate the heating assembly from the loadcell so that the loadcell always remained at room temperature.

The thin film specimens were heated while the experimental apparatus was placed under the IR camera to determine the precise temperature along the gauge section. The samples could be tested under the IR camera at slow strain rates, limited by the camera frame rate to 1 fps, or under the optical microscope at strain rates in the range of 10^{-6} to 20 s^{-1} that were supported by a camera that allowed for 100,000 fps. All experiments presented in this paper were conducted at the nominal strain rate of 9×10^{-5} s^{-1} and at 110°C.

B. Resistive Specimen Heating

The use of resistive heating to achieve high temperatures in micro and nanoscale samples presents several challenges. A serious concern, which is also one of the objectives of this paper, is the determination of the temperature profile along the specimen gauge section, which although of uniform width, it is not at uniform temperature. Furthermore the immensely larger grips act as large heat sinks, further accentuating the uneven temperature distribution along the specimen. IR images showed highly non-uniform temperature profiles in resistive heating experiments. Actually, a challenge was to select a temperature to conduct such experiments as the specimen gauge section experiences various temperatures along its length. To be consistent with the uniform heating experiments we conducted for comparison purposes, we selected the maximum temperature in the center of resistively heated specimens to be the same as that in resistively heated specimens, i.e. 110 °C, as determined from the IR images.

A practical challenge in conducting experiments with resistive specimen heating lies with the ability to ensure a conductive path between the grips and the specimen. Scientific grade high purity silver paint was used to attach the thin film specimens to a metal grip, as shown in Figure 1, to maintain electrical conductivity and to apply voltages in the order of 0.3 to 0.7 V.

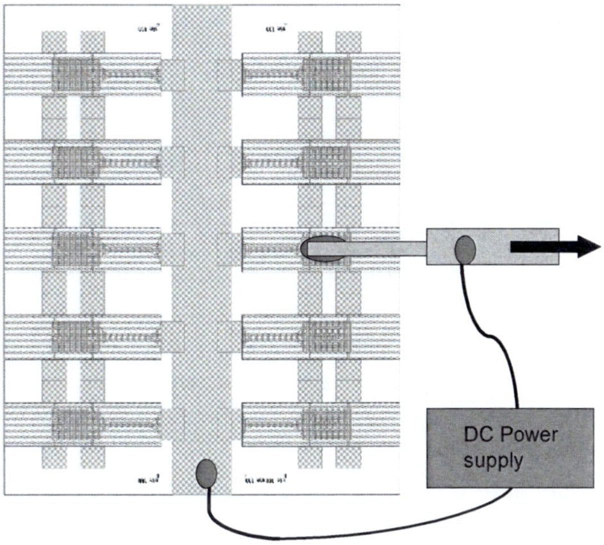

Figure 1. Schematic of resistive heating of freestanding thin film specimens. The square die carrying the 10 freestanding specimens is 1×1 cm^2. The power supply and cable dimensions are not drawn to scale.

RESULTS

A. Experiments at Uniform Gauge Temperature

Relatively low temperature gradients of about 5 °C or less were observed along the gauge of uniformly heated thin film samples as shown in Figure 2. The temperature profile along the sample section remained the same even at high engineering strains. The contrast and color distribution in IR images are affected by the angle of the specimen surface with respect to the camera plane and, therefore, reliable IR images are not always possible at high engineering strains (>10%) because of thin film out-of-plane deflection. The temperature line profiles along the centerline of the specimen gauge sections showed a rise in temperature near the right specimen grip due to the different emissivity of the silver paint applied to the grip.

The uniform temperature field resulted in uniform strain distributions as resolved from optical images by DIC. As shown in Figures 3(a,b), the axial strain distribution (line strain is shown here) along the gauge section was uniform at a various levels of strains. The axial strain was fairly uniform along the specimen length, and only at high strains specimen end effects were measurable, as seen in Figure 3(b) where a slight deviation from uniformity along the specimen gauge section is discernible. This deviation could also be due to a small gradient in the temperature distribution along the gauge section, as shown in the temperature line profile in Figure 2 due to the large external mass of the probe shown in the IR image in Figure 2. Despite these minor deviations, the DIC-derived strain distributions indicated that this method for specimen heating provides reliable high temperature experiments at the microscale.

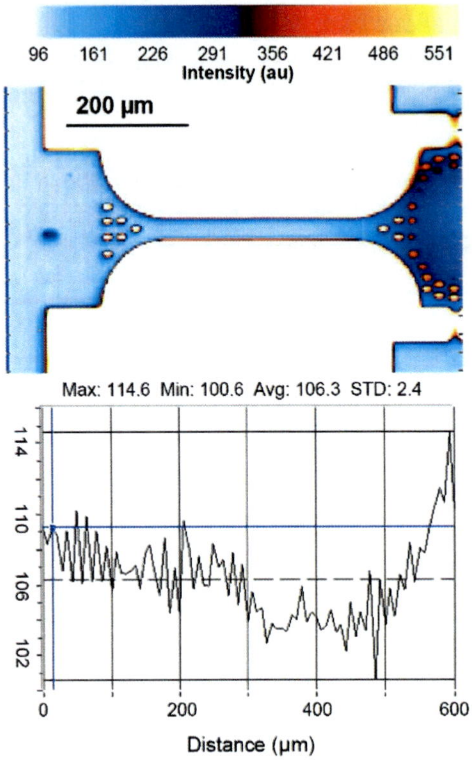

Figure 2. IR intensity image and temperature distribution along the specimen centerline for 830 nm thick Au specimen subjected to uniaxial tension at uniform temperature.

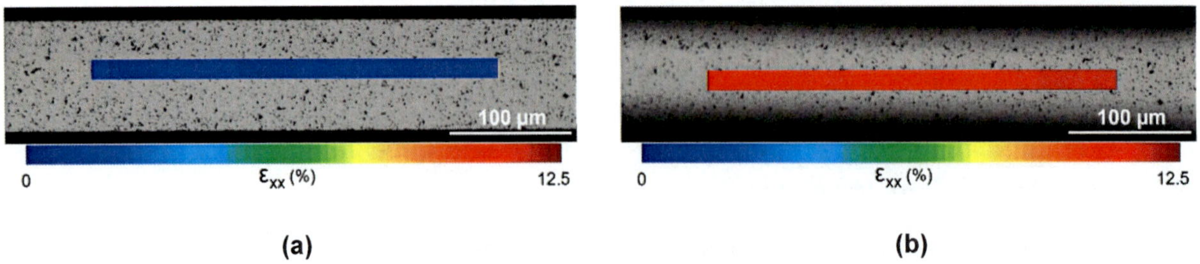

Figure 3. Full field line strains from a uniform temperature tension experiment on Au thin film samples.

B. Experiments with Resistive Heating

Unlike the uniform temperature experiments, the temperature profiles along the gauge section of resistively heated samples were highly non-uniform even at very low applied forces. This temperature non-uniformity increased during the course of an experiment due to the change in the sample resistance because of strain localization and local reduction in the mid-specimen thickness where the temperature was the highest.

As shown in Figure 4(a) the temperature profile was highly non-uniform in the beginning of an experiment with the temperature in the gauge section ranging between 50-110 °C. There is no significant change in temperature distribution during the early stages of the experiment. Thus, the small but gradual change in local electrical resistance during the course of an experiment did not affect the temperature distribution along the gauge section.

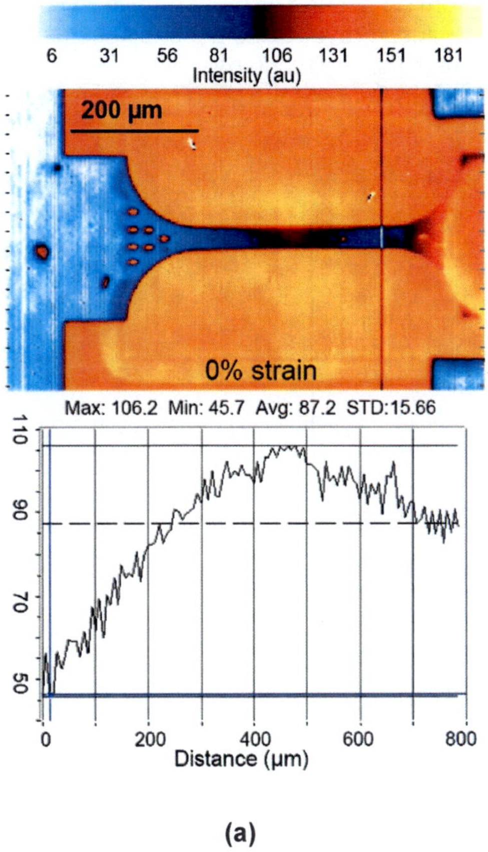

Figure 4. IR intensity images and temperature distributions along the specimen axis of symmetry for resistively heated Au thin film specimens subjected to uniaxial tension.

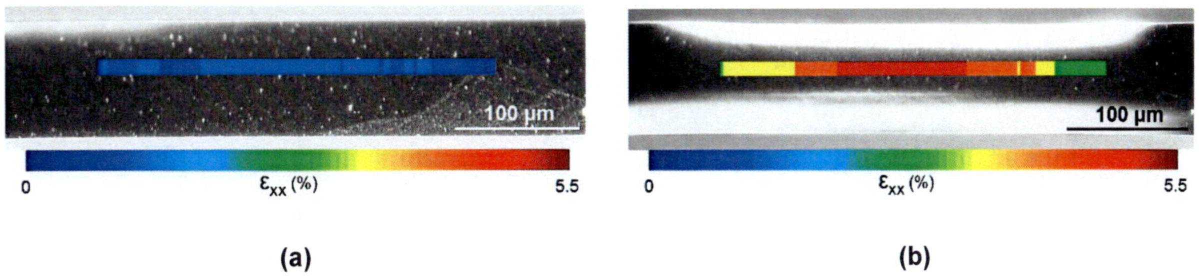

Figure 5. Axial strain distribution along a resistively heated Au thin film specimen.

Full-field strain measurements were conducted under an optical microscope at 20× magnification and were compared with the temperature distributions in Figure 4. As shown in Figure 5(a), a fairly uniform strain field was recorded along the gauge section before local yielding in the specimen center. This is expected, as the elastic modulus is temperature insensitive in this range of temperatures. However, at higher strains, the non-uniformity of the strain field increased drastically as the material underwent localized plastic deformation initiated at the location of the highest temperature. As suggested by Figure 5(b), at higher strains, the temperature in the mid-section increased from 110 ˚C to 140-150 ˚C, while the ends of the sample remained at 50 ˚C. As a result, the temperature gradient increased in the sample and consequently the strain field localized even further in the center portion of the specimen, leading to the highly inhomogeneous strain field shown in Figure 5(b).

In Figure 6, the engineering stress vs. strain responses of Au thin films tested by the two methods are compared. Although it is difficult to compare the material response from an experiment with a uniform strain field to that from a specimen with highly non-uniform strain field, this comparison reveals the uncertainties associated with the latter and the large differences in the resulting engineering stress vs. strain curves.

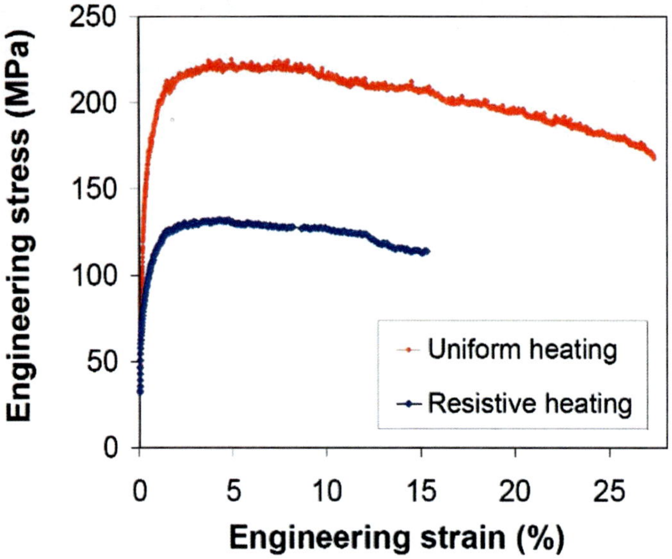

Figure 6. Comparison of engineering stress-engineering strain responses of 830 nm Au thin films at 110 ˚C by using uniform specimen heating and resistive heating.

Table 1. Comparison of properties of Au thin films tested at 110 ^{0}C derived from the stress-strain curves by the two heating methods.

Method	Elastic Limit (MPa)	Yield Strength (MPa)	Ultimate Tensile Strength (MPa)	Failure strain (%)
Uniform heating	140	165	224	27.4
Resistive heating	90	100	130	15.4

Table 1 is very instructive in this regard, as the properties calculated by the two methods differ by about 50%.The temperature gradient along the specimen potentially resulted in the formation of a neck in the specimen thickness, which, in turn, limited the load bearing capacity of the specimens loaded under resistive heating. The comparison between the two methods is further complicated by the fact that resistive heating also results in non-uniform strain rate along the specimen length and at different strain amplitudes, which is not as important in uniform heating experiments.

CONCLUSIONS

Elevated temperature microscale tension experiments with freestanding thin film specimens with thicknesses of the order of a micron were evaluated by the use of two common methods of uniform specimen heating and the resistive heating. Resistive heating provided several practical advantages such as localized heating at high temperatures and minimal effect on the measurement instruments but it resulted in highly non-uniform temperature distribution along the specimens, which in turn produced non-uniform strain fields. Uniform temperature heating on the other hand was more difficult to implement. A combination of optical and IR imaging was used to evaluate the accuracy and efficacy of the two methods by performing a quantitative comparison between the full-field strain measurements (and the associated stress-strain curves) and the full-field temperature measurements. The strain distribution from uniform heating experiments was also very uniform supporting the accuracy of this type of experiments. The corresponding IR images indicated a very mild temperature gradient that was less than 5 °C along the gauge section, only to be owed to the attachment of an external grip for specimen loading. In comparison, the resistively heated specimens had very high temperature gradients in the gauge section, ranging from 50 to 110 °C, which resulted in yield and ultimate tensile strengths that were lower by as much as 50%. Hence, the properties measured from a tension experiment conducted by resistively heating a test specimen are not representative and uniform specimen heating should be implemented when possible.

ACKNOWLEDGMENTS

The authors acknowledge the support by the National Science Foundation under grants CMS 0555787 and CMMI 0927149 ARRA.

REFERENCES

1 Emery RD, Povirk GL, "Tensile behavior of free-standing gold films Part I," Acta Materialia, 51 (7), 2067-2078, 2003.
2 Emery RD, Povirk GL, "Tensile behavior of free-standing gold films Part II," Acta Materialia, 51 (7), 2079-2087, 2003.
3 Espinosa HD, Prorok BC, "Size effects on the mechanical behavior of gold thin films," Journal of Materials Science, 38 (20), 4125-4128, 2003.
4 Jonnalagadda KN, Chasiotis I, Yagnamurthy S, Lambros J, Pulskamp J, Polcawich R, Dubey M, "Experimental Investigation of Strain Rate Dependence of Nanocrystalline Pt Films," Experimental Mechanics, 50, 25–35, 2010.
5 Jansen F, MacHonkin MA, Palmieri N, Kuhman D, "Thermal expansion and elastic properties of plasma-deposited amorphous silicon and silicon oxide films," Applied Physics Letters, 50 (16), 1059-1061, 1987.
6 Tada H, Kumpel AE, Lathrop RE, Slanina JB, Nieva P, Zavracky P, Miaoulis IN, Wong PY, "Thermal expansion coefficient of polycrystalline silicon and silicon dioxide thin films at high temperatures," Journal of Applied Physics, 87, 4189-4193, 2000.
7 Kahn H, Ballarini R, Heuer AH, "Thermal expansion of low-pressure chemical vapor deposition polysilicon films," Journal of Materials Research, 17 (7), 1855-1862, 2002.
8 Chae J-H, Lee J-Y, Kang S-W, "Measurement of thermal expansion coefficient of poly-Si using microgauge sensors," Sensors and Actuators A: Physical, 75 (3), 222-229, 1999.
9 Emery R, Simons C, Mazin B, Povirk GL, "High temperature tensile behavior of free-standing gold films," Materials Research Society Symposium - Proceedings 505:57-62, 1998.
10 Gruber PA, Olliges S, Arzt E, Spolenak R, "Temperature dependence of mechanical properties in ultrathin Au films with and without passivation," Journal of Materials Research, 23 (9), 2406-2419, 2008.

11 Oh C-S, Coles G, Sharpe Jr WN, "High temperature behavior of polysilicon," Materials Research Society Symposium – Proceedings, 741, 53-58, 2002.

12 Oh C-S, Sharpe Jr. WN, "Techniques for measuring thermal expansion and creep of polysilicon," Sensors and Actuators, A: Physical, 112 (1), 66-73, 2004.

13 Zupan M, Hayden MJ, Boehlert CJ, Hemker KJ, "Development of high-temperature microsample testing," Experimental Mechanics, 41 (3), 242-247, 2001.

14 Zupan M, Hemker KJ, "High temperature microsample tensile testing of γ-TiAl," Materials Science and Engineering A, 319-321, 810-814, 2001.

15 Haque MA, Saif MTA, "Thermo-mechanical properties of nano-scale freestanding aluminum films," Thin Solid Films, 484 (1-2), 364-368, 2005.

16 Kalkman AJ, Verbruggen AH, Janssen GCAM, "High-temperature bulge-test setup for mechanical testing of free-standing thin films," Review of Scientific Instruments, 74 (3 I), 1383-1385, 2003.

17 Teh KS, Lin L, "Time-dependent buckling phenomena of polysilicon micro beams," Microelectronics Journal, 30 (11), 1169-1172, 1999.

18 Brotzen FR, Rosenmayer CT, Cofer CG, Gale RJ, "Creep of thin metallic films," Vacuum, 41(4-6), 1287-1290, 1990.

19 Hyun S, Brown WL, Vinci RP, "Thickness and temperature dependence of stress relaxation in nanoscale aluminum films," Applied Physics Letters, 83 (21), 4411-4413, 2003.

20 Mönig R, Keller RR, Volkert CA, "Thermal fatigue testing of thin metal films," Review of Scientific Instruments, 75 (11), 4997-5004, 2004.

21 Chasiotis I., Cho SW, Jonnalagadda KN, "Fracture toughness and subcritical crack growth in polycrystalline silicon," J App. Mechanics Trans. ASME, 73 (5), 714-722, 2006.

22 Sutton MA, Li N, Joy DC, Reynolds AP, Li X, "Scanning electron microscopy for quantitative small and large deformation measurements Part I: SEM imaging at magnifications from 200 to 10,000," Experimental Mechanics, 47 (6), 775-787, 2007.

23 Sutton MA, Li N, Joy DC, Reynolds AP, Li X, "Scanning electron microscopy for quantitative small and large deformation measurements Part II: SEM imaging at magnifications from 200 to 10,000," Experimental Mechanics, 47 (6), 789-804, 2007.

24 Karanjgaokar N, Jonnalagadda K, Chasiotis I, Chee J, Mahmood A, Peroulis D, "Mechanical behavior of nanocrystalline Au films as a function of strain rate and film thickness," SEM-11th International Congress and Exhibition on Experimental and Applied Mechanics, 4, 1860-1866, 2008.

Proceedings of the SEM Annual Conference
June 7-10, 2010 Indianapolis, Indiana USA

Surface Texturing Using Gold Nanoparticles to Reduce Adhesion in MEMS

N. Ansari[a,*], K. M. Hurst[a] and W. R. Ashurst[a]
[a]Department of Chemical Engineering, Auburn University, Auburn, AL 36849 USA
*Corresponding author: ansarna@auburn.edu (N. Ansari)

Abstract

Since the advent of Micro-electromechanical systems (MEMS) technology, researchers have used surface texturing as one of the approaches to alleviate unintentional adhesion in MEMS. However, the conventional methods used for surface texturing are reported to reduce apparent in-plane adhesion only by a factor of 20. Further, the test surfaces used to-date are inherently rough, as a result of which, the effects of surface texturing could not be studied independently. We report on a novel method of texturing inherently smooth Si(100) surfaces by depositing dodecanethiol capped gold nanoparticles using a gas-expanded liquid technique. The dodecanethiol capping ligands are removed by exposing the treated surfaces to UV-Ozone atmosphere for an hour and the textured surfaces thus obtained are characterized by atomic force microscopy (AFM) and scanning electron microscopy (SEM). The textured Si(100) surfaces exhibit a significant reduction in apparent in-plane work of adhesion, which is determined using the cantilever beam array (CBA) technique, compared to untextured smooth Si(100) surfaces having only native oxide on them.

1. Introduction

Unintentional adhesion, commonly known as stiction in the MEMS community, is one of the major obstacles hindering the commercialization of a wide variety of useful MEMS [1]. Stiction is broadly classified into two categories, namely release stiction and in-use stiction. Release stiction, which limits the yield of MEMS, is encountered when microstructures are dried following their release. Exposure of adjacent MEMS surfaces to a liquid-vapor interface created by an evaporating liquid during drying initiates strong surface interactions between them, which eventually lead to permanent adhesion and device failure, if the restoring forces are not strong enough to overcome those surface interactions [1]. Common approaches used to address the issue of release stiction involve the use of release or drying techniques that eliminate the liquid-vapor interface, such as critical point drying, freeze sublimation drying, polymer support ashing, hydrofluoric acid (HF) vapor etching and hydrophobic self assembled monolayer (SAM) coating [2–6]. While all of these approaches do eleminate release stiction successfully, except SAM coating, none of them can be used to address in-use stiction [6]. In-use stiction, which limits the lifetime and hence reliability of MEMS, is encountered during storage as well as during normal operation, when adjacent MEMS surfaces come in close proximity and experience strong attractive surface forces which are greater than their restoring forces.

Surface texturing is one of the first strategies devised by the MEMS community to address in-use stiction. However, to-date it is reported to reduce in-use stiction only by a moderate factor of 20 [7]. Only recently, DelRio et. al. studied the effect of silicon carbide particles (diameter ranging between 20-50 nm) on interfacial adhesion between polysilicon surfaces and reported that nanoparticles (NPs) deposited on contacting surfaces can strongly influence adhesion between them by increasing the average separation distance between them [8]. However, surface texturing using NPs still remains a challenge due to unavailability of well established methods for depositing NPs on MEMS. The conventional drop-casting and solvent evaporation methods are not preferred since they cause deformation of microstructures, which ultimately leads to device failure. A process developed by the Robert's group at Auburn University precipitates dispersed metallic NPs from a solution and deposits them as uniform, wide area thin films using gas-expanded liquids [9]. In this paper, we investigate the applicability of this process, which is compatible with current microfabrication technologies, for surface texturing of MEMS. We also investigate the effect of surface texturing using gold NPs on in-plane apparent in-use stiction.

2. Experimental Details

CBAs are used to investigate the effects of surface texturing on the apparent in-plane adhesion. The test structures, which are optically measured to be 29.765 μm wide (design width is 30 μm) and 1.89 μm thick (design thickness is 2

DOI 10.1007/978-1-4419-8825-6_26, © The Society for Experimental Mechanics, Inc. 2011

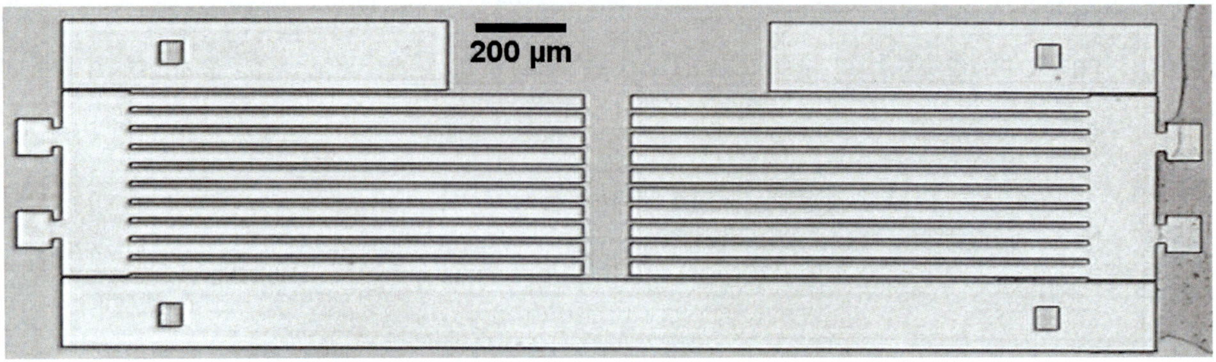

Figure 1: Optical image of a Cantilever Beam Array

μm) are fabricated using the standard surface micromachining technology on a SOI wafer. Thickness of the sacrificial layer is 2 μm. Figure 1 is an optical image of a CBA. Each CBA consists of ten cantilever beams, each of the same length. This allows estimation of 10 distinct data points from each CBA, thereby enabling a statistical analysis of the local adhesion. The bottom surface of a cantilever beam and the top surface of the substrate, which are the two in-plane test surfaces used, are both Si(100) surfaces and hence are inherently smooth. Before using the fabricated test structures for determining the apparent in-plane work of adhesion, they are released by etching them in conc. HF (49 wt.%). After the structures are completely released, the etchant is rinsed away by deionized (DI) water until the pH of the solution becomes neutral. The test structures are then transferred (without being exposed to ambient air) to hot H_2O_2 (75-80 °C) for 10 min. for oxidizing the test surfaces, after which they are rinsed with DI water followed by a displacement rinse with electronic grade isopropanol. From this point onwards, the control sample, which will not be subjected to surface texturing and the test sample, which will be subjected to surface texturing recieve different treatments. The control sample is supercritically dried, while the test sample is thoroughly rinsed with electronic grade anhydrous hexane. The test sample is then subjected to surface texturing.

A two-phase liquid arrested precipitation process similar to that reported by Sigman et. al. is used to synthesize dodecanethiol capped gold nanoparticles, which are used for texturing the test surfaces [10]. 36 ml of an aqueous solution containing 0.38 g of hydrogen tetrachloroaurate is added to an organic solution containing 2.7 g of tetraoctylammonium bromide (phase transfer catalyst) and 24.5 ml toluene. The mixture is stirred for an hour, after which the aqueous phase is separated and discarded. Gold ions present in the organic phase that is left behind are then reduced to ground state by adding 30 ml of an aqueous solution containing 0.5 g of NaBH4. The resulting mixture is stirred for 8-10 h to allow particle growth before removing and discarding the aqueous phase. 240 μl of 1-dodecanethiol is then added to the organic solution and the mixture is stirred for 4 h to cap and stabilize the gold nanoparticles. Excess thiol and NaBH4 are removed by centrifuging the dodecanethiol capped nanoparticle dispersion with equal parts of ethanol at 4500 rpm for 5 min. After repeating the centrifugation step several times, the nanoparticles are dispersed and stored in hexane. Average diameter of gold nanoparticles used for this study is determined to be 4.4 $\pm$ 1.4 nm using transmission electron microscopy.

Without exposing the test sample to ambient air, it is carefully placed upside-down on top of a stainless steel washer inside a glass vial, which is kept inside a large beaker filled with hexane. Another washer is placed on top of the test sample to hold it in place. The vial is then removed from the beaker, and hexane is carefully removed from the vial to a level just above the test sample. 25 μl of concentrated dodecanethiol capped gold nanoparticle dispersion (in hexane) is then added to the vial and carefully mixed, after which the vial is placed inside a 30 ml stainless steel high pressure vessel. Approximately 400 μl of pure hexane is also added alongside the vial to saturate the headspace of the high pressure vessel and prevent evaporation of the organic nanoparticle dispersion prior to sealing the high pressure vessel. After sealing the vessel, the chamber is pressurized with CO_2 first to approximately 23 bar at room temperature (i.e., 22 °C) using a 500 ml ISCO piston syringe pump and then upto the vapor pressure of the gas (i.e., 58 bar at 22 °C) by setting the pump flow rate to 0.4 ml/min. During pressurization, CO_2 dissolves in the organic phase, effectively expanding the volume and reducing the solvent strength of the hexane/CO_2 mixture, which results in precipitation of the suspended nanoparticles. Specifically, dodecanethiol stabilized AuNPs in the size range of 3-7 nm are reported to precipitate in the pressure range of 34.5-55.2 bar [9]. The precipitating nanoparticles are reported to conformally coat all the test surfaces [11]. Pressurization is continued until the entire chamber is filled with liquid CO_2. The mixture of liquid CO_2 and organic solvent is then heated isochorically to 40 °C to transform it into a supercritical fluid, after which the chamber is flushed with several volumes of pure supercritical CO_2 (at 40 °C and 90 bar) at a rate of approximately 1.0 ml/min to ensure complete removal of the organic solvent. Finally, the chamber is slowly depressurized (at 40 °C) to atmospheric pressure and the dry, dodecanethiol capped AuNP coated test sample is removed.

The final processing done on both the control and test samples is supercritical drying. Samples dried supercritically using CO_2 are known to have contaminated test surfaces. In order to clean the test surfaces and also to remove the dodecanethiol capping ligands before testing, both the control and test samples are exposed to an UV-Ozone atmosphere for an hour. CBAs on both the control and test samples are actuated manually by pushing the cantilever beams with a sharp tungsten probe tip until they come in contact with the substrate, after which the probe tip is retracted slowly and carefully. On releasing, the beams peel apart up to a certain characteristic length called as the crack length (s), which depends on the interfacial properties of the two contacting surfaces. The crack length of each actuated beam is determined optically using interferometry. The apparent in-plane work of adhesion ($\mathcal{W}$) between the substrate and the bottom surface of the cantilever beam is determined using Eq. 1 suggested by Hurst et. al., where E and t are the Young's modulus and thickness of the structural film respectively, s is the crack length and a_1 and a_2 are parameters extracted from the fitted experimental height profile of the actuated cantilever beam [12].

$$\mathcal{W} = \frac{3Et^2}{2s^4}\left(a_1^2 + a_1a_2 + \frac{1}{3}a_2^2\right) \tag{1}$$

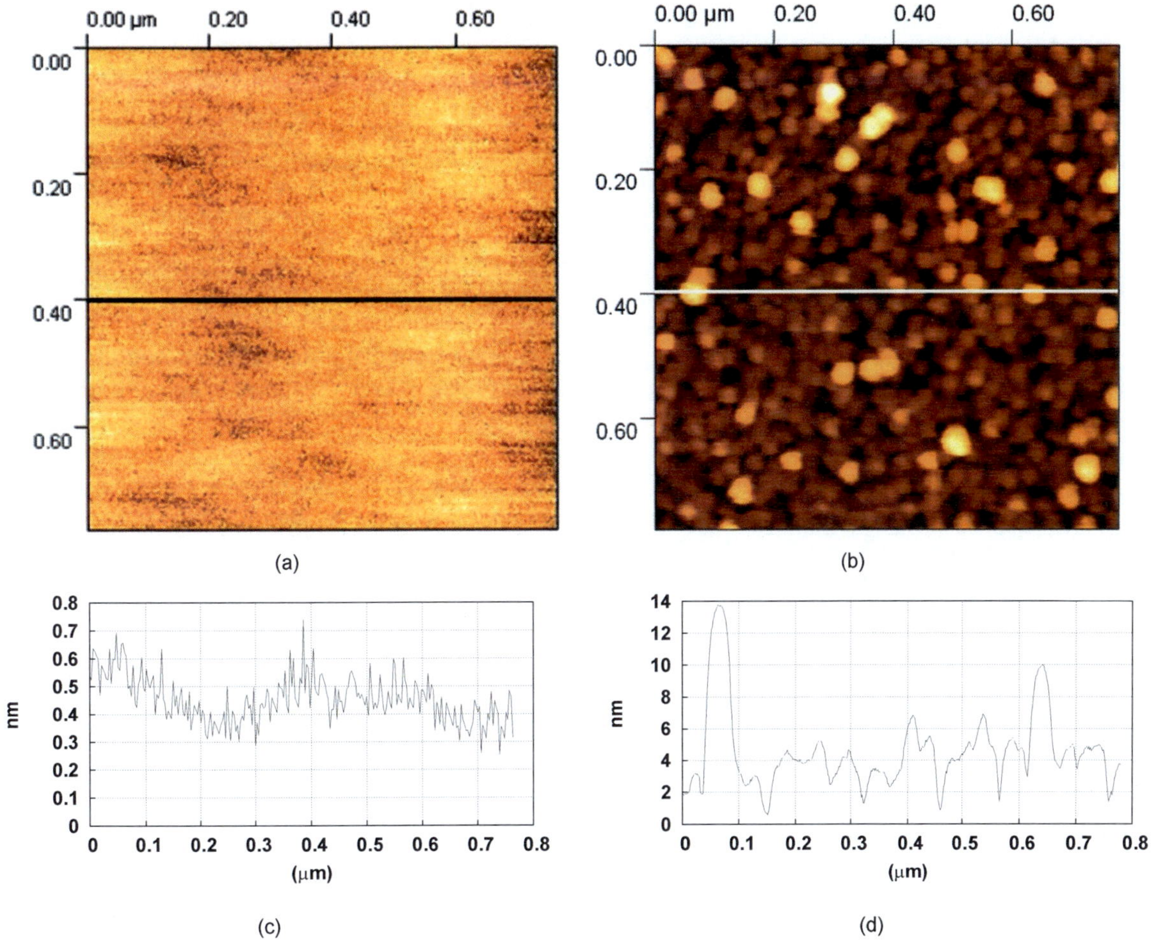

Figure 2: (a) AFM image of an untextured substrate (b) AFM image of a substrate textured using gold nanoparticles (c) Linescan across image a (d) Linescan across image b

3. Results and Discussions

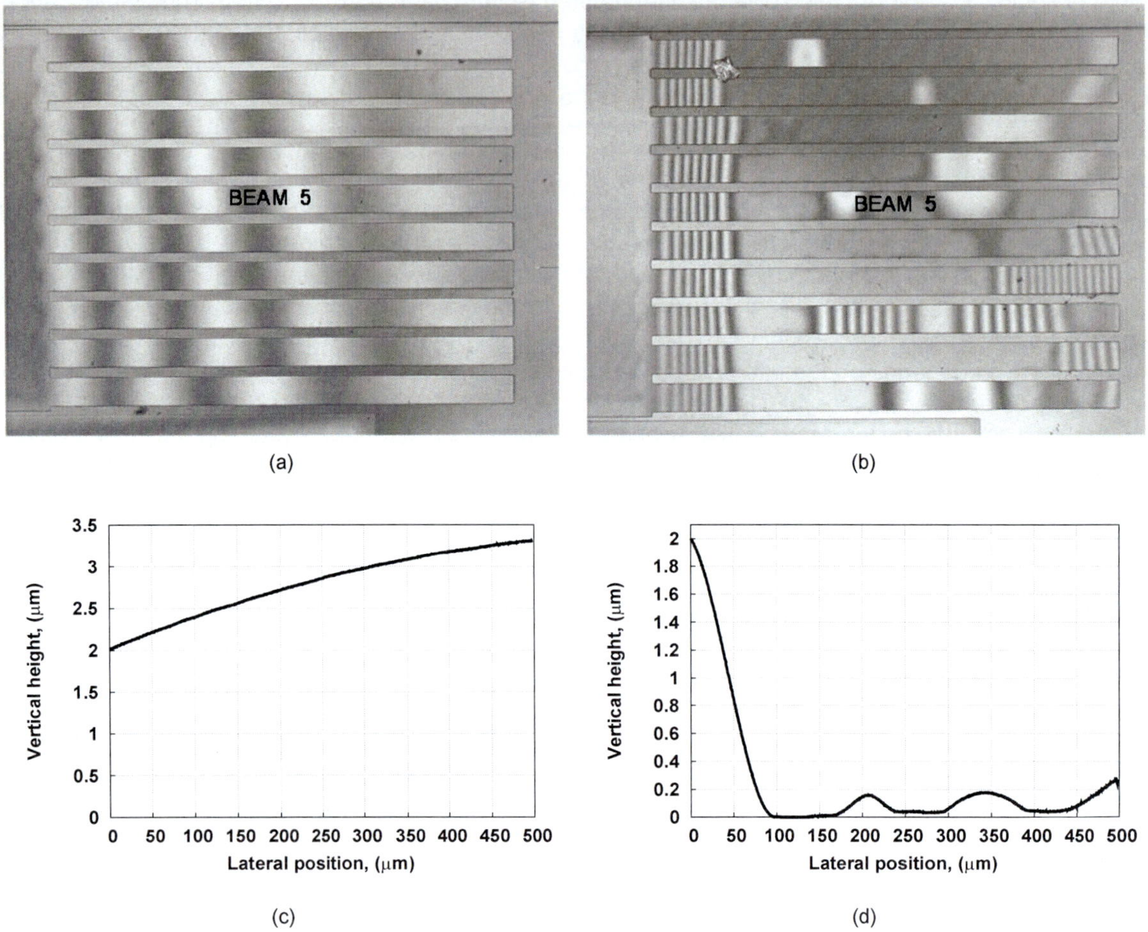

Figure 3: (a) Interferogram of an unactuated, 500 μm long CBA on the control sample (b) Interferogram of an actuated and released, 500 μm long CBA on the control sample (c) Height profile of beam 5 shown in image a (d) Height profile of beam 5 shown in image b

SEM analysis of AuNP coated surfaces indicates that AuNPs are conformally coated on every surface of the cantilever beams and the substrate. Further, SEM analysis of the underside of a cantilever beam, which is broken and turned upside down confirms the conformality of AuNP coating. AFM images of both untextured as well as a substrate textured with AuNPs are shown in Fig. 2a and Fig. 2b respectively. It is clearly evident from the AFM images that AuNPs are not deposited individually but have formed small islands, which is in accordance with the Volmer-Weber film growth model [13]. Figure 2d, which is a linescan across the AFM image of the textured substrate, yields that the AuNP islands have a height and width of 10-15 nm and 25-50 nm respectively. Formation of AuNP islands, however, is advantageous because it increases the roughness of the test surfaces significantly, thereby drastically reducing the real area of contact between them. This is substantiated by the rms roughness values of 0.097 nm and 2.06 nm for untextured and textured test surfaces determined using the AFM scans shown in Figs. 2a and 2b respectively. The significantly different peak to valley ratios evident in the linescans shown in Figs. 2c and 2d further emphasize the drastic difference in the topography of untextured and textured test surfaces.

On both the control and the test samples, four CBAs are actuated, two of which have 500 μm long cantilever beams, while the other two have 1000 μm long cantilever beams. All the experiments are conducted in laboratory air at room temperature (i.e., 22.5 °C) and at relative humidity of 25%. An interferogram of an unactuated, 500 μm long CBA on the control sample is shown in Fig. 3a. All the beams shown in Fig. 3a are free standing, which is clearly evident from the height profile shown in Fig. 3c. The interferogram collected after actuating and releasing the CBA shown in Fig. 3a is shown in Fig. 3b, which indicates that the control sample exhibits uniform local adhesion. Height profile of

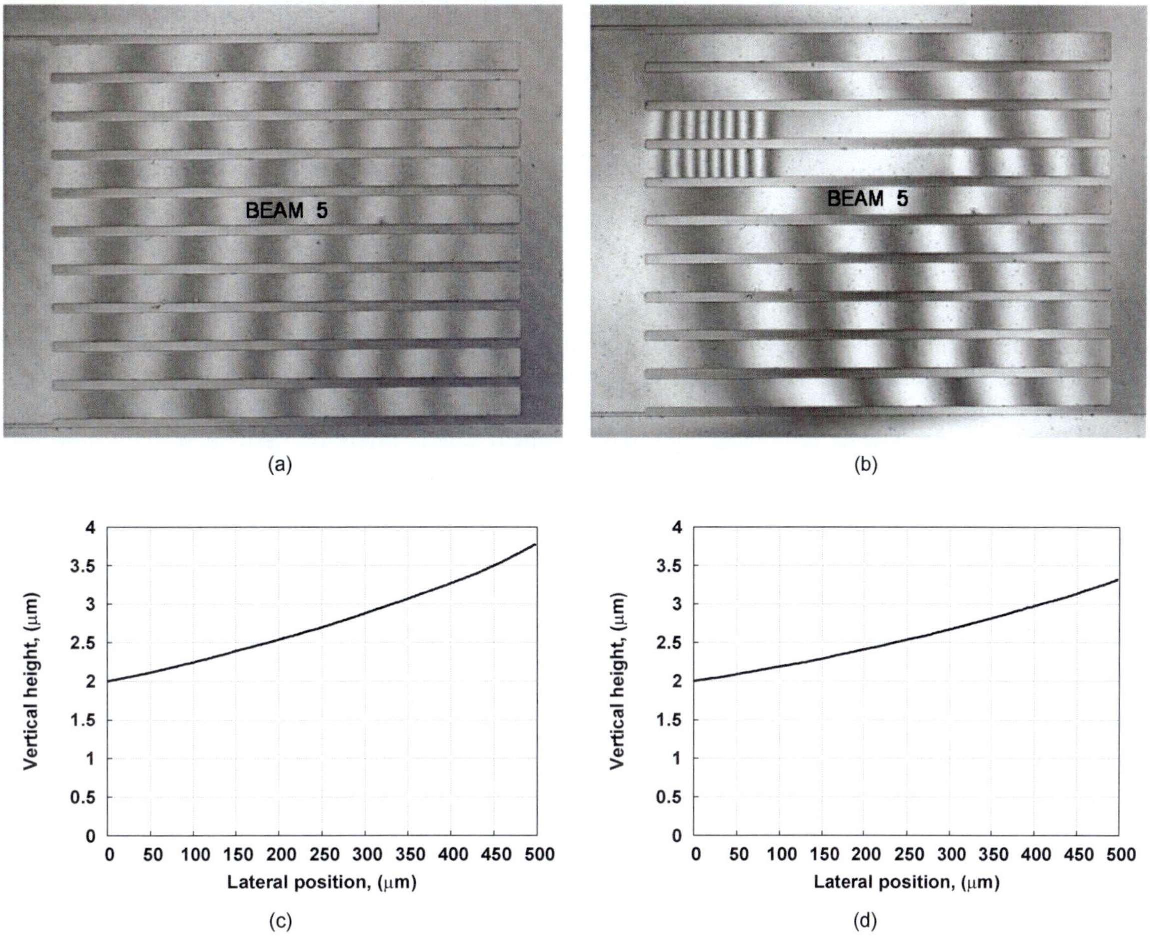

Figure 4: (a) Interferogram of an unactuated, 500 μm long CBA on the test sample (b) Interferogram of an actuated and released, 500 μm long CBA on the test sample (c) Height profile of beam 5 shown in image a (d) Height profile of beam 5 shown in image b

an actuated and released, 500 μm long cantilever beam shown in Fig. 3d indicates that the smooth in-plane surfaces on the control sample exhibit a crack length (s) of 96 $\pm$ 5 μm. Figure 4a shows the interferogram of an unactuated, 500 μm long CBA on the test sample. Again, as evident from the height profile shown in Fig. 4c, all the cantilever beams shown in Fig. 4a are free standing before actuation. The interferogram shown in Fig. 4b indicates that most of the 500 μm long cantilever beams on the test sample do not stick to the substrate on actuation. The height profile shown in Fig. 4d further substantiates that the height profile of an actuated, 500 μm long cantilever beam on the test sample is the same as that of an unactuated cantilever beam shown in Fig. 4c. Since the 500 μm long cantilever beams actuated on the test sample do not stick at all, they could not be used to determine the apparent adhesion between in-plane textured surfaces. However, the 1000 μm long cantilever beams actuated on the test sample do exhibit S-shaped height profiles similar to that exhibited by the 500 μm long cantilever beams actuated on the control sample. A crack length of 475 $\pm$ 30 μmis determined using the 1000 μm long CBAs on the test sample. Also, the local adhesion between the in-plane textured surfaces of the test sample is less uniform than that between the in-plane untextured surfaces of the control sample.

Table 1 lists the crack lengths observed along with the corresponding calculated apparent work of adhesion between in-plane surfaces on the control and the test samples. The textured in-plane surfaces of the test sample exhibit a significantly reduced apparent work of adhesion as compared to that exhibited by the untextured in-plane surfaces of the control sample. The factor of 300 reduction in the apparent work of adhesion corelates well with the increased roughness exhibited by the textured in-plane surfaces of the test sample as compared to the untextured in-plane surfaces of the control sample. The increased roughness increases the average seperation distance between contacting

Sample	Crack Length $s \pm$ std.dev. (μm)	Apparent In-plane Work of Adhesion $\mathcal{W} \pm$ std.dev. (mJ/m^2)
Control	96 ± 5	30.5 ± 6
Test	475 ± 30	0.11 ± 0.03

Table 1: Apparent work of adhesion exhibited by untextured and textured in-plane test surfaces.

test surfaces.

4. Conclusions

We have successfully demonstrated a novel method of texturing inherently smooth Si(100) test surfaces by depositing dodecanethiol capped gold nanoparticles from CO_2-expanded hexane using a gas-expanded liquid technique coupled with supercritical drying. The textured cantilever beams coated with gold nanoparticles exhibit a significantly reduced in-plane apparent work of adhesion of 0.11 ± 0.03 mJ/m^2, which is two orders of magnitude lower than that exhibited by untextured cantilever beams. This reduction in apparent in-plane adhesion is a result of increase in surface roughness and commensurate reduction in the real area of contact between test surfaces caused by texturing. The surface texturing technique demonstrated in this paper is not only compatible with the commonly used microfabrication techniques but also has a conformal effect on all the test surfaces. Although, for this study the dodecanethiol capping ligands are removed before testing, the novel surface texturing technique reported in this paper provides an opportunity to control not only the surface topography but also the surface composition of contacting test surfaces, which is not possible with traditional surface texturing techniques used.

5. Acknowledments

The authors wish to thank the Electrical & Computer Engineering department at Auburn university for allowing them to use the AMSTC facilities to fabricate the test strucrures. The authors also gratefully acknowledge the financial support of Auburn University.

References

[1] R. Maboudian, Surface Processes in MEMS Technology, Surface Science Reports, 30, 207–269, 1998.

[2] G. T. Mulhern, D. S. Soane, R. T. Howe, Super-critical Carbondioxide Drying of Microstructures, In Proceedings of International Conference on Solid State Sensors and Actuators, Transducers'93, 296–299, 1993.

[3] K. Petersen, Silicon as a Mechanical material, Proceedings of the IEEE, 70 (5), 420–457, 1982.

[4] C. H. Mastrangelo, G. Saloka, A dry-release method based on polymer columns for microstructure fabrication, In Proceedings of the IEEE MicroelectroMechanical Systems Workshop, 77–81, 1993.

[5] J. Anguita, F. Briones, HF/H_2O vapor etching of SiO_2 sacrificial layer for large-area surface-micromachined membranes, Sensors and Actuators A-Physical, 64 (3), 247–251, 1998.

[6] M. R. Houston, R. T. Howe, R. Maboudian, Self-assembled monolayer films as durable anti-stiction coatings for polysilicon microstructures, In Proceedings of the 1996 Solid-State Sensor and Actuator Workshop, Hilton Head, 42–47, 1996.

[7] Y. Yee, K. Chun, J. D. Lee, C. J. Kim, Polysilicon surface-modification technique to reduce sticking of microstructures, Sensors And Actuators A-Physical, 52 (1-3), 145–150, 1996.

[8] F. W. DelRio, M. L. Dunn, B. L. Boyce, d. M. P. Corwin, A. D., The effect of nanoparticles on rough surface adhesion, Journal of Applied Physics, 99, 104304(pp. 9), 2006.

[9] M. C. McLeod, M. Anand, C. L. Kitchens, C. B. Roberts, Precise and Rapid Size Selection and Targeted Deposition of Nanoparticle Populations Using CO_2 Gas Expanded Liquids, Nano Letters, 5 (3), 461–465, 2005.

[10] M. B. Sigman, A. E. Saunders, B. A. Korgel, Metal Nanocrystal Superlattice Nucleation and Growth, Langmuir, 20 (3), 978–983, 2004.

[11] K. M. Hurst, C. B. Roberts, W. R. Ashurst, A gas-expanded liquid nanoparticle deposition technique for reducing the adhesion of silicon microstructures, Nanotechnology, 20 (18), 185303(pp. 9), 2009.

[12] K. M. Hurst, C. B. Roberts, W. Ashurst, A New Method to Determine Adhesion of Cantilever Beams Using Beam Height Experimental Data, Tribology Letters, 35, 9–15, 2009.

[13] M. Ohring, Materials Science of Thin Films: Deposition & Structure, Academic Press, San Diego, CA, 2001.

Proceedings of the SEM Annual Conference
June 7-10, 2010 Indianapolis, Indiana USA

Design of Microswitch Systems Avoiding Stiction due to Surface Contact

Ling WU [1,2], L. NOELS [1], V. ROCHUS [1],M. PUSTAN [1,3], J.C. GOLINVAL [1]

[1] University of Liege, Aerospace and Mechanical Engineering Department, Belgium
L.Wu@ulg.ac.be
[2] Northwestern Polytechnical University, School of Aeronautics, 710072 Xi'an, China
[3]Technical University of Cluj-Napoca, Cluj-Napoca, Romania

ABSTRACT

Stiction which results from contact between surfaces is a major failure mode in micro electro-mechanical systems (MEMS). Increasing restoring forces using high spring constant allows avoiding stiction but leads to an increase of the actuation voltage so that the switch's efficiency is threatened. A statistical rough surfaces interaction model, based on Maugis' and Kim's formulations is applied to estimate the adhesive forces in MEMS switches. Based on the knowledge of these forces, the proper design range of the equivalent spring constant, which is the main factor of restoring force in MEMS switches, can be determined. The upper limit of equivalent spring constant depends mainly on the expected actuator voltage and on the geometric parameters, such as initial gap size and thickness of dielectric layer. The lower limit is assessed on the value of adhesive forces between the two contacting rough surfaces. It mainly depends on the adhesive work of contact surfaces and on the surfaces' roughness. In order to study more complicated structures, this framework will be used in a multiscale model: resulting unloading micro adhesive contact-distance curves of two rough surfaces will be used as contact forces in a finite-element model. In this paper the extraction of these curves for the particular case of gold to gold micro-switches is pursued.

1. INTRODUCTION

Micro electro-mechanical systems (MEMS) are becoming more and more popular due to their miniature sizes, low power requirements and manufacturing cost reductions. Despite those advantages, the inherent characters of MEMS such as the large surface area-to-volume ratio, smooth surfaces, small interfacial gaps and small restoring forces, make them particularly vulnerable to the surface forces that could lead to permanent adhesion, commonly called stiction, which can severely affect the reliability of the devices.

In-use stiction cannot be completely eliminated due to the existence of the van der Waals forces between materials. Therefore, models are required in order to predict and avoid this kind of failure. Toward these ends, a multi-scale model is required: at the lower scale the adhesive contact forces of two rough surfaces are predicted, while at the higher scale a finite element analysis is considered, in which the adhesive contact-distance curve is

DOI 10.1007/978-1-4419-8825-6_27, © The Society for Experimental Mechanics, Inc. 2011

integrated on the surface of the finite elements as a contact law.

In order to model the adhesive contact forces, it has been revealed that, in the dry environment where the capillary forces can be neglected, main governing parameters are the surface energies and other parameters related surface forces (e.g. van der Waals (VDW) force) [1]. In the presence of van der Waals (VDW) force, two earlier theories of adhesive contact between an elastic sphere and a flat surface have been developed: the Johnson-Kendall-Roberts (JKR) model [2] and the Derjaguin-Muller- Toporov (DMT) model [3]. The JKR model is ideal for soft materials with a larger contact curvature and a high surface energy. The DMT theory, in contrast, is well suited for hard materials with a smaller contact curvature and a lower surface energy. Maugis [4] provided a transition solution for intermediate cases between JKR and DMT regimes. In this model the transition between the two regimes is characterized by the Maugis' transition parameter λ, which involves surface and material properties.

Owing to the rough property of real micro surfaces, investigation of the adhesion problems based on rough surfaces is mandatory for MEMS. One approach is the statistical framework introduced by Greenwood and Williamson (GW) [5], for which the rough surfaces are simulated by the configurations of multi-asperities with random height [6].

This GW asperity model has been combined to Maugis transition model by the authors in order to study the minimal restoring force requires by a 1D model in order to avoid stiction between two rough micro surfaces [7]. However in order to study more complex structures, a multi-scale approach should be developed. In such a framework, the higher scale is constituted by a finite element analysis of structures entering into contact, while the adhesion contact model previously developed as the lower scale [7], and based on Maugis' formulations, is directly integrated on the surfaces of the finite-elements [8].

Micro-scale model developed in [7] will result in the definition of unloading adhesive contact-distance curves of two rough interacting surfaces. As said, this model is established from a combination Greenwood-Williamson asperity model and Maugis transition's theory, and is dependant on the material and surface properties as well as on the rough surfaces' topography parameters.

In this paper it is proposed to extract these curves for gold to gold micro-switches, which in turns will be used in a multi-scale model [8].

2 GREENWOOD – WILLAMSON MODEL and ADHESIVE CONTACT THEORY

Greenwood and Williamson's 'asperity-based model' [5] can be applied to simulate the rough surface/flat plane contact, with the rough surface being described by a collection of spherical asperities with identical end radii, whose height h have a statistical distribution. The contact of two rough surfaces can be represented by the contact between an equivalent rough surface and a smooth plane [6]: if the two initial contacting rough surfaces have respectively the asperities' end radii of R_1 and R_2 and the standard deviation in asperity heights σ_1 and σ_2, the equivalent rough surface is defined by asperities' end radii $R = R_1R_2/(R_1+R_2)$ with the standard deviation in asperity heights $\sigma_s = (\sigma_1^2 + \sigma_2^2)^{1/2}$. A Gaussian distribution can be used [6] to describe the asperity heights distribution of the equivalent rough surface

$$\varphi(h) = \frac{1}{\sigma_s\sqrt{2\pi}}\exp(\frac{-h^2}{2\sigma_s^2}) \tag{1}$$

The total contact force F_{nT}, between the equivalent rough surface and a smooth plane, can be expressed as the sum of the contact force F_{ni}, (i=1,2,...N) on each asperity (see Fig. 1). This total contact forces can be expressed as a function of the surfaces' distance d, which is originally defined as the separation between the two rough surfaces' mean planes of asperity heights, and which is replaced by the distance from the equivalent rough surface's mean plane of asperity heights to the smooth surface in the equivalent representation. One obviously has

$$F_{\mathrm{nT}} = \sum_{i=1}^{N} F_{\mathrm{n}i} \tag{2}$$

The contact forces F_{ni}, which include the Hertz contact forces due to the elastic deformation of the asperity at micro contacts and the adhesion forces due to van der Waals attractive forces, can be calculated with the Maugis' adhesive contact theory combining Kim's Expansion [4, 9], as it has been developed in [7]. Maugis theory [4] is an analytical theory based on Dugdale assumption of inter-atomic attractions characterized by a critical value of separation z_0 of two surfaces within which a constant force per unit area σ_0 is acting. Maugis' transition parameter defined as

$$\lambda = \frac{2\sigma_0}{(\pi\varpi K^2 / R)^{1/3}} \tag{3}$$

characterizes the importance of the adhesive traction to the Hertz elastic deformation pressure. In Eq. (3), R is the effective radius of the equivalent rough surface and ϖ is the adhesive energy. As Dugdale assumption is made, the adhesive energy satisfies the relation $\varpi=\sigma_0 z_0$. The reduced elastic modulus K is defined as $K=4/3 \cdot[(1-\nu_1^2)/E_1+(1-\nu_2^2)/E_2]^{-1}$, where E_1 and E_2 represent the Young's modulus and where ν_1 and ν_2 represent the Poisson's ratio for the respective materials of the two interacting bodies.

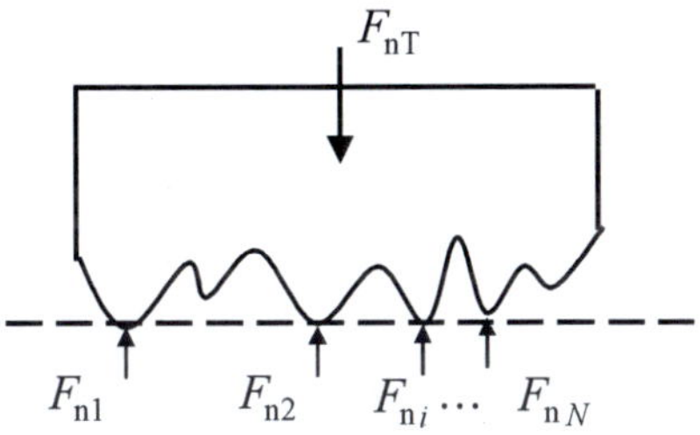

Fig. 1: Schematic view of the contact forces between the equivalent rough surface and the smooth plane represented by the dashed line.

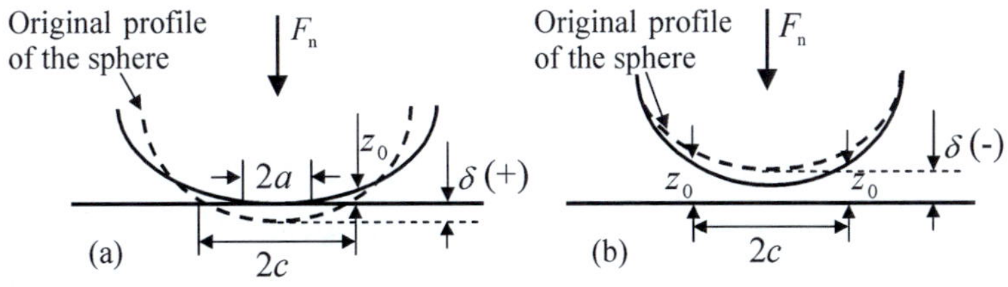

Fig. 2: Illustration of physical parameters. (a) Contact radius a and adhesive contact radius c, interference δ is a positive value for original Maugis theory. (b) In Kim *et al* extension, interference δ is a negative value.

Maugis' theory [4] allows expressing the contact force F_n of an asperity with a flat surface when the intimate contact radius $a\neq 0$, see Fig. 2(a). In this figure, c is the adhesive interaction radius within which adhesion forces are non-zero. Kim et al [9] extend the solution of Maugis-Dugdale to the non-contact regime of a=0 and $c\neq 0$ (see Fig.

2(b)) by the adjustment of Maugis' governing equations, see [7] for details.

Resulting combination of Maugis theory and Kim's extension allows rewriting the contact forces (2) per unit area in a dimensionless way

$$\overline{F}_{\mathrm{nT}}(\overline{d}) = \frac{N}{\overline{\sigma}_s\sqrt{2\pi}}\int_{\Delta_1}^{\infty}\overline{F}_{\mathrm{n}}(\Delta)\exp(-\frac{(\Delta+\overline{d})^2}{2\overline{\sigma}_s^2})d\Delta \tag{4}$$

in terms of the dimensionless distance. In Eq. (4), N is the number of asperities per unit area of surface, $\overline{\sigma}_s$ is the standard deviation of asperities heights normalized by $(\pi^2\varpi^2 R/K^2)^{1/3}$, and term Δ_1 limits the integration range in order to account for asperities subjected to adhesive-contact forces only [7].

From Eq. (4), it can be found that the dimensionless contact force $\overline{F}_{\mathrm{nT}}$ per unit area at a certain contact distance d depends on the Maugis' transition parameter λ and on the rough surface's topography parameters: the area density of asperities N and the standard deviation of asperities heights σ_s.

3. GOLD TO GOLD INTERACTION

In order to exploit the micro adhesive-contact forces (4) in a multiscale approach [8], the parameters required (Maugis' transition parameter λ and the rough surface's topography parameters) have to be identified for the materials and surfaces the micro-switches are made of. In this paper gold-on-gold electric contact, which has been typically used in MEMS switches due to its excellent electrical conductivity and corrosion resistance, is studied. The low stiffness of gold leads to severe adhesion problems in MEMS switches, which justifies its study in the present paper.

3.1 Adhesive parameters

According to the descriptions of Maugis' theory and Dugdale assumption, the adhesive energy per unit area ϖ is the energy per unit area necessary to separate two bodies (1 and 2) of perfectly flat surface in contact, which in case of two identical bodies is twice the surface energy γ of the material. This value can be obtained theoretically for crystals by considering a cleavage in the lattice

$$2\gamma = \frac{N_{\mathrm{rupture}}\varepsilon}{S_{\mathrm{ref}}} \tag{5}$$

where N_{rupture} is the number of broken bonds during the separation of the plane, which can be calculated as 2×1/2×5 (center atoms)+2×1/8×1 (corner atoms) for <1 1 0> cleavage of FCC lattice, where S_{ref} is the area of the separation surface, which is equal to $2D_0 \times 2D_0/2^{1/2}$ for plane <1 1 0> of FCC lattic, where D_0 is the equilibrium separation between two atoms, and where ε is the potential energy between two atoms at equilibrium distance. In case of gold, D_0 = 2.88 Å [10], ε equalș 8.55×10^{-20} J [11], and the surface energy can be evaluated as γ = 0.96 J/m^2. This is a typical value for clean pure metals and agrees with numerically computed results of the literature [10], which give values around 1.0 J/m^2 for different lattice planes. Accordingly, the adhesive energy ϖ is set to be 1.92 J/m^2 for gold surfaces.

In the definition of Dugdale assumption, the critical value of separation z_0 satisfies $\varpi = \sigma_0 z_0$. Therefore, for usual potential description of gold [11] it can be deduced that $z_0 = 0.97D_0$, and we have $z_0 = 2.80$ Å for the gold to gold contact.

3.2 Statistical characters of surface topography

Three important parameters in the Greenwood-Williamson (GW) model, are the assumed constant radius of the spherical asperities R, the standard deviation of the asperity heights σ_s (assumed to be a Gaussian distribution), and the area density of asperities N. In the study of Poon and Bhushan [12], the standard deviation of the asperity heights is referred to as the root-mean-square (RMS) roughness R_q when the reference plane is the same as the mean plane contrarily to the standard derivation of asperities heights σ_s, which considers only the tip of asperities heights accordingly to GW model. According to the research of McCool [13], these three parameters can be extracted from the three spectral moments of the surface topography $z(x, y)$. The three spectral moments are the variance of heights m_0, slope m_2 and curvature m_4.

The three statistical surface parameters in GW model can be calculated by the expressions as follow [13]:

$$R = \frac{3\sqrt{\pi}}{8\sqrt{m_4}}, \qquad N = \frac{1}{6\sqrt{3}\pi}\frac{m_4}{m_2}, \qquad \sigma_s = \sqrt{m_0 - \frac{3.717 \times 10^{-4}}{N^2 R^2}} \tag{6}$$

Practically, these values are obtained from the atomic force microscopy (AFM) measurements. In this paper, a gold sample was measured, which consists in the bottom electrode of the MEMS device fabricated by the Laboratory of Analysis and Architecture of Systems (LAAS) in Toulouse France. On the surface of the gold sample, a scan size of 27.0×27.0 µm was measured with AFM at 256 × 256 sample data size. Then, in order to account for the scale effect on the surface properties [8], a square of 10×10 µm was cut from the measured area and was applied to analysis to extract the statistical characters of surface. Fig. 3 shows the AFM images of the surface. Based on the AFM measurements, the statistical characters of the surface are obtained according to Eqs. (6) and are presented in Tab.1. In order to consider the adhesive contact of two rough surfaces, which have the same statistical characters, the statistical characters of the equivalent rough surface can also be computed, and are reported in Tab.1.

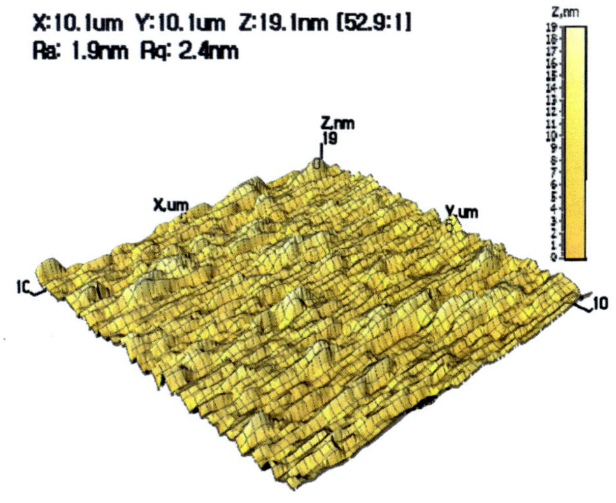

Tab. 1. Statistical characters and the mechanical properties of the Au rough surface.

Measured rough surface	R_q (nm)	σ_s (nm)	R (μm)	N (μm^{-2})	E (GPa) [14]	ν [14]
	2.4	2.11	9.02	1.86	112	0.44
Equivalent rough surface	R_q (nm)	σ_s (nm)	R (μm)	N (μm^{-2})	K (GPa)	
	3.39	2.98	4.51	2.63	92.59	

3.3 Adhesive contact force between Au-Au rough surfaces

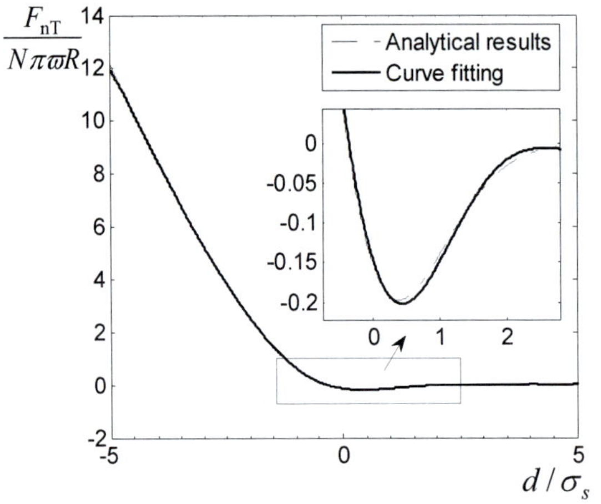

Fig. 4: The normalized adhesive contact force vs. the normalized distance between the rough surfaces for λ=6.08 and σ_s=2.98 nm.

The adhesive-contact vs distance curve (4) of gold-to-gold contact can be established for the parameters of the equivalent rough surface (Tab.1). Eq. (3) is used to evaluate the Maugis transition parameter, where ϖ = 1.92 J/m^2 and z_0 = 2.80 Å, leading to a value of λ = 6.08 for the Maugis' transition parameter. The resulting normalized adhesive contact force vs. the normalized distance is presented in Fig. 4, with the standard derivation of asperity heights σ_s equals 2.98 nm.

In Fig.4, positive forces represent compression and negative forces represent adhesion between the two interacting surfaces. In the curves it clearly appears that when the surfaces are in contact, Hertz forces lead to a resulting positive compressive force. When the distance increases, the contact compressive forces are progressively released but adhesive forces are still acting, and for a defined distance range, the resulting force is negative, which corresponds to surface attraction. Finally if the distance keeps increasing, the adhesion forces are decreasing and the resulting force tends toward zero.

4. CONCLUSIONS

In order to predict stiction in MEMS structures, a possible approach is to consider a multi-scale framework. If at the higher scale a finite element analysis can be considered, it required an adhesive-contact law to be integrated on the interacting surfaces.

The definition of this adhesive-contact law constitutes the micro-scale. In this paper, this adhesive contact-distance curve of two rough surfaces interacting was established from a combination of

Greenwood-Williamson asperity model and Maugis transition's theory.

As the computed unloading adhesive contact-distance curves are dependant on the material and surface properties as well as on the rough surfaces' topography parameters, parameters required to define the micro adhesive contact curves were identified for gold-to-gold interaction. These parameters were either evaluated from theoretical models or from AFM measurement.

ACKNOWLEDGEMENT

The work of the first author has been supported by the project COROMIS, "First Postdoc Project 2007" funded by the Walloon Region of Belgium. The third author acknowledges the financial support of the Belgian National Fund for Scientific Research.

REFERENCE

[1] A. Hariri, J. W. Zu and R. Ben Mrad, Modeling of dry stiction in micro electro-mechanical systems (MEMS), J. Micromech. Microeng. 16 (2006) 1195–1206.

[2] Johnson K. L., Kendall, K. and Roberts, A. D., (1971), Surface Energy and the Contact of Elastic Solids, Proc. R. Soc. Lond, A 324, pp. 301-313.

[3] Derjaguin B. V., Muller V. M. and Toporov Y. P., (1975), Effect of contact deformation on the adhesion of elastic solids, J. Colloid Interface Sci. 53, pp. 314–26.

[4] Maugis D., Adhesion of Spheres: The JKR-DMT Transition Using a Dugdale Model, J. Colloid and Interface Science, vol.150 (1992), pp. 243-269.

[5] Greenwood J. A. and Williamson J. B. P., Contact of nominally flat surfaces Proc. R. Soc. Lond., A vol. 295, pp. 300–19 1966.

[6] Greenwood J A and Tripp J H 1971 The contact of two nominally flat rough surfaces Proc. Instn Mech. Eng. 185 625–33.

[7] L. Wu, V. Rochus, L. Noels, and J. C. Golinval, Influence of adhesive rough surface contact on microswitches, J. Appl. Phys. 106, 113502 (2009).

[8] L. Wu, L. Noels V. Rochus, M. Pustan, and J. C. Golinval, A Multiscale Approach to Overcome Stiction due to Surface Contact in Micro electro-mechanical Systems (In preparation).

[9] Kim K. S., McMeeking R. M. and Johnson K. L. J., Adhesion, Slip, Cohesive Zones and Energy Fluxes for Elastic Spheres in Contact ,Mech. Phys. Solids 46 243–66, 1998.

[10] R.J. Stokes, D.F. Evans, Fundamentals of Interfacial Engineering, Wiley–VCH, New York, 1997.

[11] Ning Yu, Andreas A. Polycarpou, Adhesive contact based on the Lennard–Jones potential: a correction to the value of the equilibrium distance as used in the potential, Journal of Colloid and Interface Science 278 (2004) 428–435.

[12] Poon C. Y. and Bhushan B., Comparison of surface roughness measurements by stylus profile AFM and non-contact optical profiler, Wear, 1995, vol. 190, pp. 76–88.

[13] McCool, J. I., 1986, "Predicting Microfracture in Ceramics Via a Microcontact Model," ASME J. Tribol., 108, pp. 380–386.

[14] Zong Zong, Yifang Cao, Nima Rahbar, and Wole Soboyejo, Nano- and microscale adhesion energy measurement for Au–Au contacts in microswitch structures, Journal of Applied Physics 100, 104313 (2006).

Proceedings of the SEM Annual Conference
June 7-10, 2010 Indianapolis, Indiana USA

Measurement of adhesive force between two mica surfaces with multiple beam interferometry

T. Y. Chen and J. C. Jung

Department of Mechanical Engineering
National Cheng-Kung University
Tainan, Taiwan, 70101, ROC

ABSTRACT

Surface forces play a crucial role in the contact behavior of micro-components as well as the application of MEMS products. In this study, a microscopic measurement system based on multiple beam interferometry is developed to measure the adhesive force between two mica thin films. The contact area on the mica can be determined from the FECO fringes. A double cantilever spring is used to measure the adhesive (pull-off) forces between the two mica thin films. The usefulness of the adhesive force measurement system is validated by comparing with the result from JKR contact theory.

Keywords：Multiple beam interferometry, Mica thin film, Adhesive force, JKR contact theory

I. INTRODUCTION

In nano-scale, physical and mechanical properties of materials will show a significantly different characteristics and phenomena. The establishment and use of precision measurement technology and system become an important part to understand and to effectively control the materials, structures or installations in nano-scale. Adhesion force plays an important role in the contact behaviour of micro-components and the application of MEMS products. It is necessary to develop a measurement system to understand and control the material characteristics.

Tabor and Winterton [1] developed an apparatus to measure the force between two mica sheets in vacuum and air. Multiple beam interferometry (MBI) is regularly used for measuring film thickness and refractive index. Based on the designs of Tabor and Winterton, Israelachvili and Adams [2] developed a well-known surface force apparatus which allows the forces to be measured as a function of separation between the two mica sheets in contact. The separation between the contact surfaces is determined by using multiple beam interferometry, and the force is determined by measuring the deflection of a weak spring. Parker et.al.[3] developed an improved device for measuring the force and separation between two surfaces. One of the surfaces is mounted on a double-cantilever force measuring spring which is supported at the end of a shaft. The upper surface is mounted in the center of a piezoelectric tube.

II. METHOD AND THEORY

1 Multiple beam interferometry

A schematic setup of multiple beam interferometry system is shown in Fig. 1. The white light source inject into the two cylindrical mica sheets perpendicularly. Two cylindrical axes of the mica sheets are orthogonal, and each piece of mica should be coated with a 4~6 μm layer of silver film. When the light has interference in the two highly reflective silver films by multiple reflection, the wavelength of transmitted light is related to the thickness and refractive index of the material passed through. After the light passes through two mica sheets, it injects into the slit by a right-angle mirror and sampled by a spectrometer. In the spectrograph, as shown in Fig. 2, one can see the bright fringes which are called the fringe of equal chromatic order (FECO). The vertical pixel in FECO image corresponds to the actual location of the lateral position of the mica sheets. As the mercury light passing through the spectrometry, its spectral lines have a fixed wavelength and can be used to correct the horizontal direction out of FECO images of the relationship between pixel location and wavelength. In general, the film thickness and the refractive index can be obtained by analyzing the FECO images using fast spectral correlation method. For friction study, the contact length or area can be measured from the FECO image after calibration of the pixel distance.

DOI 10.1007/978-1-4419-8825-6_28, © The Society for Experimental Mechanics, Inc. 2011

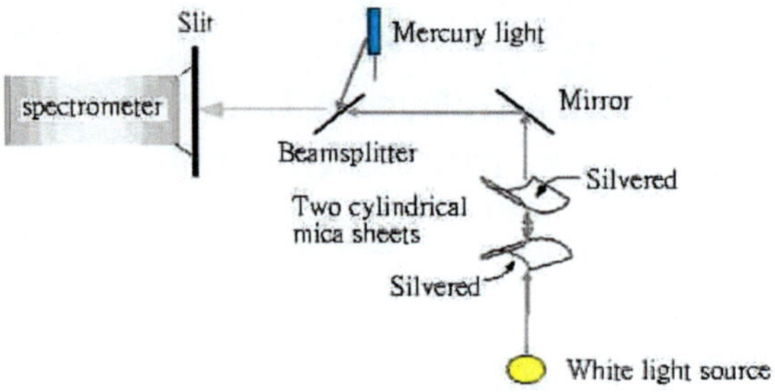

Fig. 1. Schematic setup of multi-beam interferometry system.

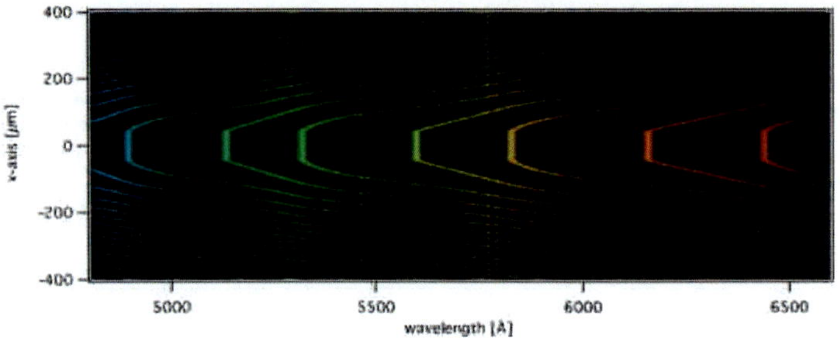

Fig. 2. Typical image of fringe of equal chromatic order (FECO).

2 Theory of contact

Generally, the mechanics of two initially curved bodies in contact can be regarded as Hertzian or non-Hertzian. In Hertzian type, the contact surfaces of two elastic bodies, as shown in Fig. 3, are considered molecularly smooth, and would separate freely at zero contact area and load. With no adhesion between the contact bodies, Hertz showed that

$$a_0^3 = \frac{RP_0}{K} \tag{1}$$

Where a_0 is radius of the contact circle, P_0 is the applied load, and

$$R = \frac{R_1 R_2}{R_1 + R_2}, \quad K = \frac{4}{3}\left(\frac{1-v_1^2}{E_1} + \frac{1-v_2^2}{E_2}\right), \tag{2}$$

where R_1 and R_2 are radii, E_1 and E_2 are Young's modulus, and v_1 and v_2 are Poisson's ratios of the two sphere bodies.

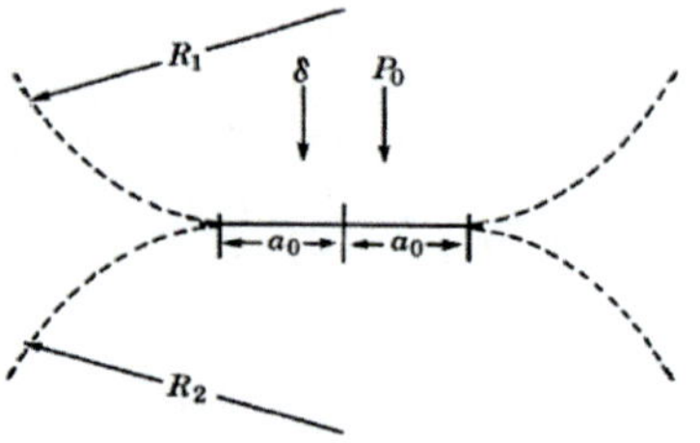

Fig. 3 Two spheres in contact.

Johnson et. al.[4] extended the Hertz theory to two adhering solids by including surface energies in their formulation. The load between the contact surfaces P_1 is modified as

$$P_1 = P_0 + 3\gamma\pi R + \sqrt{6\gamma\pi R P_0 + (3\gamma\pi R)^2} \quad (3)$$

where γ is the surface energy per unit area between the contact surfaces. Thus the radius of the contact area can be expressed as

$$a^3 = \frac{R}{K}\left(P + 3\gamma\pi R + \sqrt{6\gamma\pi RP + (3\gamma\pi RP)^2}\right) \quad (4)$$

It can be derived from eq. (4) that the radius is finite under zero load. Further more, when the contact area is zero, .there is a negative load P (called pull-off force) given as

$$P = -\frac{3}{2}\gamma\pi R \quad (5)$$

III. SYSTEM CONFIGURATION

The experimental system constructed to measure the frictional force is shown in Fig. 4. The system contains a white light source, a multi-beam interference system based on two mica sheets, a spectrometer to generate spectrograph and a PC-based image processing systems for data processing. The multi-beam interference system consists of the structure ontology, motorized translation stage and piezoelectric translator for coarse and fine separation of mica sheets adhered on the surface of two silica lens. The two mica sheets were placed in two lens holder and form symmetric three-layer interference (mica-air layer under test-mica). The white light passing through an optical access and is launched into the multi-beam interference system with air between two mica sheets. The light emerging from the multi-beam interference system is guided into a spectrometer and the produced FECO image was grabbed by the image processing system for further processing to determine the contact area.

The adhesion force measuring device is enlarged as shown in Fig. 4. The top mica sheet holder is fixed at the bottom of two thin spring plates which are connected to a U-shaped mounting frame. The mounting frame is fixed on a precision motorized translation stage having a measurement range of 80 mm and resolution of 0.16 m. The bottom mica sheet was fixed at one end of a double cantilever spring. The other end of the spring is fixed on a piezoelectric translator having a resolution of 0.3nm in a range of 15 m. The spring constant of the double cantilever spring was analyzed by using ANSYS and calibrated by a displacement sensor using dead weight. The experimental data compared very well to the numerical ones. The spring constant was determined as 376 N/m. The separation or contacting of the mica sheets is initiated by the piezoelectric translator. The deflection of the spring is measure and the adhesive (pull-off) force is calculated by multiplying the spring constant.

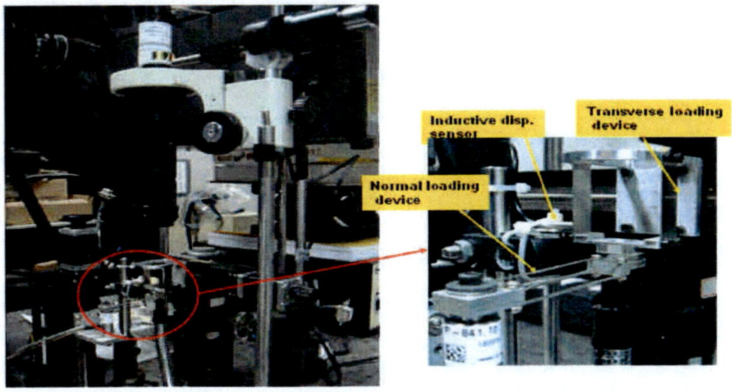

Fig. 4. The experimental system constructed to measure the friction force.

IV. EXPERIMENTAL RESULTS AND DISCUSSION

Fig. 5(a) shows the FECO images of the two mica sheets separated by 70 nm. During the force measurement experiment, the bottom mica sheet was moved slowly toward the top mica sheet by the piezoelectric translator. When the mica surfaces are brought together, the surface force attraction causes them to adhere. Fig. 5(b) shows the FECO image at that moment. Further movement of the piezoelectric translator, Fig. 5(c) shows the FECO images of the two mica sheets in contact due to adhesion force and a normal force of 15mN, respectively. Then the piezoelectric translator was moved backward and the contact areas were measured as given in Fig. 6. The pull-off force determined is 6.1 mN. The calculated surface energy is 60.5 mJm2, which agrees with the values determined in Ref [5]. Comparison of the experimental data to JKR curve shows that the trend agrees well with each other. Further examining the loading and unloading behaviour of the experimental data, adhesion hysteresis was also observed.

(a)Separated by 70 nm (b) Contact under adhesion force (c) Contact under a 15 mN normal load

Fig. 5. Experimental FECO images of two mica sheets under various contacts.

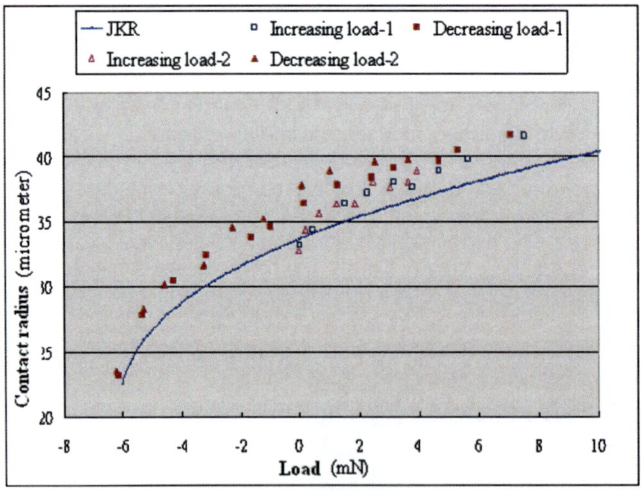

Fig. 6. Contact areas measured under various loadings.

V CONCLUSION

A microscopic measurement system based on multiple beam interferometry is developed to measure the adhesion force between two mica thin films. The surface behaviour due to adhesion between two mica sheets is report. The evaluated surface energy between two mica sheets agrees well to the existing data. Adhesion hysteresis was also observed from the loading and unloading data. The applicability the system developed is validated.

Acknowledgement

The authors are grateful to National science council, Republic of China, for supporting this work under the contract no. NSC95-2221-E-006-037-MY2.

References

[1] 1. Tabor, D. and Winterton, R. H. S., "Surface forces: direct measurement of normal and retarded van der Waals forces", Proc. R. SOC. A 312, 435, 1969.

[2] Israelachvili, J. N. and Adams, G. E., "Measurement of forces between two mica surfaces in aqueous potassium nitrate solutions", J. Chem. Soc. Faraday Trans 1, 74, 975, 1978.

[3] Parker J.L. ,Christenson H.K. ,Ninham B.W. , "Device for measuring the force and separation between two surfaces down to molecular separations", Rev. Sci. Instrum .60, 3135,1989.

[4] Johnson, K. L., Kendall , K., and Roberts, A. D., Proc. R. Soc. London A 324, 301(1971)

[5] Horn, R.G. , Israelachvili , J. N. ,and Pribac , F. , "Measurement of the deformation and adhension of solids in contact" , J. colloid and interface science, vol. 115 ,1986

Proceedings of the SEM Annual Conference
June 7-10, 2010 Indianapolis, Indiana USA

Performance Studies of A Prototypical MEMS Thermal Actuator

Wei-Yang Lu[1], Ernest J. Garcia[2], Helena Jin[3], Bo Song[3]
[1,3] Sandia National Laboratories, MS9042, Livermore, CA 94551-0969
[2] Sandia National Laboratories, MS1064, Albuquerque, NM 87123
[1] wlu@sandia.gov

ABSTRACT

This paper presents the performance of a thermal actuator made from single crystal silicon. The thermal actuator, shaped like a chevron, has four beams on each side and a shuttle in the middle. The shuttle's motion is caused by thermal expansion and bending of the beams. The power curve, temperature profiles, in-plane and out-of-plane displacements, frequency response, as well as load-displacement curves of this device were measured at various operating voltages at the ambient condition. Experimental setup and results are described.

INTRODUCTION

Thermal actuators, which can provide relatively high forces and permit high positioning accuracy, have found many applications in Microsystems, e.g. relays, switches, valves, displays, etc. For in-plane linear motion, one popular design is the chevron-shaped bent-beam actuator. It is driven by passing current through beams at the two ends. The beams function as structural elements as well as resistors. The thermal expansion of the beams due to resistance heating pushes the tip outward. The angle of the chevron increases as heat is applied and decrease as heat is removed.

In this paper the basic mechanical and thermal behaviors of a prototypical bent-beam actuator, shown in Figure 1, are investigated. The actuator has four parallel beams on each side and a shuttle in the middle. All eight beams have the same dimension. The nominal length, width and depth are 7000, 65 and 250 um, respectively. The distance between the adjacent beams is 250 um from center to center. The dimension of the shuttle is approximately 1500x100x250 um. A line is marked on the shuttle and vernier scales are printed on a fixed island next to it. The displacement of the shuttle can be detected. The top-right insert in Figure 1 shows the initial (or reset) position, where the marked line is aligned with 0 um.

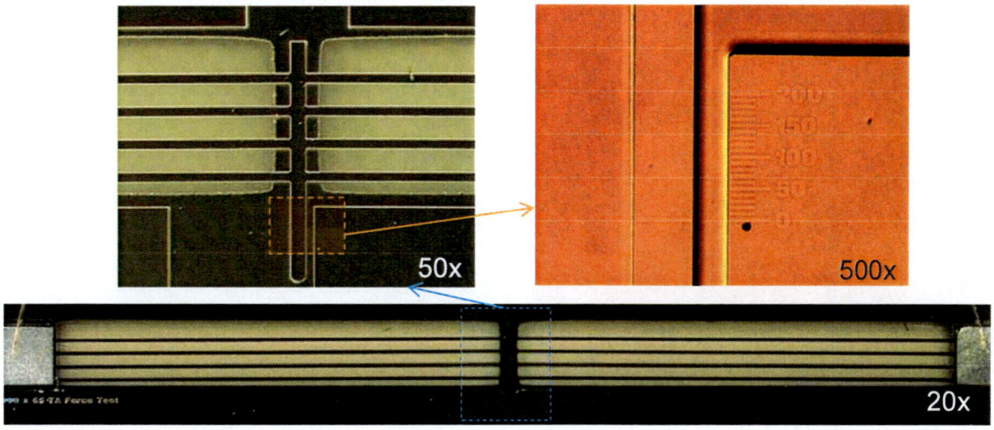

Figure 1 Chevron-shaped bent-beam thermal actuator.

T. Proulx (ed.), *MEMS and Nanotechnology, Volume 2*, Conference Proceedings of the Society for Experimental Mechanics Series 2,
DOI 10.1007/978-1-4419-8825-6_29, © The Society for Experimental Mechanics, Inc. 2011

The thermal actuator was fabricated using a silicon-on-insulator (SOI) wafer. With this structure, the device wafer could be etched by means of the deep reactive ion etch (DRIE) process, also known as the Bosch Process, to the precision required while using single crystal Si to establish the force required for device functionality.

Validated models can be used to predict response to various environments and loadings [1]. Experimental characterization of this particular thermal actuator was performed to provide data for FEM model validation. Experiments to obtain the power curve, temperature profiles, in-plane and out-of-plane displacements, time and frequency responses, and load-displacement curves are described in the following.

POWER CURVE

To activate the thermal actuator, a circuit was connected as shown in Figure 2. The thermal actuator and a resistor R = 1 KΩ were in series. By slowly increasing or decreasing the excitation V_o step by step, voltages V_1 and V_2 were recorded. V_1 was the voltage applied to the actuator. The current and power consumption of the actuator were $i = V_2/R$ and $P = i \cdot V_1$, respectively. Figure 3 shows the power curve, P versus V_1.

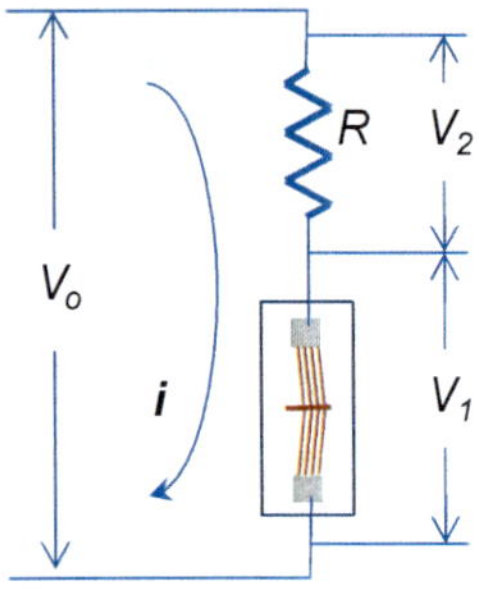

Figure 2 Actuation circuit.

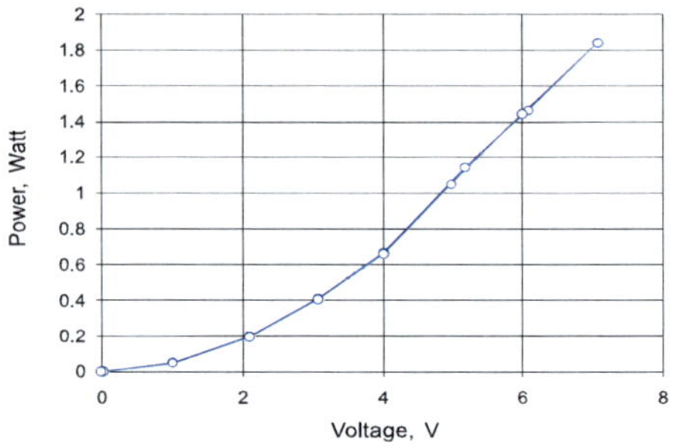

Figure 3 Power curve of the thermal actuator.

DISPLACEMENT

In-plane

Notice the beams have a high aspect ratio, depth / width ≈ 4. The shuttle's motion is designed to be in-plane. There are several ways to measure the in-plane displacement of the shutter. One is to estimate it from the vernier scale. As shown in Figure 4(a), the shuttle moves about 130 um when the excitation is 5 V. It can be done by using a measuring microscope too, which has a better resolution. During the power measurement experiment described in the previous section, the displacement was also recorded at each step. The displacement versus voltage curve is shown in Figure 4(b). The relation is nonlinear. The shuttle displacement reaches d = 200 um when $V_1 \approx 6.7$ V.

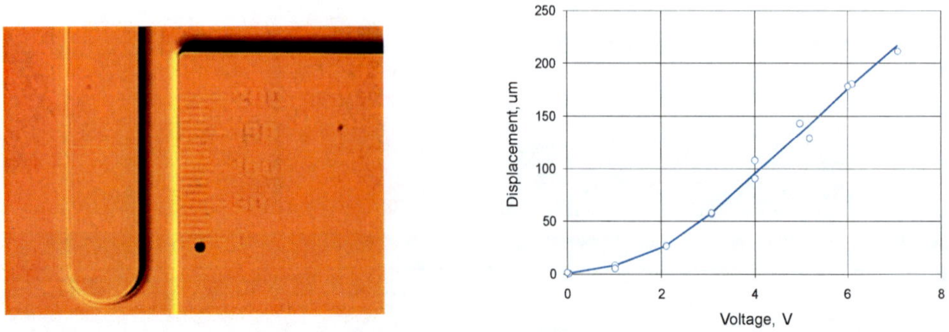

(a) (b)

Figure 4 (a) Shuttle displacement at 5 V, and (b) the displacement-voltage curve of the actuator.

Out-of-plane

Although Bosch process can fabricate 90° (truly vertical) walls, but often the walls are slightly tapered, which may cause the shuttle to have an out-of-plane component of displacement during actuation. The insulator (oxide) is 2 um thick. If the out-of-plane component is large enough, the shuttle may hit the bottom handling wafer or packaging structures on the top. Using a Veeco interferometer, the displacement of the shuttle was determined. As shown in Figure 5, the out-of-plane and in-plane components of displacement are approximately 0.5 and 200 um, respectively, at 7 V excitation.

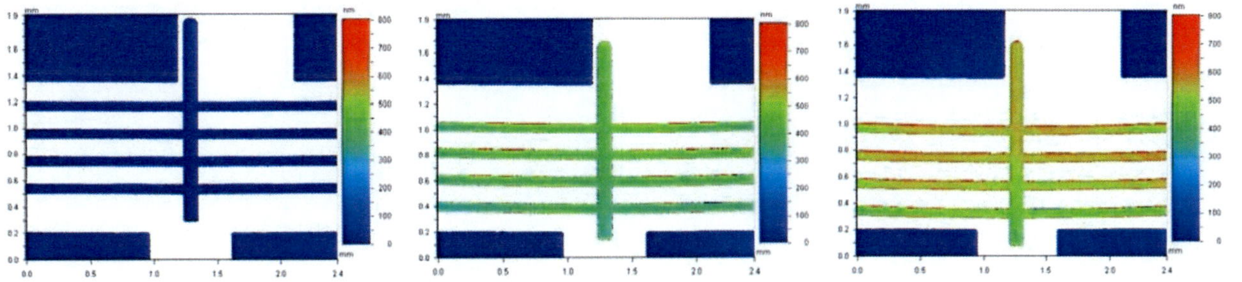

(a) 0 V, $d_{out\text{-}of\text{-}plane}$ = 0 um (b) 5 V, $d_{out\text{-}of\text{-}plane}$ = 0.4 um (c) 7 V, $d_{out\text{-}of\text{-}plane}$ = 0.5 um

Figure 5 In-plane and out-of-plane components of the shuttle displacement of at various levels of excitation.

TEMPERATURE

Calibration

An Indigo thermal imaging system was used to characterize the temperature profiles of the actuator. The intensity or ADC (Analog to Digital Converter) count of each pixel does not provide the temperature reading directly. It represents the total thermal or photon energy (flux) detected during the integration time. The count number has to be converted to temperature through a calibration curve. The calibration was done by placing a die on a hot plate, as shown on Figure 6(a). A thermal couple was attached on the top surface of the die. The infrared camera was focused on the surface of the die at an area near the thermal couple, Figure 6(b). Another thermal couple was attached to the hot plate to control its temperature. When both thermal couple readings were stabilized, the temperature of the die surface and the corresponding averaged ADC count were recorded. The procedure was repeated at several discrete temperatures to obtain a calibration curve. Since the integration time could be adjusted to improve the sensitivity of measurement for a given temperature range, a number of calibration curves related to different integration times were acquired. Three calibration curves with integration times of 0.02, 0.05 and 0.5 ms are displayed in Figure 6(c).

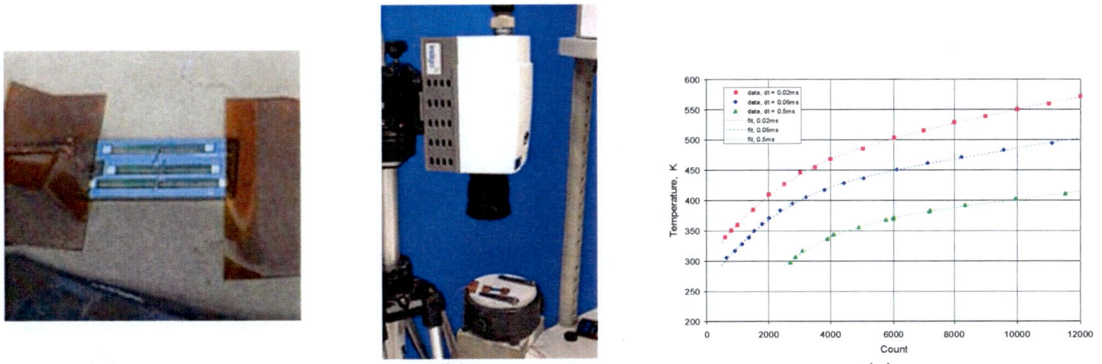

(a) (b) (c)

Figure 6 Calibration of a thermal imaging system

Measurement

Due to its structural symmetry, the viewing range of the thermal camera was adjusted to cover half of the actuator for better spatial resolution. Thermal images were acquired at various states of actuation and then post-processed using a Matlab routine to display temperature fields according to the assigned colormap. Examples of the temperature fields at 3, 5 and 7 V excitation are shown in Figure 7. Detailed temperature profiles along the shuttle and beams of the 7 V excitation are plotted in Figure 7(c). The temperatures of two inner beams (#2 and 3) are generally the same and higher than the temperatures of two outside beams (#1 and 4). The hottest region of #2 (or #3) is about 2.5 to 3.0 mm from the shuttle, which is not exactly at the midpoint of the beam. The maximum temperature is close to 600 °K.

(a) 3 V

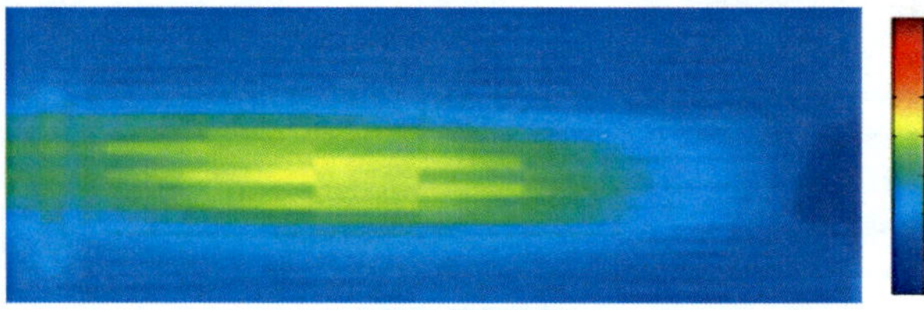

(b) 5 V

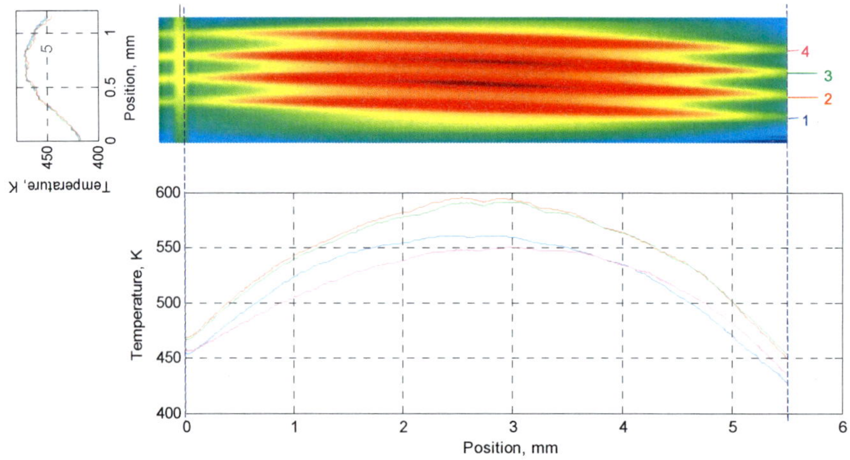

(c) 7 V

Figure 7 Temperature (unit: °K) profiles of the bent beams and shuttle at (a) 3 V, (b) 5 V and (c) 7 V.

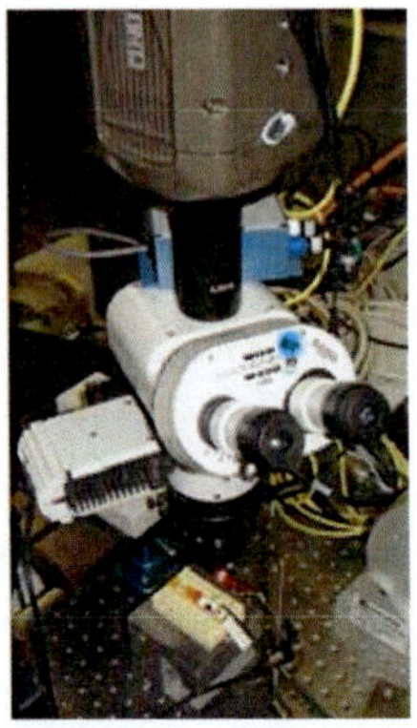

Figure 8 High speed imaging and data acquisition

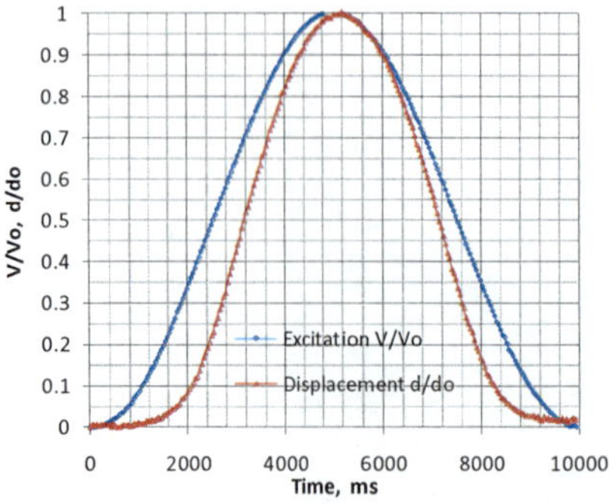

(a)

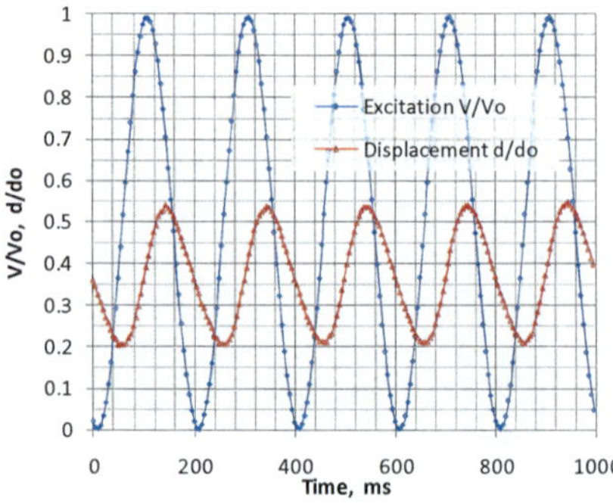

(b)

Figure 9 Normalized excitation and displacement at (a) 0.1 and (b) 5.0 Hz. (V_o = 4.0 V, d_o = 95 um).

FREQUENCE RESPONSE

The operating frequency of an actuator may depend on beam's geometry/size, thermal boundary conditions, etc. It has been reported that micro thermal actuators could operate at high frequencies, possibly exceeding 1 KHz [2]. To characterize the frequency response of the actuator accurately, the displacement of the shuttle and the excitation of the actuator needed to be sampled at a frequency at least twenty times faster than the excitation frequency. A Phantom high speed camera system with A/D input channels was integrated with a microscope. The displacement of the shuttle in high resolution videos and the excitation signal in discrete numbers were recorded in sync. The quantitative displacement data were determined from the videos in post-experiment image analyses. The integrated system is shown in Figure 8.

The excitations were Haversine functions $V = V_o(1-\cos(2\pi f))/2$, where V_o was the amplitude of the input voltage and f was the frequency. Figure 9 shows the normalized steady-state excitation V/V_o and displacement d/d_o of the actuator at two different frequencies, $f = 0.1$ and 5.0 Hz. The value of d_o was the static displacement corresponding to V_o, which was determined from the characteristic displacement-voltage curve of the actuator shown in Figure 4(b). The image and data acquisition rate was 24 fps for 0.1 Hz input and 200 fps for 5.0 Hz input. For the examples shown in Figure 9, $V_o = 4.0$ volt and $d_o = 95$ um. At the frequency $f = 0.1$ Hz, shown in Figure 9(a), $(d/d_o)_{max} = 1$ and $(d/d_o)_{min} = 0$. The shuttle has a full range of stroke, the same as that of static loading, but with a small amount of phase lag. At the frequency $f = 5.0$ Hz, $(d/d_o)_{max} = 0.54$ and $(d/d_o)_{min} = 0.20$, shown in Figure 9(b). The stroke is significantly reduced, only 34% of the stroke of static loading; it also has a large phase delay of about 75 degree.

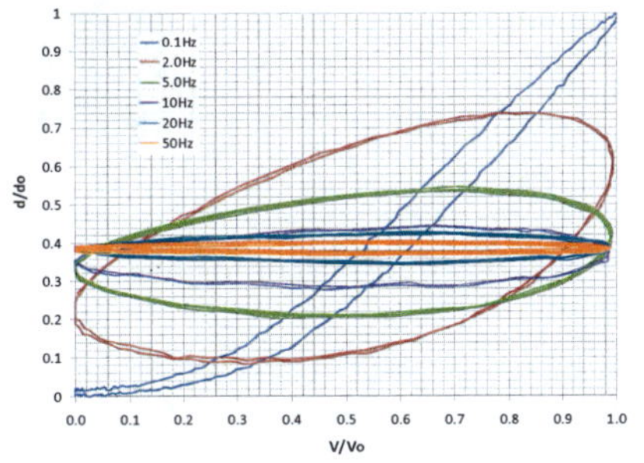

Figure 10 Normalized displacement-voltage curves for $V_o = 4.0$ V at various frequencies.

A number of normalized displacement-voltage curves of different frequencies, but same $V_o = 4.0$ V, are plotted in Figure 10. As the frequency gets higher, the range of the stroke becomes smaller and the stroke approaches the asymptote $d/d_o = 0.38$ or $d = 36$ um. The high frequency response of the actuator to excitation $V_o = 4.0$ V is equivalent to the static loading at 2.4 V.

LOAD-DISPLACEMENT CURVE

To measure the force of the actuator, an external load cell was utilized. Shown in Figure 11(a), the load cell is mounted on a micro manipulator with a probe attached on the other side. The end of probe is bent about 90°, so it can reach the shuttle and keep the load cell and shuttle aligned on the same axis, Figure 11(b). Under the microscope, the probe can be positioned and aligned at the end of the shuttle by controlling x, y, z and up-down of the manipulator, Figure 11(c). Then the probe is ready to load the actuator using the x-control. Since the bent probe is compliant, the displacement of the manipulator is not the same as the shuttle. The displacement of the shuttle is determined from the digital microscope.

The load-displacement curve of the actuator was characterized under the constant excitation voltage. During experiment, the actuator was first activated by a constant excitation V_o and the shuttle moved freely to a displacement d_o. Then in each following loading steps, the probe was moved slowly to push the shuttle for a distance δ. The shuttle's total displacement became $d = do - \delta$. The values of the force F and displacement d were recorded. It was repeated until snap-thru occurred or being stopped manually.

(a) (b) (c)

Figure 11 Experimental setup to measure the force of the actuator.

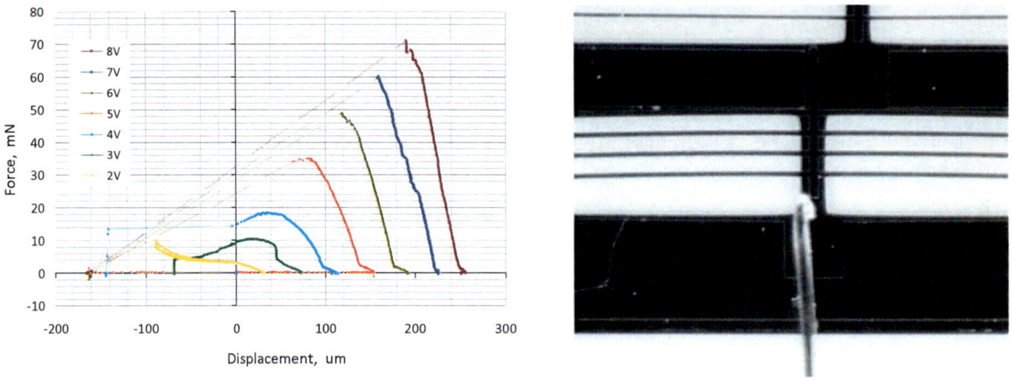

(a) (b)

Figure 12 (a) Load-displacement curves of the actuator at various level of actuation; (b) snap-thru.

Figure 12(a) shows a series of load-displacement (*F-d*) curves of the actuator. In general, the curve of higher excitation voltage has a stiffer slope and a higher snap-thru load. Data show that for most cases the shuttle ends up the same position after snap-thru. The reason is that two actuator beams are rest against fixed walls as shown in Figure 12(b).

SUMMARY

The basic mechanical and thermal behaviors of a thermal actuator were characterized. The power, temperature, in-plane and out-of-plane components of displacement, and load-displacement curves as functions of the actuation voltages were obtained under quasi-static loading at the ambient condition. The relations are all nonlinear. The frequency response of the actuator was also investigated. When the operating frequency is greater than 1 Hz, the range of stroke starts to decrease and becomes negligible at about 50 Hz.

ACKNOWLEDGEMENT

Sandia is a multiprogram laboratory operated by Sandia Corporation, a Lockheed Martin Company, for the United States Department of Energy under contract DE-AC04-94-AL85000.

REFERENCES

[1] J.R. Torczynski, M.A. Gallis, E.S. Piekos, J.R. Serrano, L.M. Phinney, and A.D. Gorby, "Validation of Thermal Models for a Prototypical MEMS Thermal Actuator," SAND2008-5749, September 2008.

[2] L.L. Chu, L. Que, A.D. Oliver, and Y.B. Gianchandani, "Lifetime Studies of Electrothermal Bent-Beam Actuators in Single-Crystal Silicon and Polysilicon" Journal Of Microelectromechanical Systems, Vol. 15, No. 3, June 2006.

Proceedings of the SEM Annual Conference
June 7-10, 2010 Indianapolis, Indiana USA

A New Electrothermal Microactuator with Z-shaped Beams

Changhong Guan and Yong Zhu
North Carolina State University, Mechanical and Aerospace Engineering
2601 Stinson Dr., Raleigh, NC 27695-7910
E-mail: yong_zhu@ncsu.edu
http://www.mae.ncsu.edu/homepages/zhu/index.html

Abstract

A new class of thermal microactuators, Z-shaped thermal actuator, is introduced in comparison with the well-established V-shaped thermal actuator. Though they share many features in common, Z-shaped thermal actuator offers several advantages: compatibility with anisotropic etching, smaller feature size, larger displacement, and larger variety of stiffness and output force. While the Z-shaped thermal actuator was modeled analytically and verified by multiphysics finite element analysis (FEA), the beam width and length of the central beam were identified as the major design parameters in tuning the device displacement, stiffness, stability and output force. Experimental measurements were taken on three arrays of Z-shaped thermal actuator with variable parameters. Results agreed well with the finite element analysis. The development of Z-shaped thermal actuator is applicable in simultaneous sensing and actuating applications. During the quasi-static test of individual Z-shaped thermal actuator, the average temperature in the device structure was estimated based on electric resistivity at each actuation voltage.

Nomenclature

f_{ij}	Compliance coefficients
F_x	Internal (reaction) axial force
P	Virtual unit force
M	Internal (reaction) moment
α	Thermal expansion coefficient
U	Deflection at the tip
L	Length of the long arm beam
l	Length of the central beam
w	Beam width
h	Beam thickness
E	Young's modulus of silicon
I	Beam moment of inertia ($=w^3h/12$)
k	Stiffness of a Z-shaped beam
k_p	Thermal conductivity of silicon
V	Applied voltage across a Z-shaped beam
ρ	Resistivity of silicon

Introduction

Electrostatic actuator [1] and electrothermal actuator [2-4] are the two major in-plane actuators in micro-electromechnical systems (MEMS). Electrostatic actuators, also known as comb drive actuators, have an output force typically on the order of 1 μN when actuation voltage is more than 30 V [5], while thermal actuators can easily generate a force of 1 mN at an actuation voltage around 5-10 V [6]. Since comb drive actuators require high actuation voltage (> 30 V), which is not compatible with microelectronic power, and large area of comb structures, thermal actuators have attracted significant attentions in recent years, as they are demonstrated as a compact, stable and high-force actuation apparatus [4].

Thermal expansion is the operating principle of all kinds of thermal actuators. As for in-plane thermal actuator, U-shaped and V-shaped thermal actuators have been explored and implemented for a few years. Former, also known as thermal actuator [8-10], employs asymmetrical thermal beams with different cross-sectional areas (cold arm and hot arm). The locus of motion of single U-shaped thermal actuator is an arc, which a pair of U-shaped thermal actuator could translate linear motion. The latter, also known as bent-beam thermal actuator [6,7,11-13], utilizes thermal expansion of symmetric, slanted beams to generate rectilinear displacement of the central shuttle. The V-shaped thermal actuators, especially, have been implemented in many applications including linear and rotary microengines [6], nanoscale material testing systems [11,12], and nanopositioners [14].

The advantages of V-shaped thermal actuators are large force (on the order of mN), huge stiffness, low actuation voltage and small features size. However, the slanted beams in V-shaped actuator usually are not oriented along a crystalline orientation, so that anisotropic etching cannot be used for fabricating these structures, which largely limits the available fabrication methods as well as materials. Also, the slanted beams pose challenges for fabricating small features, which deteriorates as the beam width gets close to the resolution of photolithography (typically ~ 2 μm). Though large stiffness makes the V-shaped thermal actuator a perfect displacement controlled actuator, it cannot be used as a simultaneous sensor and actuator as the electrostatic devices. Thus an additional load sensor is always required for such applications as nanomechanical testing [12] and nanomanufacturing [15].

The Z-shaped thermal actuator introduced in this paper is a new class of thermal actuators with Z-shaped beams for in-plane motion. It offers a large range of stiffness and output force that is complementary to the comb drives and V-

T. Proulx (ed.), *MEMS and Nanotechnology, Volume 2*, Conference Proceedings of the Society for Experimental Mechanics Series 2,
DOI 10.1007/978-1-4419-8825-6_30, © The Society for Experimental Mechanics, Inc. 2011

shaped actuators. Similar to V-shaped actuators, the structure and operating principle of the Z-shaped actuators is first described and modeled analytically. Finite element multiphysics simulations were then performed to verify the experimental results that were measured in vacuum. The devices were fabricated using the SOI-MUMPs (silicon-on-insulator multi-user MEMS process) (MEMSCAP, Durham, NC) with 10 μm thick silicon as the structural layer.

Concept and Modeling

A. Structure

Schematic of the Z-shaped and V-shaped thermal actuator is shown in Fig. 1 for comparison. It is seen that the basic unit of a Z-shaped actuator is a pair of Z- shaped beams and a shuttle in the middle. Despite of principle similarity, Z-shaped actuators rely on bending of the symmetric Z-shaped beams induced by thermal expansion to achieve rectilinear displacement of the central shuttle. Due to Joule heating, the device is heated up when a current is passed through. The temperature rise leads to thermal expansion of all the beams especially the long beams; the long beams cannot expand straight due to symmetry constraint of the structure, rather they bend to accommodate the length expansion. As a result the shuttle is pushed forward. Figure 1(c) shows a scanning electron microscopy (SEM) image of a Z-shaped thermal actuator with two pairs of Z-shaped beams.

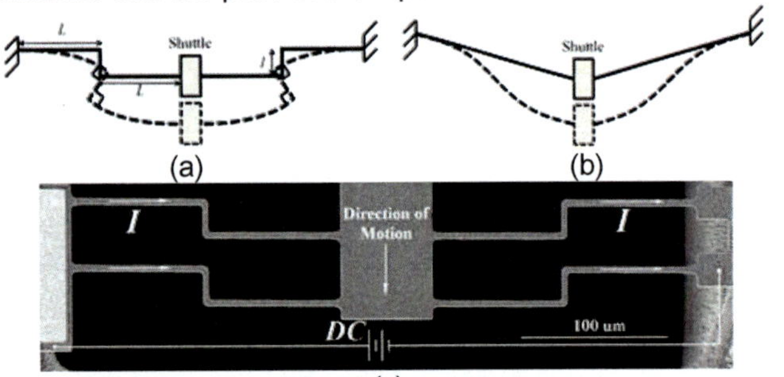

(c)

Fig.1. Schematics of (a) a Z-shaped thermal actuator and (b) a V-shaped thermal actuator before and after motion. Drawn not to scale. (c) SEM image of the Z-shaped thermal actuator. The black area is an etched hole underneath. I is the current passing through thermal beams, while DC is the power source.

B. Mathematical Modeling

Following assumptions have been made for analytical derivation: central shuttle is rigid and its thermal expansion is neglected; thermal expansion of short beam (with length *l* in Fig.1) is neglected; small strains and displacements are considered; average temperature rise in a Z-shaped beam is given [11].

Mechanical response of the structure in Fig.1 can be equivalently modeled by considering half of the structure without the shuttle, as shown in Fig. 2. In energy method, three reaction forces, axial force F_x, virtual force *P* and moment *M*, can be obtained by solving the following set of equations [8]

$$\begin{bmatrix} f_{11} & f_{12} & f_{13} \\ f_{21} & f_{22} & f_{23} \\ f_{31} & f_{32} & f_{33} \end{bmatrix} \begin{bmatrix} F_x \\ P \\ M \end{bmatrix} = \begin{bmatrix} 2\alpha\Delta TL \\ U \\ 0 \end{bmatrix} \qquad (1)$$

where

$$f_{11} = \frac{2L}{EA} + \frac{l^3}{3EI} + \frac{Ll^2}{EI} \quad f_{12} = \frac{3L^2 l}{2EI} + \frac{Ll^2}{2EI} \quad f_{13} = -\frac{l^2}{2EI} - \frac{Ll}{EI}$$

$$f_{21} = \frac{Ll^2}{2EI} + \frac{3L^2 l}{2EI} \quad f_{22} = \frac{l}{EA} + \frac{L^2 l}{EI} + \frac{8L^3}{3EI} \quad f_{23} = -\frac{2L^2}{EI} - \frac{Ll}{EI}$$

$$f_{31} = -\frac{l^2}{2EI} - \frac{Ll}{EI} \quad f_{32} = -\frac{2L^2}{EI} - \frac{Ll}{EI} \quad f_{33} = \frac{2L}{EI} + \frac{l}{EI}$$

Set virtual force *P* equal to zero, deflection in the *y* direction is derived as

$$U = \frac{12\alpha\Delta TL^3}{l^2 + 6L\left(l + \frac{w^2}{3l}\right)} \qquad (2)$$

The stiffness of a Z-shaped beam is given by

$$k = \frac{Ew^3h\left(l^3 + 2Lw^2 + 6Ll^2\right)}{\left(8L^3l^3 + w^2l^4 + 16w^2L^4 + 2w^4Ll + 12L^4l^2 + 6w^2Ll^3\right)} \qquad (3)$$

The internal force is given by

$$F_x = \frac{24\alpha\Delta TEIL}{l^3 + Ll^2 + 2Lw^2} \qquad (4)$$

The output force *f* is given by the product of displacement and stiffness, *f=kU*. The stiffness of a Z-shaped actuator with a pair of beams is 2*k*, thus the output force is *2f* accordingly.

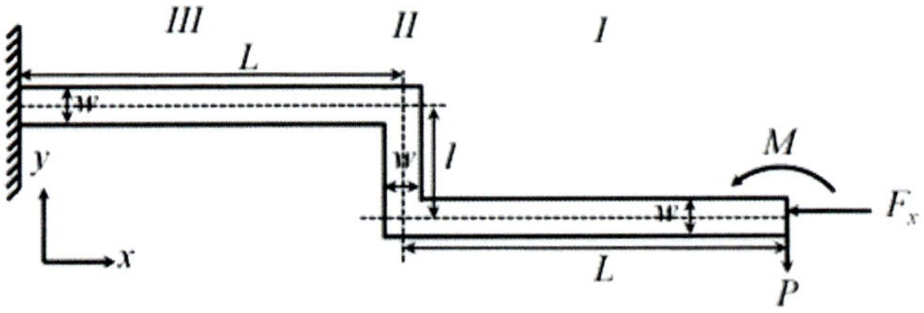

Fig.2. Free-body diagram of a single Z-shaped beam.

There are three possible modes of heat transfer: conduction, convection, and radiation. Convection and radiation are generally neglected in MEMS structures; conduction through the air layer between the device and the substrate is a major heat transfer mechanism for surface micromachined devices, since the air layer is typically very thin (on the order of a few μm) [8,10-12]. But in SOI devices, the only heat transfer mechanism is heat conduction to the anchors across the beams, since the underneath silicon substrate is totally etched as shown in Fig.1(c).

The performance of Z-shape thermal actuator is geometry dependent. The peak displacement of one single Z-shaped beam at a given temperature can be increased by simply increasing the length of the long arm (*L*). The device thickness (*h*) is not related to the displacement, but affects the stiffness and loading force in direct proportion. In our design, all the long beams (*L*) length was 88 μm and structural thickness is 10 μm as specified in the SOI-MUMPs. Widths of all the long beams and central beams are the same for simplicity. The central beam length (*l*) and beam width (*w*) are variables for parametric study.

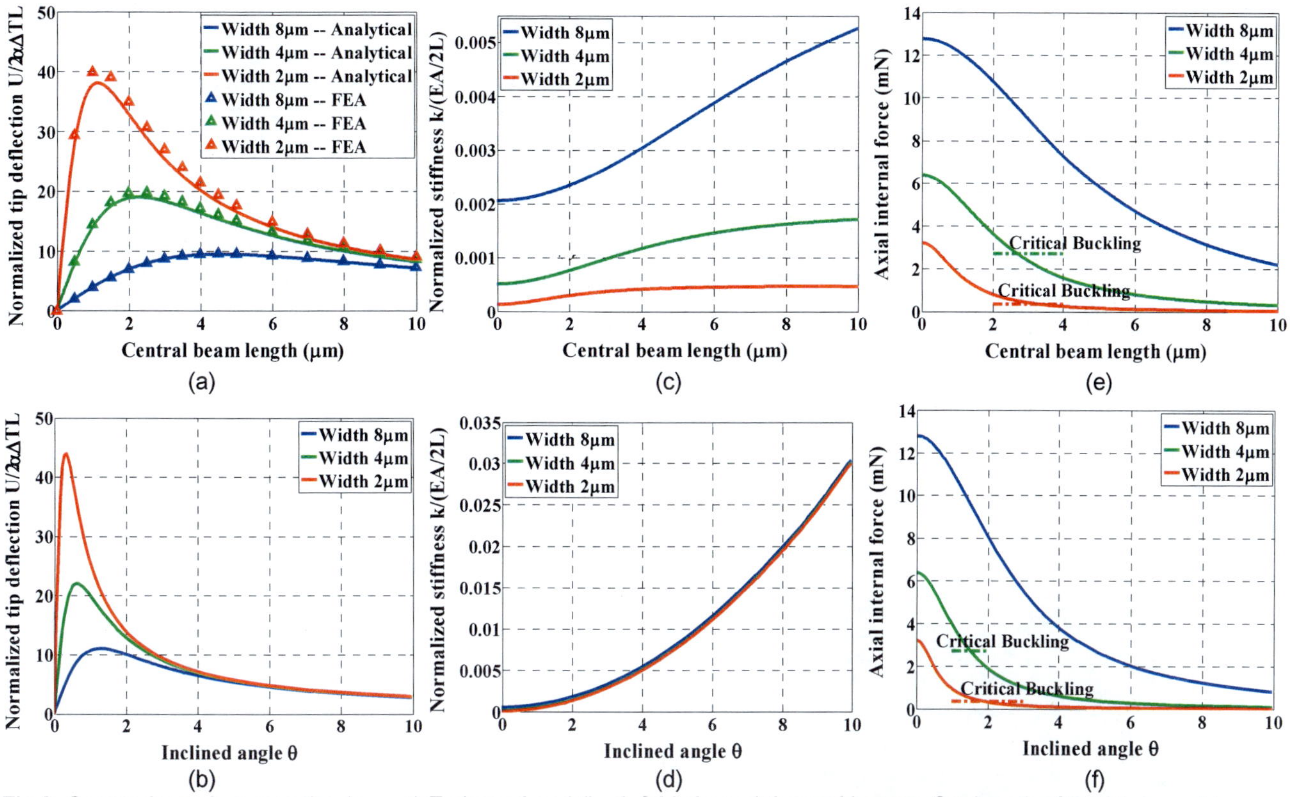

Fig 3. Comparisons between the (a, c, e) Z-shaped and (b, d, f) V-shaped thermal beams. Set length of the V-shaped beam equal to 2*L* as in the Z-shaped beam. For given beam widths, the variable for Z-shaped beam is the central beam length, while for V-shaped beam is the inclined angle. (a) and (b) Dimensionless displacement, in which L = 88 μm and $2\alpha\Delta TL$ = 0.176 μm; (c) and (d) Dimensionless stiffness, in which $EA/2L$ = 7.27×10^{4} N/m (for beam width of 8 μm); (e) and (f) internal axial force with critical buckling criterion.

C. *Multiphysics FEA Simulation*

Thermomechanical finite element analysis (FEA) was performed to verify the mathematical modeling. A 2D multi-field plane element PLANE223 was used in the FEA (ANSYS v11.0) simulation, which involves electric, thermal and mechanical fields. A constant temperature increase of 400 K was applied to the entire Z-shaped beam as shown in Fig. 2. Note that, for verification of the mathematical model, all thermal properties of single crystalline silicon (SCS) are constants at room temperature; for comparison with the experimental results, however, all thermal properties of SCS are temperature dependent as listed in Table 1 in the end of the paper.

D. *Comparison between Z-shape and V-shape*

A systematic comparison between the Z-shaped thermal actuators and the V-shaped thermal actuators was carried out to further illustrate their characteristics, as shown in Fig. 3. Fig. 3(a) shows an excellent agreement for displacements between the FEA and analytical solution, which confirms the validity of the analytical model. Figures 3(a) and (b) together show that, for both actuators, the smaller the beam width, the larger the displacement. However, fabricating a small beam width (especially ≤2 mm) is more challenging for inclined beams (V-shape). In this regard, the Z-shaped actuators can achieve relatively larger displacement.

Fig. 3(c) and (d) show that the stiffness of the Z-shaped actuators is about one order of magnitude smaller than that of the V-shaped actuators, when the beam width ranges from 2 to 8 μm. The stiffness of V-shaped actuators does not change with the beam width; by contrast, that of Z-shaped scales approximately with square of beam width, because V-shaped actuators are mainly based on beam extension while Z-shaped actuators are mainly on beam bending. Since the stiffness of bending beam is proportional to the cube of the beam width, Z-shaped actuators possess a large stiffness range for different beam width For some applications that requires simultaneous sensing and actuating functions, Z-shaped thermal actuator alone could also be consider as senor at the same time, while V-shaped thermal actuator should combine with a certain type of sensor.

Fig. 3(e) and (f) show that both actuators share the same column effective length factor, which means they have the same critical buckling force and possess similar level of stability. The output force is in the range of 30 to 490 μN, calculated by $f=kU$. Apparently, the output force of the Z-shaped actuators is smaller than that of the V-shaped thermal actuators [6,11].

Quasi-static Results

The Z-shaped thermal actuators were fabricated by the SOI-MUMPs process in run 27. All the Z-shaped thermal devices in our design have the same anchor-anchor distance (412μm); central beam length and width are two design parameters. Three arrays of Z-shaped thermal actuators with different beam widths (2 μm, 4 μm and 8 μm) were fabricated for parametric study and device optimization. Within each array, the length of the central beam varies from 1 to 20 μm. One such array of the Z-shaped thermal actuators is shown in Fig.4. The displacement was measured using an optical microscope with the edge detection method [16]. Resolution is calibrated as 81.5 nm/pixel. Fig.5 shows the displacement of the Z-shaped thermal actuators of two arrays. The array with 2 μm and 8 μm beam width were actuated under 2 V and 3 V, respectively. Multiphysics FEA results for both arrays with the corresponding applied voltages are also plotted in the Figure. The displacement results agreed quite well between experiments and FEA, which are also in line with the analytical modeling as shown in Figure 3(a).

Fig.4 An array of Z-shaped thermal actuator with the same beam width of 4 mm.

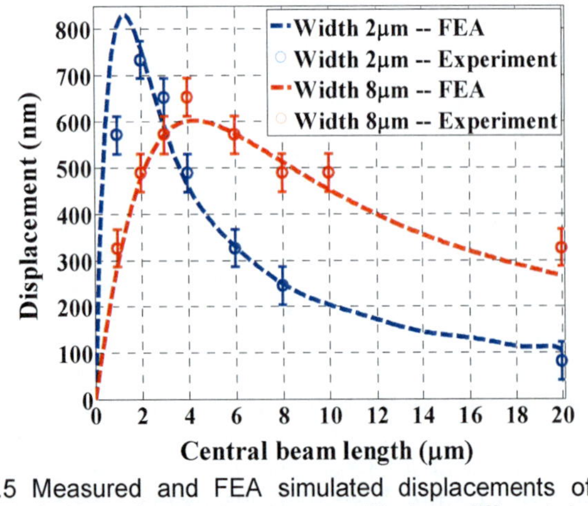

Fig.5 Measured and FEA simulated displacements of Z-shaped thermal actuator arrays with two different beam widths (2 mm and 8 mm). The actuation voltage on the array with 2 μm beam width is 2 V and that on the array with 8 mm beam width is 3 V.

The following tests were carried out inside an SEM (JEOL 6400F) on a particular Z-shaped thermal actuator with 4μm beam width and 20 μm central beam length. Displacement was measured with the actuation voltage from 0 to 6 V. The measured displacement is plotted with respect to the input current, as shown in Fig. 6(a). The stiffness of the structure was calculated to be 273.4 N/m based on the measured dimensions. Figure 6(b) shows the dependence of electric resistance of the structure on the input current. Assuming the linear dependency of resistivity on temperature as listed in Table 1, the average temperature in the device was estimated [8,9], and also plotted in Fig.6(b).

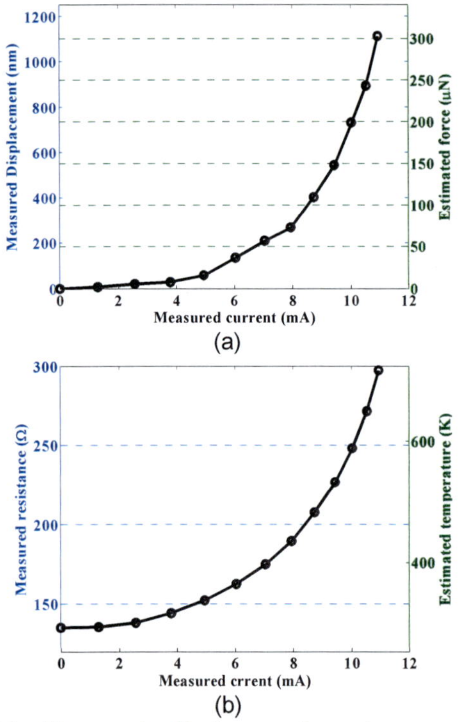

Fig.6 (a) Measured displacement and corresponding (calculated) force as functions of input current. (b) Measured resistance and corresponding (estimated) average temperature change as functions of input current.

Conclusions

A new electrothermal actuator with symmetric Z-shaped beams was developed in this paper. Compared to V-shaped one, it offers some unique advantages such as compatibility with anisotropic etching, larger displacement, and smaller stiffness and output force. The variable stiffness and force of Z-shaped actuators fill the gap between those of the comb drives and V-shaped thermal actuators. Additionally a Z-shaped actuator with smaller stiffness could be used as a simultaneous load sensor. Among all of the design parameters, the length of the central beam and beam width were identified as the major ones in tuning the device features. The quasi-static experimental measurements of three arrays of Z-shaped thermal actuators agreed well with the FEA predictions.

References

1. Tang W.C., Nguyen T.C.H., Judy M.W. and Howe R.T., "Electrostatic-comb drive of lateral polysilicon resonators", *Sensors and Actuators A* **21,** 328-31, (1990).
2. Guckel H., Klein J., Christen T., Skrobis K., Landon M. and Lovell E.G., "Thermo-magnetic metal flexure actuators", *Tech. Digest IEEE Solid State Sensor and Actuator Workshop* 73–5, (1992).
3. Comtois J.H., Bright V.M. and Phipps M.W., "Thermal microactuators for surface-micromachining processes", *Proc. SPIE* 2642 10, (1995).
4. Geisberger A.A, Sarkar N., Ellis M. and Skidmore G., "Electrothermal properties and modeling of polysilicon microthermal actuators", *Journal of Microelectromech. System* **12** 513–23, (2003).
5. Legtenberg R., Groeneveld A.W. and Elwenspoek M., "Comb-drive actuators for large displacements", *Journal of Micromechanics and Microengineering* 6 320-9, (1996).
6. Que L., Park J.S. and Gianchandani Y.B., "Bent-beam electrothermal actuators—Part I: Single beam and cascaded devices", *Journal of Microelectromechanical System* **10** 247–54, (2001).
7. Park J.S., Chu L.L., Oliver A.D. and Gianchandani Y.B., "Bent-beam electrothermal actuators—Part II: linear and rotary microengines", *Journal of Microelectromechanical System* **10** 255–62, (2001).
8. Huang Q. and Lee N., "Analysis and design of polysilicon thermal flexure actuator" *Journal of Micromechanics and Microengineering* **9** 64–70, (1999).
9. Hickey R., Kujath M. and Hubbard T., "Heat transfer analysis and optimization of two-beam microelectromechanical thermal actuators", *Journal of Vacuum Science and Technology A* **3** 971-4, (2003).
10. Mankame N.D. and Ananthasuresh G.K., "Comprehensive thermal modelling and characterization of an electro-thermal-compliant microactuator", *Journal of Micromechanics and Microengineering* **11** 452-62, (2001).
11. Zhu Y., Corigliano A. and Espinosa H.D., "A thermal actuator for nanoscale in situ microscopy testing: design and characterization", *Journal of Micromechanics and Microengineering* **16** 242-53, (2008).
12. Zhu Y. and Espinosa H.D. "An electro-mechanical material testing system for in-situ electron microscopy and applications", *Proceedings of the National Academy of Sciences* USA **102** 14503-8, (2008).
13. Lott C.D, McLain T.W, Harbb J.N and Howell L.L, "Modeling the thermal behavior of a surface-micromachined linear-displacement thermomechanical microactuator" *Sensors and Actuators A* **101** 239-50, (2002).
14. Chu L.L. and Gianchandani Y.B., "A micromachined 2D positioner with electrothermal actuation and sub-nanometer capacitive sensing", *Journal of Micromechanics and Microengineering* **13** 279–85, (2003).
15. Dong J. and Ferreira P.M., "Simultaneous actuation and displacement sensing for electrostatic drives", *Journal of Micromechanics and Microengineering* **18** 035011(10pp), (2008).
16. Zhu Y., Moldovan N. and Espinosa H.D., "A microelectromechanical load sensor used for in-situ electron and x-ray microscopy tensile testing of nanostructures", *Applied Physics Letter* **86**, 013506(3pp), (2005).
17. Li L. and Uttamchandani D., "Dynamic response modelling and characterization of a vertical electrothermal actuator", *Journal of Micromechanics and Microengineering.* **19** 075014(9pp), (2009).

Table 1. Silicon properties used in simulations of Z-shaped thermal actuators

Material properties	Unit	Value	Reference
Young's modulus	GPa	160	[17]
Poisson's ratio	-	0.28	[17]
Thermal conductivity (constant)	$Wm^{-1}K^{-1}$	146	[4]
Thermal conductivity (temperature dependent)	$Wm^{-1}K^{-1}$	$k_t(T)=210658T^{-1.2747}$	[17]
Resistivity (constant)	Ωm	5.1×10^{-5}	Measured
Resistivity (temperature dependent)	Ωm	$\rho(T)=5.1\times10^{-5}[1+3\times10^{-3}(T-273)]$	[17]
Thermal expansion coefficient (constant)	K^{-1}	2.5×10^{-6}	[4]
Thermal expansion coefficient (temperature dependent)	K^{-1}	$\alpha(T)=-4\times10^{-12}T^2+8\times10^{-9}T+4\times10^{-7}$	[4]

Proceedings of the SEM Annual Conference
June 7-10, 2010 Indianapolis, Indiana USA

Electrothermal Actuators for Integrated MEMS Safe and Arming Devices

Robert A. Lake, LaVern A. Starman and Ronald A. Coutu Jr.,
Air Force Institute of Technology
Wright Patterson Air Force Base, OH, USA
Ronald.Coutu@afit.edu

***Abstract* - The use of electrothermal actuators to achieve the necessary motion of a MEMS based safe and arming device was thoroughly explored. Multiple variants of thermal actuators were designed, modeled, fabricated, and tested in order to gain a better understanding of their specific characteristics. Design variations included both single and double hot arm actuators as well as bent beam thermal actuators. Studies were performed to analyze and compare the displacement and output force of these actuators both as standalone devices as well as multiple actuators joined together. Detailed analysis of the results of the modeling and testing demonstrated the advantages and disadvantages of each style of thermal actuator. Furthermore, the specific variant of electrothermal actuator that is best suited for implementation into MEMS based safe and arming devices can be effectively determined. Finally, detailed analysis of the performance of electrothermal actuators integrated into a functioning MEMS safe and arm device will be presented. Methods in which these actuators are incorporated to best take advantage of their particular characteristics is shown as well as methods that were incorporated in order to overcome some of the shortcomings inherent with these actuators in order to provide the overall safe and arming device with reliable and efficient performance.**

I. INTRODUCTION

In all forms of munitions, the fuze is one of the most critical components. The recent trends in weapon design to make smaller, more sophisticated munitions coupled with the fact that the fuze industry has diminished greatly since the Cold War has resulted in a growing interest in the miniaturization of the fuze components, particularly the safe and arming feature.

Smart munitions, specifically air dropped precision guided munitions (PGM's) are the weapon of choice for the immediate future [1]. The increasing need for smarter weapons combined with the current trend towards smaller munitions creates the problem of decreasing room within the weapon for all the necessary components such as the explosive, guidance and navigation system, and the fuze.

This problem leaves manufacturers with trade-offs that must be considered in order to produce accurate and effective weapons. In the case of a larger weapon with much higher explosive yield, precision accuracy is not as critical of a consideration to a manufacturer because even a miss, within the general vicinity of the intended target will still likely destroy it. In urban settings such as Iraq where we find many of our modern day conflicts taking place, smaller weapons with a much lower explosive yield are more desirable in order to reduce unwanted collateral damage. The drawback being that these smaller weapons have less chance of successfully destroying the target in the event of a near miss, thus making the need for very accurate guidance systems of greater importance.

II. SAFE AND ARMING DEVICES

The way in which many modern fuzes operate is on the principle of an explosive train (Figure 1.) The way in which these work is that when certain conditions such as altitude or target proximity have been met, a small, highly sensitive initiator is triggered. This initiator then causes the detonation of an additional charge of higher output energy, which in turn detonates an even more powerful explosive. The reason for this is primarily safety. The main charge has very large explosive energy, but has very low sensitivity. In other words, it requires significant input energy in order to detonate. Additionally, while the initiator is much more sensitive than the main charge, its output energy is low enough so that it may be blocked by a safe and arming device.

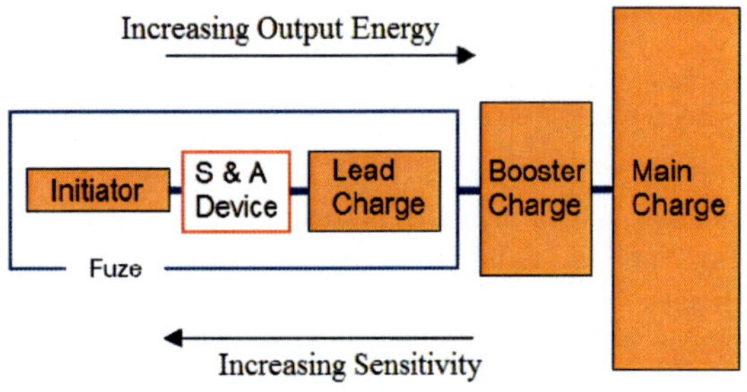

Figure 1: Illustration of the explosive train of a fuze. The initiator is the first stage in the train, having the highest sensitivity and lowest output energy. The detonation of this initiator leads to the detonation of subsequent charges which result in the final detonation of the main charge [2].

T. Proulx (ed.), *MEMS and Nanotechnology, Volume 2*, Conference Proceedings of the Society for Experimental Mechanics Series 2,
DOI 10.1007/978-1-4419-8825-6_31, © The Society for Experimental Mechanics, Inc. 2011

The safe and arming device works on the principle of an interrupted explosive train. Simply put, a mechanical barrier exists between the initiator and the subsequent charge in the explosive train. The safe and arming device of the fuze is generally the largest component of a fuze and is mechanical in nature [3]. As a result, it is an obvious target for miniaturization with MEMS components.

Since the device is mechanical in nature, and requires the implementation of a movable barrier, some sort of actuation method is required in the design. Additionally, because of the nature of the device, mechanical safety locks must also be incorporated. According to the Department of Defense's Safety Criteria for Fuze Design, MIL-STD-1316E [4], S&A device must contain two individual safety mechanisms.

III. Fabrication

The devices were fabricated by two different methods, PolyMUMPS and SUMMiT V. Both of these processes use deposition and selective etching of layers of polysilicon and silicon oxide in order to create MEMS structures.

PolyMUMPS, (Polysilicon Multi User MEMS Process,) is a MEMS fabrication process that allows for up to two releasable, structural layers to be incorporated into a design [7]. The advantages of this method are that it is simple, has a rapid turn-around time (two months,) is relatively inexpensive, and allows for a certain degree of flexibility with its design rules. Some disadvantages of this process include large minimum feature and spacing size requirements, the designer is restricted to two structural layers, and all layers are conformal to the underlying topology. A cross sectional view of all the layers of the PolyMUMPS process is illustrated in Figure 2. Table 1 provides a description of each of these layers.

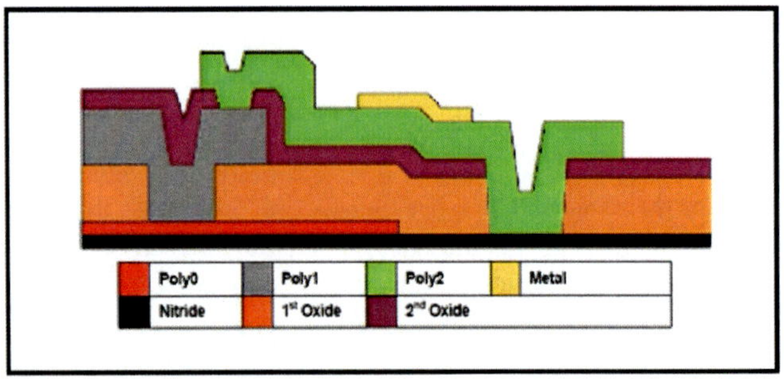

Figure 2: Cross sectional view of all seven layers of the PolyMUMPS process (Not to scale.) The two oxide layers are sacrificial layers, and the nitride, or isolation layer, is the surface which provides the foundation for the device. It is grown on a <100> silicon wafer [7].

SUMMiT V, (Sandia Ultra-planar Multi-level MEMS Technology,) is a MEMS fabrication process invented at Sandia National Laboratories (Figure 3.) It offers several advantages over the PolyMUMPS process such as tighter tolerances, smaller feature size and spacing, (typically one half to one micron,) and an additional two releasable structural

Table 1: PolyMUMPS process layer break down [7].

Material Layer	Layer Thickness	Layer Description
Nitride	0.6 μm	Provides electrical isolation from wafer
Poly0	0.5 μm	Electrical layer for ground plane
1st Oxide	2.0 μm	First sacrificial layer providing gap between Poly0 and Poly1
Poly1	2.0 μm	First mechanical layer
2nd Oxide	0.75 μm	Second sacrificial layer providing gap between Poly1 and Poly2
Poly2	1.5 μm	Second mechanical layer
Metal	0.5 μm	Electrical contact layer or optically reflective layer

layers which allow for more complex designs that would not be possible to fabricate using the PolyMUMPS process [8]. In addition, the two top layers of this process are planarized by using chemical mechanical polishing. This removes all the conformality of its underlying layres which provides a tremendous advantage when creating structures on top of structures such as sliding plates. If the two plates are completely flat, there is no risk of features getting caught on each other as they would if the plates were conformal to everything underneath it as with PolyMUMPS. Additional benefits include the *in situ* doping of its polysilicon layers and the lack of a metal layer, which result in minimal residual stresses in the devices upon release.

There are however some disadvantates of the SUMMiT process. For example, the turn around time on a SUMMiT run is typically four to five months, and is considerably more expensive than a PolyMUMPS die site. Also, because o the added complexity of designs possible with the SUMMiT V process, the layout and design requires considerably more effort and meticulous care, especially when it is taken into accound that the SUMMiT V process allows for only mimial flexibility when it comes to its design rules.

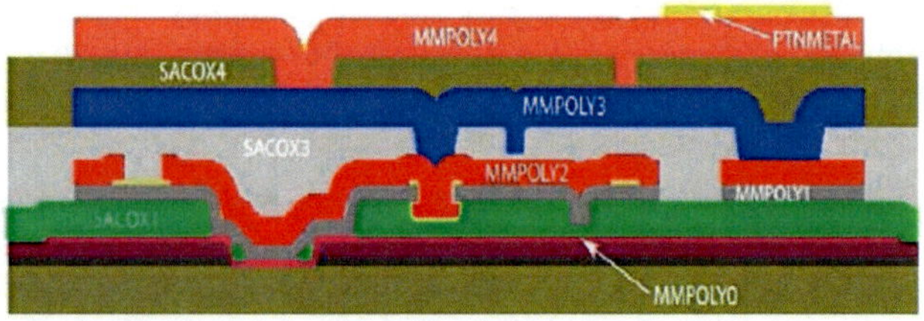

Figure 3: Cross sectional view of the layers used in the SUMMiT V fabrication process (Not to scale.) SACOX1 through SACOX4 are all sacrificial oxide layers. Note the planarization of the top two layers.

IV. Electrothermal Actuators

Electrothermal actuation takes advantage of the phenomena of thermal expansion of materials. When materials heat up, they expand. The coefficient of thermal expansion is a property of any given material which dictates how much that material will expand relative to a change in its temperature as illustrated by Figure 4.

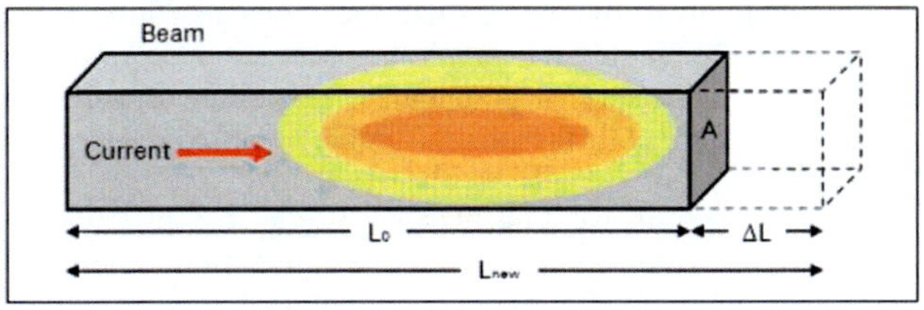

Figure 4: Illustration of thermal expansion. As the beam of length L_0 is heated, in this case by electrical current, it expands by a displacement of ΔL [2].

The new length of the beam, L_{new}, is given as

$$L_{new} = L_o + \alpha_L[L_o(T_{avg} - T_o)] \quad (1)$$

where L_0 is the original length of the beam, α_L is the coefficient of thermal expansion, T_{avg} is the average temperature of the beam after the expansion, and T_0 is the initial temperature of the beam [2].

The MEMS based electrothermal actuators operated based on the asymmetrical thermal expansion of two arms made out of the same conductive material [5]. One arm known as the cold arm is very wide (20 μm) in comparison to the other arm (3 μm) known as the hot arm as illustrated in Figure 5. The wider cold arm has a much lower resistance than the narrower hot arm due to its larger cross sectional area in correspondence with Equation 2.

$$R = \frac{\rho l}{A}\ (\Omega), \quad (2)$$

where R is the resistance, ρ is the resistivity of the material, l is the length of the arm, and A is the cross sectional area. The higher resistance of the hot arm will result in a larger temperature increase localized in the hot arm relative to that of the cold arm because of a higher power dissipation according to Equation 3.

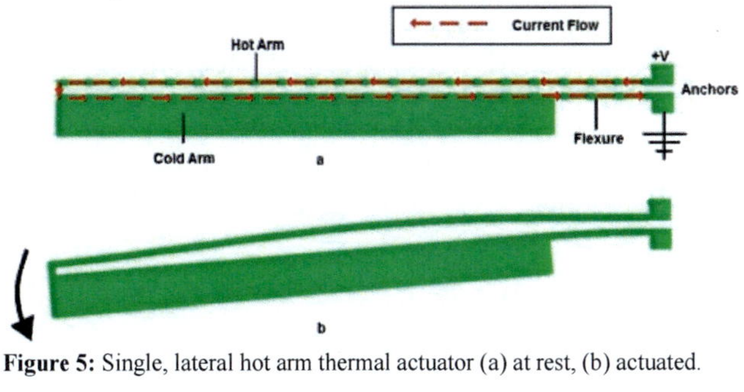

Figure 5: Single, lateral hot arm thermal actuator (a) at rest, (b) actuated.

$$P = I^2 R\ (\text{W}), \quad (3)$$

where P is the dissipated power, I is the current, and R is the resistance of the arm. As the temperature of the hot arm increases, it causes it to expand more when compared to the cold arm. As a result, it tends to push the tip of the actuator in an arcing motion in the direction of the cold arm as illustrated in Figure 3(b).

A different configuration known as a double hot arm thermal actuator works on the same principles as the single hot arm actuator as illustrated by Figure 6. The major difference is that instead of one narrow hot arm, there are two narrow hot arms in which the current is confined. Like the single hot arm thermal actuator, the two hot arms expand much more relative to that of the cold arm, which in this case undergoes minimal thermal expansion, causing the actuator to deflect in the direction of the cold arm.

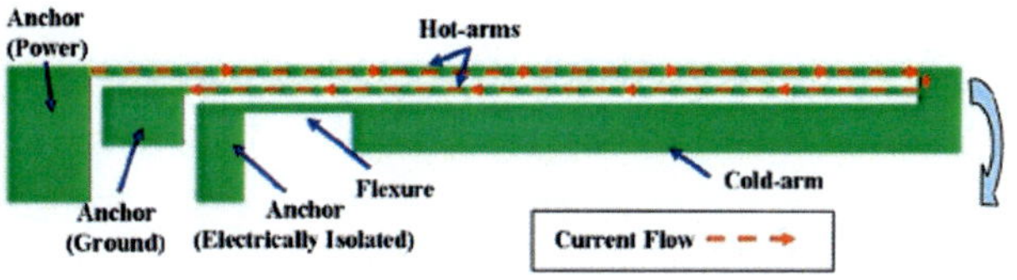

Figure 6: Diagram of the major components of a double hot arm thermal actuator.

An additional configuration arises when multiple thermal actuators are grouped together. This can be accomplished by connecting the tips of the actuators together with a coupling yoke as seen in Figure 7.

V. FEM Modeling

Multiple configurations of single and double hot arm thermal actuators as well as multiple configurations of grouped thermal actuators were modeled for analysis using the Coventorware 2008 software suite. These models were built upon the PolyMUMPS fabrication process. Each model was built to represent the stacked layers of the 2.0 μm thick Poly1 and the 1.5 μm thick Poly2 polysilicon layers deposited in this fabrication process. A more detailed explanation of this fabrication process is provided in the following section.

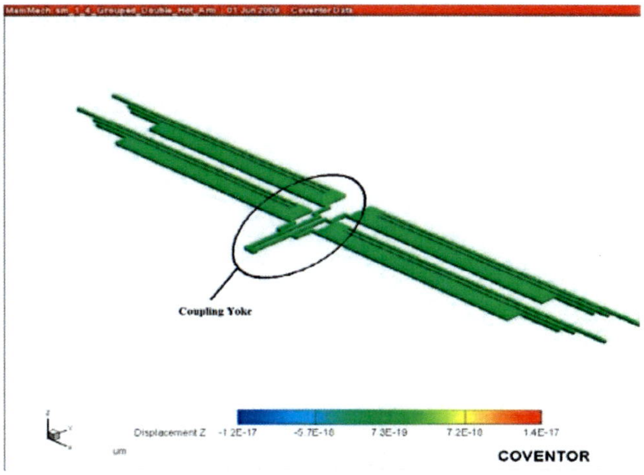

Figure 7: An illustration of an array of four thermal actuators connected together at their tips with a coupling yoke.

Three models of single hot arm actuators were analyzed as well as three models of double hot arm actuators. In each case, some parameters were the same. For example, all the cold arms are 20 μm wide, all of the hot arms are 3 μm wide and each flexure is 50 μm long. The remaining variable parameters are summarized in Table 2 and Table 3.

Table 2: Modeled length dimensions of three single hot arm thermal actuators

	Device A	Device B	Device C
Cold Arm Length	250 μm	300 μm	350 μm
Hot Arm Length	300 μm	350 μm	400 μm

Table 3: Modeled length dimensions of three double hot-arm thermal actuators

	Device D	Device E	Device F
Cold Arm Length	250 μm	300 μm	350 μm
Inner Hot Arm Length	313 μm	363 μm	413 μm
Outer Hot Arm Length	346 μm	396 μm	446 μm

In addition to these six models of varying sizes, four models consisting of thermal actuators all of the same size in groups of two, four, six, and eight were analyzed.

For each of the models, a varying potential from zero to twelve volts was applied at the input terminals. In the case of the single hot arm actuators, the potential was applied between the end of the hot arm, and the end of the cold arm as shown in Figure 5(a). For the double hot arm actuators, the potential was applied between the two hot arms, with the end of the cold arm left electrically isolated as shown in Figure 6.

The characteristics of particular interest for this analysis were the displacement and maximum output force of the actuators. The results indicate that as the applied potential increases, the displacement of the actuators increases. It is also revealed that the longer the thermal actuator is, the further it deflects for the same applied power for both the single and double hot arm actuators.

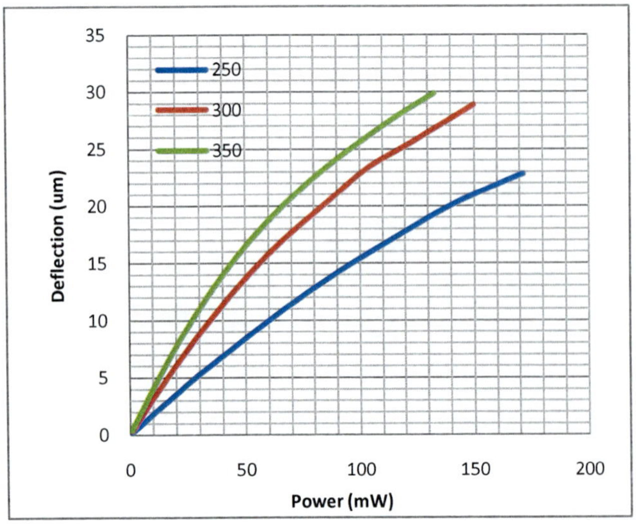

Figure 8: Graph illustrating the displacement versus applied power of a 250, 300, and 350 μm thermal actuator. It can be seen that the longer thermal actuators deflect further for the same applied power.

A comparison of the displacement versus the applied voltage for the single and double hot arm actuators indicates that for the same applied potential, a single hot arm thermal actuator will deflect further than a double hot arm actuator with the same cold arm length as illustrated by Figure 9. These results are deceiving at first glance. While it is true that the single hot arms deflect further for the same applied potential, when it is taken into consideration that at the same applied potential, more power is dissipated by the single hot arm thermal actuator due to the lower resistance encountered by the current path. When a comparison is made between the single and double hot arm actuators in terms of displacement versus dissipated power, it can be seen that the single hot arm actuator and double hot arm actuator of the same cold arm length deflect approximately the same distance for the same applied power as illustrated in Figure 10.

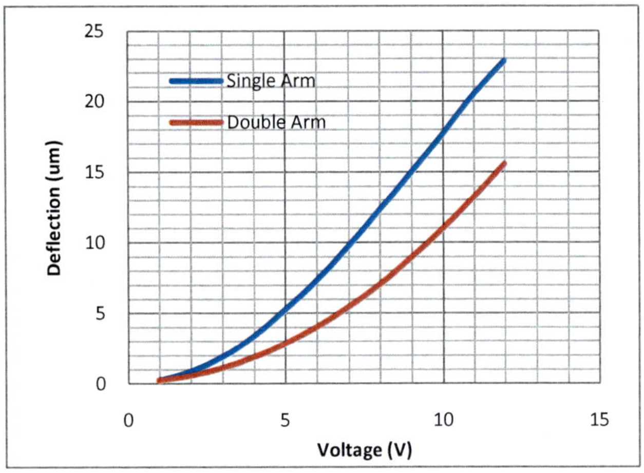

Figure 9: Graph illustrating the FEM modeling results of the deflection of a single hot arm and double hot arm thermal actuator versus the applied potential

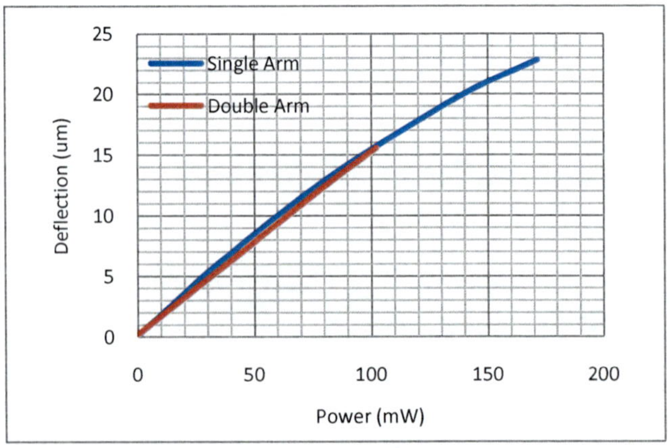

Figure 10: Graph illustrating the FEM modeling results of the deflection versus dissipated power of a single hot arm and double hot arm thermal actuator.

An analysis of the grouped thermal actuators indicates that regardless of the number of actuators that are grouped together, the displacement remains close to constant as illustrated in Figure 11.

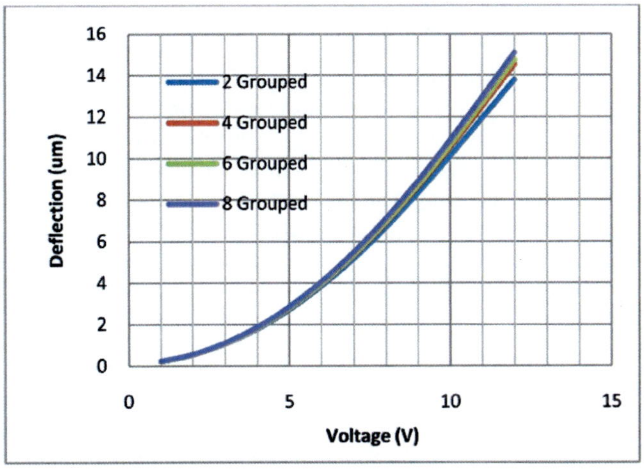

Figure 11: Graph illustrating the grouped actuator deflection for arrays of two, four, six, and eight actuators. Note that the deflection remains very close to constant regardless of how many actuators are coupled together.

In order to determine the force exerted by the actuators, a special configuration is used as shown in Figure 12. This configuration consists of a thermal actuator whose tip contacts a beam which is anchored at its other end. By knowing the parameters of this beam, treating it as a fixed-free beam and then measuring how far the beam tip is displaced by the actuator, the maximum force generated by the thermal actuator can be determined from Equation 4 [6].

$$F = \frac{Etw^3}{4l^3} x \text{ (N)}, \tag{4}$$

where F is the generated force, E is the Young's Modulus of the material, t is the thickness of the beam, w is the width of the beam, l is the length of the beam, and x is the distance that the beam is displaced.

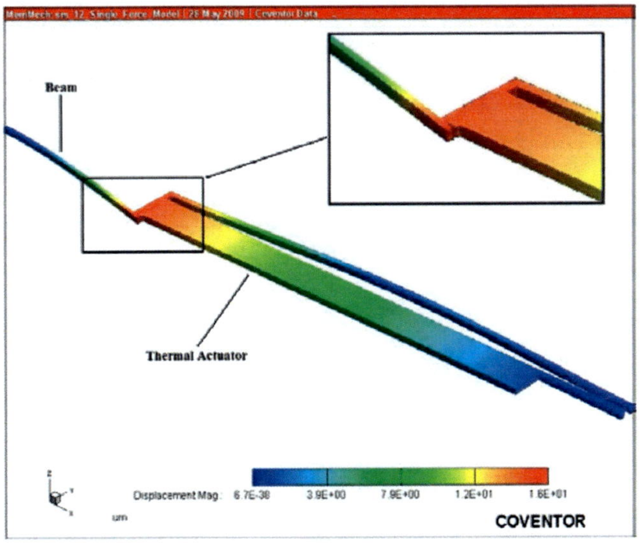

Figure 12: Special configuration used to measure the maximum output force of a thermal actuator. The tip of the actuator contacts the tip of a beam. By measuring the lateral displacement of the beam, the applied force can be determined.

The modeling results of the force characteristics of both the single and double hot arm thermal actuators demonstrate several interesting points that must be taken into consideration when designing devices with these actuators.

For the single hot arm thermal actuators, the results of the force analysis indicate that as the applied power to the actuator increases, the maximum output force generated by the actuator increase. More interestingly, there is a cross over point for the three different lengths of actuators as seen in Figure 13. At lower applied power, the longer actuators are capable of generating a higher output force, but as the applied power is increased, a point is reached where the shorter actuators are capable of generating a higher output force. This can be attributed to the fact that as the actuators deflect, they move in an arcing path. The longer ones will naturally arc more than the shorter actuators, resulting in more of the force being directed in another direction, and only a portion of its total output force being directed in the desired direction.

These results reveal a trade-off that must be considered when designing with thermal actuators. In cases where larger deflections are desired, and force is not as important of a factor, the longer actuator would be ideal. In cases where larger forces are required, it is important to consider the input power that will be available and how much deflection is required in order to choose which actuator is best suited.

Analysis of the double hot arm thermal actuators yields similar results. While the crossover points are not as pronounced, it can be seen that at lower applied powers, the output forces are very similar, regardless of actuator size. As the applied power increases, the output force of all the actuators increase and the shorter actuators are capable of generating higher output forces than the shorter actuators.

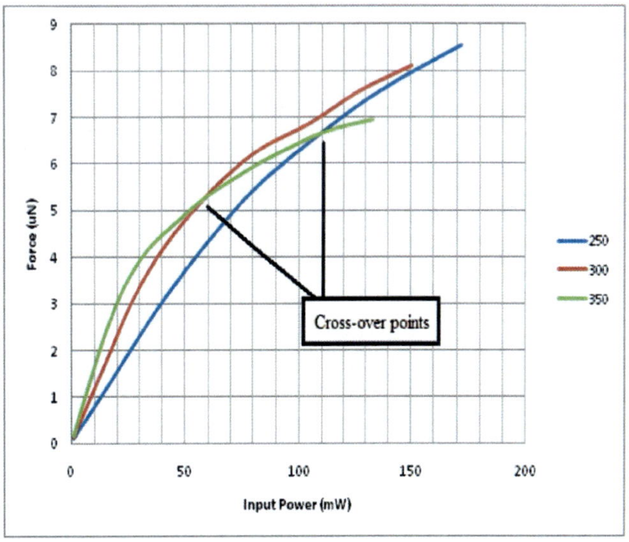

Figure 13: Graph illustrating the FEM modeling results of the output versus applied power of a 250 μm, 300 μm, and 350 μm single hot arm thermal actuator. Note the cross over points at which the shorter actuators begin to generate a higher output force.

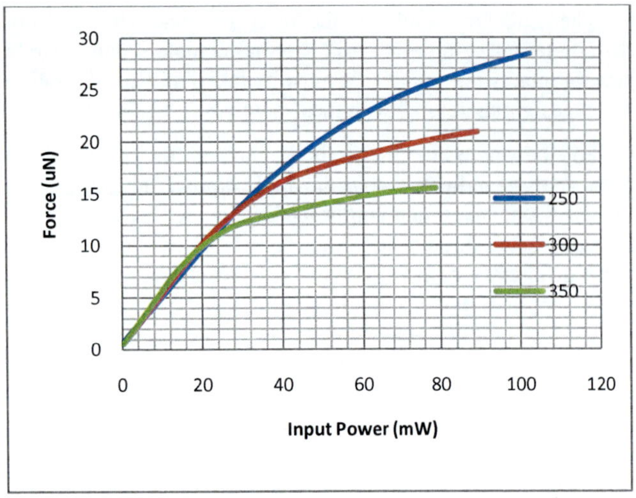

Figure 14: Graph illustrating the FEM modeling results of the output force versus applied power of a 250 μm, 300 μm , and 350 μm double hot arm thermal actuator.

A comparison of the force versus applied power characteristics of the single and double hot arm indicates that the double hot arm thermal actuators posses much greater force generating capabilities illustrated by Figure 15. This clearly demonstrates that for situations in which a higher output force is required, the double hot arm thermal actuators are the more suitable choice.

Modeling results of grouped thermal actuators demonstrate two characteristics that are important to consider when designing with them. First, it can be seen that when actuators are grouped together, their deflection is not affected by the number of actuators in the group (Figure 14.) This is a logical result in that all the actuators grouped together are the same size, and as a result would have the same deflection characteristics, and therefore not hinder or enhance each other when grouped together.

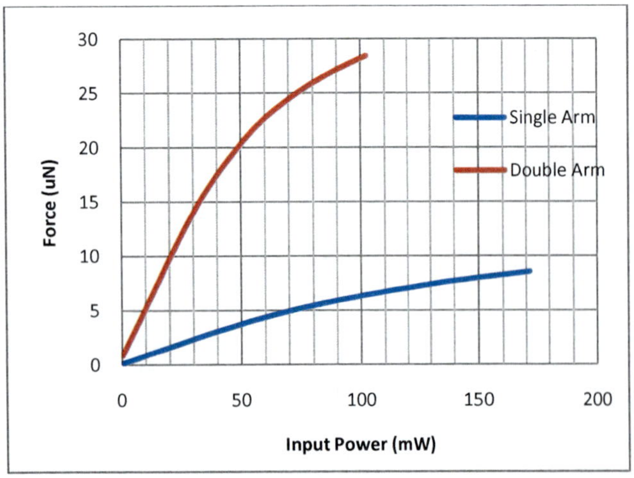

Figure 15: Graph illustrating the FEM modeling results of the output force versus dissipated power of a single hot arm thermal actuator and a double hot arm thermal actuator.

Results of the force analysis for grouped thermal actuators reveal that there is a linear dependence of the output force on the number of thermal actuators that are grouped together as seen in Figure 17. It is important to note that the increase of power consumption of these arrayed devices is directly proportional to the number of thermal actuators in the array. For example, an array with eight actuators will consume eight times the power of a single thermal actuator of the same configuration and dimensions. In situations where power consumption is limited, this would be a critical design parameter.

The results of modeling have provided important characteristics of different configurations of thermal actuators. With a better understanding of how they work and their performance characteristics, a choice as to which configurations are best suited for a particular application can more easily and accurately be made.

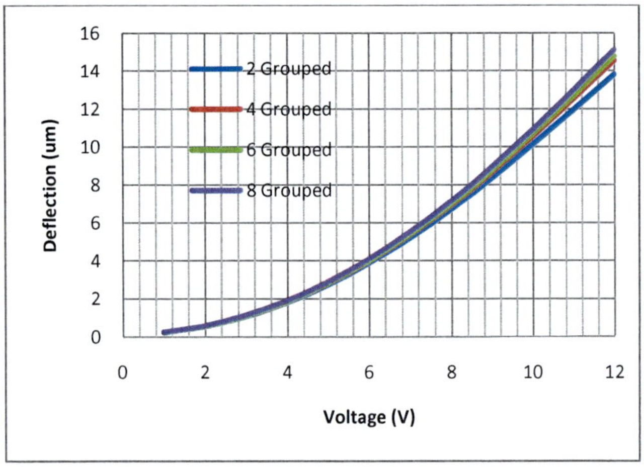

Figure 16: Graph illustrating the grouped actuator deflection for arrays of two, four, six, and eight actuators. Note that the deflection remains very close to constant regardless of how many actuators are coupled together.

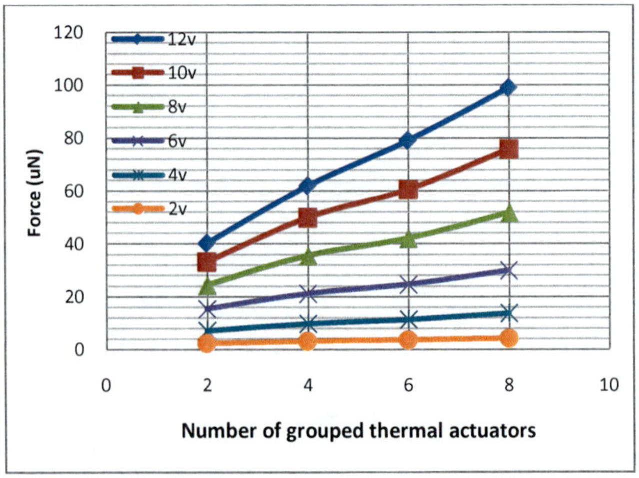

Figure 17: Graph illustrating the linear dependence of the output force upon the number of thermal actuators within the array operating from 2 – 12 volts.

VI. Integration and Operation

Thermal actuators are used exclusively in the S&A device as it is currently designed. Some considerations must be taken into account to effectively integrate these actuators. The moving barriers that interrupt the explosive are formed by rotating wheels. In order for the device to be placed from a safe state into an armed state, the outer edge of the wheel must be driven approximately 785 μm. This length is far beyond the maximum displacement capability of any electrothermal actuator.

In order to convert the relatively small displacement of the thermal actuators into a larger usable displacement, a configuration known as a pawl is incorporated. The way in which this works is that two banks of actuators are driven perpendicularly to one another, ninety degrees out of phase. The first actuator array forces a drive head into contact with the object to be moved (Figure 18(a)). The second array then actuates, pushing the object in the same direction (Figure 18(b)). Once this movement has occurred, the first actuator array disengages, breaking the contact between the drive head and the moving object (Figure 18(c)). Finally, the second actuator array is allowed to return to its original position (Figure 18(d)). This process is repeated rapidly until the barrier has been positioned into an armed mode.

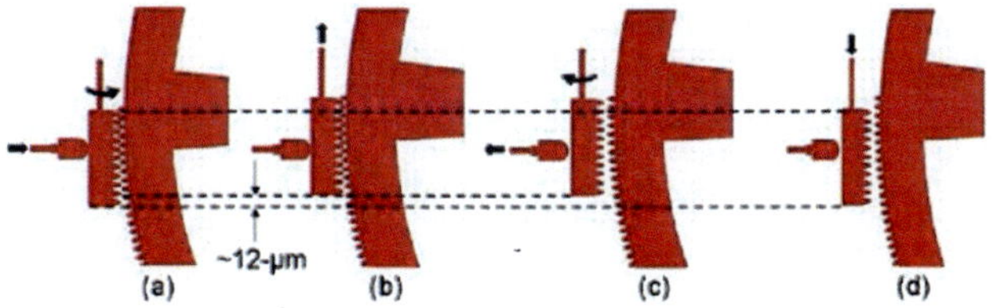

Figure 18: Pawl Operation: (a) Drive head is forced into contact with the wheel. (b) The drive head is pulled by the second array of actuators. (c) The drive head is disengaged from the wheel. (d) The drive head is returned to its original position [3].

The time to open the wheel barrier can easily be predicted. The radius of the wheel is approximately 500 μm, which results in a distance to be traveled at the outer radius of the wheel to

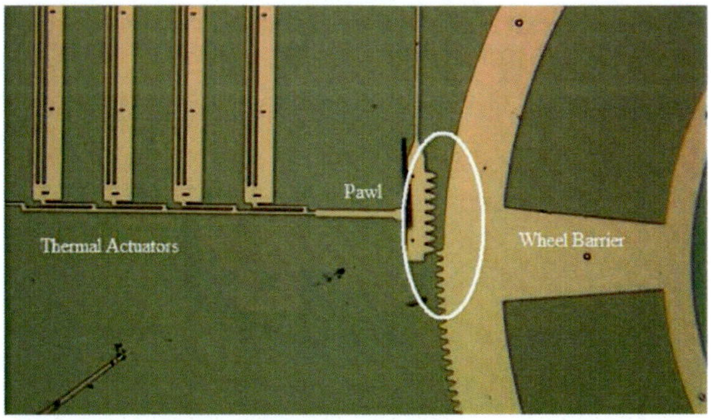

Figure 19 10x Magnification of a pawl mechanism incorporated to rotate the wheel barrier of a MEMS based S&A device.

rotate one quarter turn to be placed in the armed mode. The distance between the teeth on the wheel is approximately 10 μm. Because the thermal actuators can produce a reliable displacement of 10 to 15 μm, it can be assumed that only one tooth is engaged per cycle, giving the wheel 10 μm of travel per cycle.

The opening time of the wheel can thus be predicted by Equation 5.

$$T_{open} = \frac{78.5}{f_{eff}}, \tag{5}$$

where T_{open} is the time for the wheel to travel one quarter turn and f_{eff} is the effective frequency of the drive signal in hertz. Because the drive signal is two pulse trains 90 degrees out of phase as illustrated by Figure 20, the frequency at which the wheel is driven is not the same frequency that is applied to each individual thermal actuator array. The effective period of the composite drive signal is from the beginning of the leading signal to the end of the lagging signal, essentially adding one quarter of an individual period to the effective period. The effective frequency is given as

$$f_{eff} = \frac{f}{1.25}, \tag{6}$$

where f is the frequency of the individual drive signals.

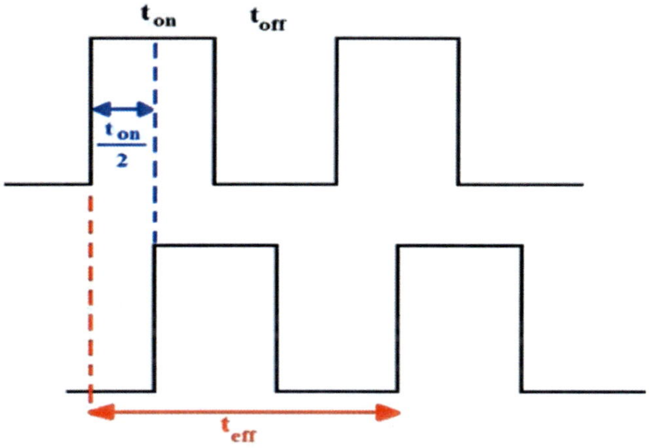

Figure 20: Timing diagram of the drive signals to drive the MEMS devices.

Table 3 provides a comparison between predicted opening times and measured opening times. The discrepancies between the predicted and measured opening times can be attributed to the fact that the pawl head may intermittently engage two teeth instead of one per cycle, causing it to open faster than predicted.

Table 4: Comparison between predicted and measured opening times

Frequency (Hz) (individual)	Predicted Opening Time (s)	Measured Opening Time (s)
2.5	39.25	37.5
25	3.925	3.60
250	0.3925	0.325

Additional thermal actuators have been incorporated to serve as safety mechanisms as required by MIL-STD-1316E [4]. These safeties simply prevent the barriers from moving when they are engaged. By applying a potential to these actuators, they disengage, allowing the barriers to move.

Safety mechanisms of this type are incorporated into designs fabricated by the PolyMUMPS™ and SUMMiT V™ fabrication processes. The safeties on the PolyMUMPS™ device fully actuate the required 10μm when a potential of 12 volts is applied, as predicted by the modeling. The device fabricated in SUMMiT however requires a potential of 25 volts in order to actuate the same distance. This is due to the fact that as polysilicon is deposited in the SUMMiT V™ process, it is doped *in situ*, causing it to have a higher resistance than polysilicon that is deposited using the PolyMUMPS™ fabrication process.

VII. CONCLUSIONS

The characteristics of MEMS electro thermal actuators have been thoroughly examined in an effort to provide a means for determining which particular configuration is best suited for a given application. Comparisons between single and double hot arm configurations, as well as the effect of their dimensions have been provided. A discussion as to how these devices are integrated into a MEMS based safe and arming device was given as well as the results and characterization of these integrated devices.

Disclaimer:
The views expressed in this document are those of the author and do not reflect the official policy or position of the United States Air Force, Department of Defense, or the U.S. Government

VIII. REFERENCES

[1] S. Carlton *et al.*, "Munitions: An Industry in Peril," tech. rep., The Industrial College of the Armed Forces, June 2002.

[2] S. Mink, "Microelectromechanical Systems (MEMS) Interrupter for Safe and Arm Devices," Master's Thesis, Air Force Institute of Technology, Mar 20006

[3] R. Platteborze, "Microelectromechanical Systems (MEMS) Safe-and-Arm Barrier for Low-Energy Exploding Foil Initiators (LEEFI)," Master's Thesis, Air Force Institute of Technology, Mar 2008.

[4] MIL-STD-1316E, "Safety Criteria for Fuze Design." Department of Defense Design Criteria Standard, Jul 1998.

[5] Carter, J. Cowen, A. Busbee, H, *et al,* "PolyMUMPS Design Handbook," Revision 11.0, MEMSCAP Inc., Copyright 1992-2005

[6] Sandia National Laboratories, SUMMiT V – Five Level Surface Micromachining Technology Design Manual, April 2008

[7] Liu, Chang, *Foundations of MEMS*, Prentice Hall, 2006

[8] E. Kolesar *et al*, "Single- and double-hot-arm asymmetrical polysilicon micromachined electrothermal actuators applied to realize a microengine," Thin Solid Films, Volumes 420 – 421, Proceedings of the 29th International Conference on Metallurgic Coatings and Thin Films, 2 December 2002, Pages 530 – 538

Proceedings of the SEM Annual Conference
June 7-10, 2010 Indianapolis, Indiana USA

Contrast Reversal on Surface Plasmon Resonance Reflectivity in Nickel and Nickel Alloy Films

A. Horvath, M. Roddy, M. Syed, A. Siahmakoun,
Rose-Hulman Institute of Technology, Terre Haute, IN 47083, U.S.A
Azad.Siahmakoun@rose-hulman.edu

ABSTRACT

We have performed surface plasmon resonance (SPR) experiments in the Kretchmann configuration on prisms coated with metal and alloy films. The experiment is performed at various wavelengths that include 1320 nm and 1550 nm wavelengths that are important for optoelectronic applications. The metal films of 20nm/50nm thickness are grown by magnetron sputtering and are binary alloy films of Nickel and Chromium (Nichrome). The aim of this study is two-fold. Our results would show SPR behavior as a sensitive function of film composition and film thickness. Our results also show two interesting thickness regimes where the reflectance is dominated by different processes that take place at the interface between the metal and the dielectric (fused silica prism in our study). Our measurements reveal a contrast reversal in plasmonic signal as we change the alloy thickness from 50 nm to 20 nm for the shorter wavelength.

INTRODUCTION

SPR effect relies on the coupling of a totally internally reflected (TIR) photon's energy and momentum with the free electrons of a metal that are excited as a two dimensional charge density wave on the interface between a metal and a dielectric. SPR effects in thin metal films have led to a renewed and sustained interest in plasmonics applications and devices that range from optoelectronic switches to biochemical sensors [1]. The materials of choice are typically gold and silver due to their desirable properties like long plasmon propagation distances, chemically stable surfaces (critical for repeated use as sensors) and very strong plasmon absorption over an extremely narrow range of angles of incidence. In addition to device applications, SPR has been investigated as a fundamental phenomenon in its own right where dimensionality effects produce interesting departures from bulk optical properties of a given material [2]. The present investigation focuses on alloy films of metals that are not typically used as plasmon supporting systems. In order to model the reflectivity of these films, information regarding the dielectric function of the metal layer is critical. We employed spectroscopic ellipsometry to extract the dielectric function of the metal layer system from companion slides grown under the same condition as a given film on the prism. This dielectric function was then used as an input for the reflectivity model that relied on the dielectric properties of the metal film and the two dielectric media surrounding it (prism and air, in this study). Surface roughness measurements were performed using a white light phase-shifting interferometer. For the samples in this study, the roughness values (RMS) were in the range of 20-30 nm.

THEORY

The model used to account for the behavior of the thin films reflectivity has been discussed by Simon et al [3]. It is a three layer model with the metal layer sandwiched between two dielectrics; these two layers are fused silica prism and air in our case. The main feature of the model is that it allows for a propagating solution at one (or both) of the metal-dielectric interfaces. This solution decays evanescently away from the metal layer in both directions and propagates along the interface with a characteristic decay length in this direction as well. Reflectivity is given by the following expression.

$$R = \left| \frac{r_{12} + r_{23}\exp(-2kd)}{1 + r_{12}r_{23}\exp(-2kd)} \right|^2 \tag{1}$$

T. Proulx (ed.), *MEMS and Nanotechnology, Volume 2*, Conference Proceedings of the Society for Experimental Mechanics Series 2,
DOI 10.1007/978-1-4419-8825-6_32, © The Society for Experimental Mechanics, Inc. 2011

Here r_{12} and r_{23} are the Fresnel reflection amplitudes at the dielectric-metal and the metal-air interfaces respectively, d is the film's thickness while k, the exponential decay due to the absorption by the film, is given by the following expression.

$$k = -i\left(\frac{\omega}{c}\right)\left(\varepsilon - n^2 \sin^2 \theta_1\right) \tag{2}$$

Where n is the index of refraction for the dielectric, ε is the complex dielectric function for the metal film, and $\boldsymbol{\theta_1}$ is the angle of incident at the dielectric-metal interface.

EXPERIMENT

Thin films of 80/20 Nichrome were grown using magnetron sputtering. A fused-silica prism was coated with a film thicknesses of 200±20 Å and a BK7 prism is coated with a film of 500±20 Å. For every prism coated with metal films, companion layers of the same nominal thickness and under the same growth conditions were also grown on glass substrate that were later used to carry out spectroscopic ellipsometry measurements. The experimental setup used to observe the SPR effects includes laser source of a chosen wavelength wavelengths followed by a polarization rotator and a linear polarizer as shown in figure 1. The coated prism was mounted on a rotational stage with 0.1 arc minute resolution which was computer controlled. Optical power meter was mounted on an independent rotational stage with 0.5 arc minute resolution. Each measurement was carried out with the light polarized in TM mode and then repeated in transverse electric (TE) mode. Both the reflected light as well as transmitted light through the film were measured. The ellipsometric measurements were carried out on Woollam spectroscopic ellipsometer for the determination of the ellipsometric parameters Ψ and Δ over a wavelength range from 400 – 900 nm.

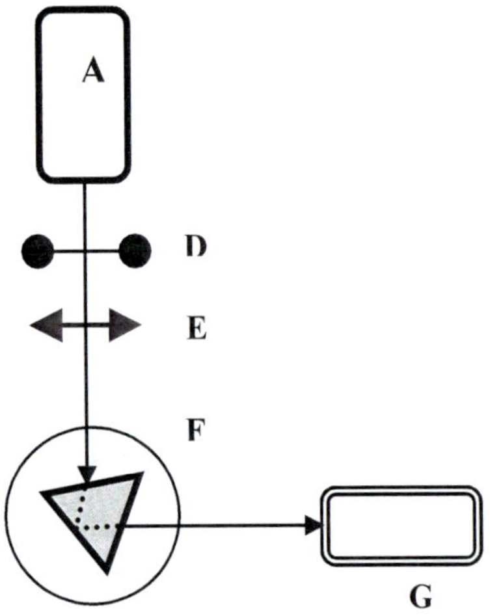

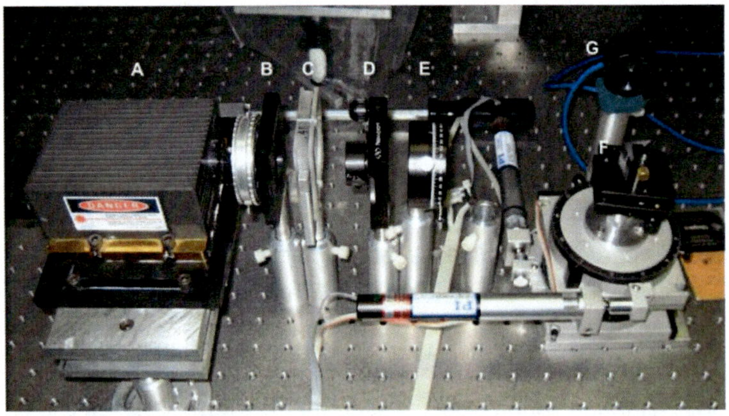

Figure 1. Schematic of experimental setup (left): (A) laser-diode, (D) polarization rotater, (E) linear polarizer, (F) coated prism on rotation stage and, (G) optical power meter.

To model the actual experimental data one has to account for the fact that the metal film was deposited on the hypotenuse of a right angle face of the prism and light was refracted onto the glass side of the prism-metal interface in the TIR geometry and after reflection exited the prism. Fresnel losses at the first and last face of the prism were also accounted for along with the reflection (Eq. 1) that took place at the hypotenuse of the right angled prism.

DISCUSSION

Sample reflectivity results are shown for various wavelengths as a function of the angle of incidence on the dielectric-metal interface, for the two types of film thicknesses (50 nm and 20 nm). The modeling procedure would involve getting the dielectric function from ellipsometry and getting roughness estimates from ellipsometric model and from measurements using the phase-shifting interferometer setup. Only *p*-polarization data is presented in detail as the *s*-polarization data does not reveal any absorption or reflection features. To illustrate the point, one sample *s*-polarization curve is shown that exhibits a fairly featureless reflectivity curves (nearly monotonic) as predicted by a simple application of Fresnel's equations (see figure 2). In other words, no plasmon solution is expected for the *s*-polarization, and no corresponding feature is ever seen in the *s*-polarization data. As clear from the figures, the theoretical model does a reasonably good job of explaining the data. While figure 2 shows the reflectivity for the 20 nm film with its high reflectance, figure 3 show the results of *p*-polarization reflectivity for the 50 nm Nichrome film with the SPR feature (now absorption) present prominently in at an angle of about 41.5°. Theoretical calculations for the plasmon angle with the appropriate choice of the dielectric functions result in an angle of 41.9° [3]. While this is close to the observed value, the difference is not insignificant. It is hoped that a more detailed investigation into the nature of the roughness and the dielectric function will result in better agreement. It should also be noted that Nickel does not show any such feature for the *p*-polarization geometry for the corresponding 50 nm single metal films. The theroetical model in general, predicts a slightly higher reflectivity but that is mostly due to losses on the slightly inferior surface of the BK7 prism (as opposed to the fused silica prisms). On the other hand the case for chromium is more complicated. While it shows no corresponding features for the shorter wavelength (633 nm) used in the experiment, a weak absorption feature is present for the 1550 nm excitation wavelength. Attempts to meodel this feature using the dielectric fnction for bulk chromium or companion layers has, to date, not met with success. It is strongly suspected that this complication is a result of a thin oxide layer that may be present on the metal surface. Given that this is a native oxide, attempts to etch it and then do the experiment are not likely to succeed, unless the experiment is carried out in an environment that can inhibit regrowth of the oxide layer. Presently we are trying to systematically study this layer and incorporate its optical properties into the model, and hope to deal with the case of chromium in detail in aletr publication.

It is worth noting that the 20 nm film shows fundamentally different behavior for the case of 633 nm excitation wavelength. The primary absorption feature is now replaced by a strong reflection peak. In order to model this data, Nichrome dielectric function values for 633 nm are obtained from the ellipsometric measurements. The starting thickness is assumed to be 50 nm but once reasonable agreement is found between experimental data and the corresponding model, thickness is allowed to vary by a small amount and surface roughness is built into the model. The real and imaginary parts of the dielectric function are 2.06 and 25.98, respectively. Typically, for the surface plasmon to be excited the real part of the dielectric function for the metal should be negative. Moreover, for metals like silver and gold that show very large SPR response with reflectivity varying over 98% in the neighborhood of the plasmon angle (compared to ~18% for Nichrome), the imaginary part of the dielectric function is very small. That is also not the case for Nichrome. The relatively large imaginary part of the dielectric function is easier to interpret than the positive real part. The large imaginary part of the dielectric function implies strong damping of the plasmon and that is evident from the relatively shallow absorption feature associated with the plasmon absorption in Nichrome. The positive real part is more difficult to explain. Another complication is the nature of roughness. The interferometer results suggest a roughness that is on the order of the thickness of the thinner of the two sets of films. This really suggests that the surface profile for both types of films may be fairly corrugated and the model should really be refined to make the interface more of a textured surface with some kind of an effective medium approximation. This would also modify our present approach as regards to the dielectric function of the two nichrome layers (50 nm and 20 nm). Presently, the companion layer analysis suggests that the same dielectric function is used for both layers. However, if an effective medium approximation is to be used then clearly, the 20 nm layer incorporates greater void fraction than the 50 nm layer. Indeed, it is surprising that the present model explains the results, as well as it does for all the wavelengths examined.

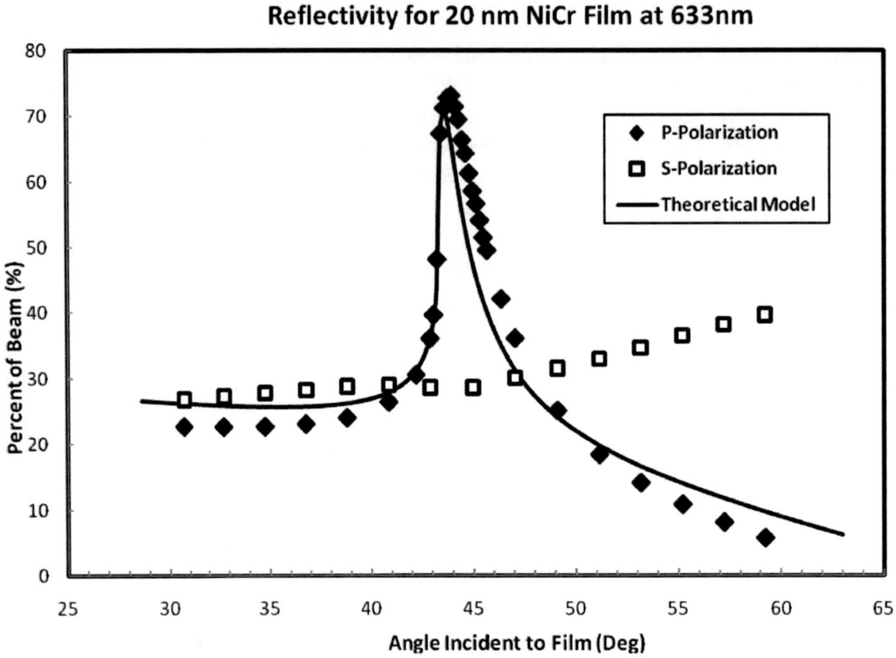

Figure 2. Reflectivity data for 20 nm Nichrome film at 633 nm wavelength.

The next two figures show the results for the reflectivity at 1550 nm wavelengths for the two thickness values. In both cases, the data looks very similar for the 1320 nm wavelengths. Dielectric functions for the nichrome film at 1320 nm and 1550 nm wavelengths are **3.2 + 41*i*** and **3.5 + 45*i***, respectively. While the 50 nm film shows an absorption feature for 633 nm excitation wavelength, it displays a reflection maximum at longer excitation wavelengths, albeit less dramatic than the reflection peak for the 20 nm film.

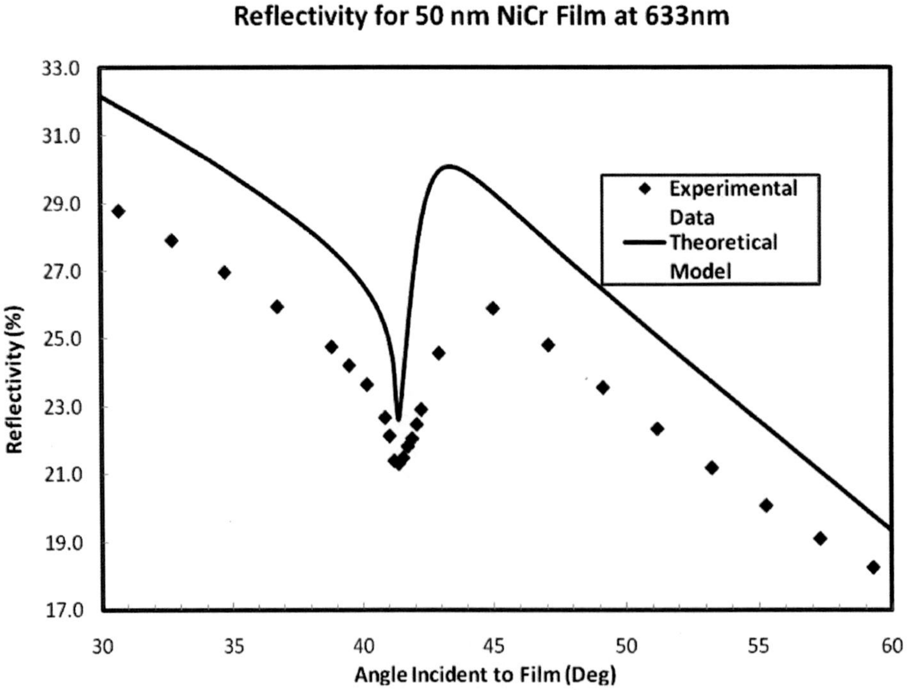

Figure 3. Reflectivity data for 50 nm Nichrome film at 633 nm wavelength.

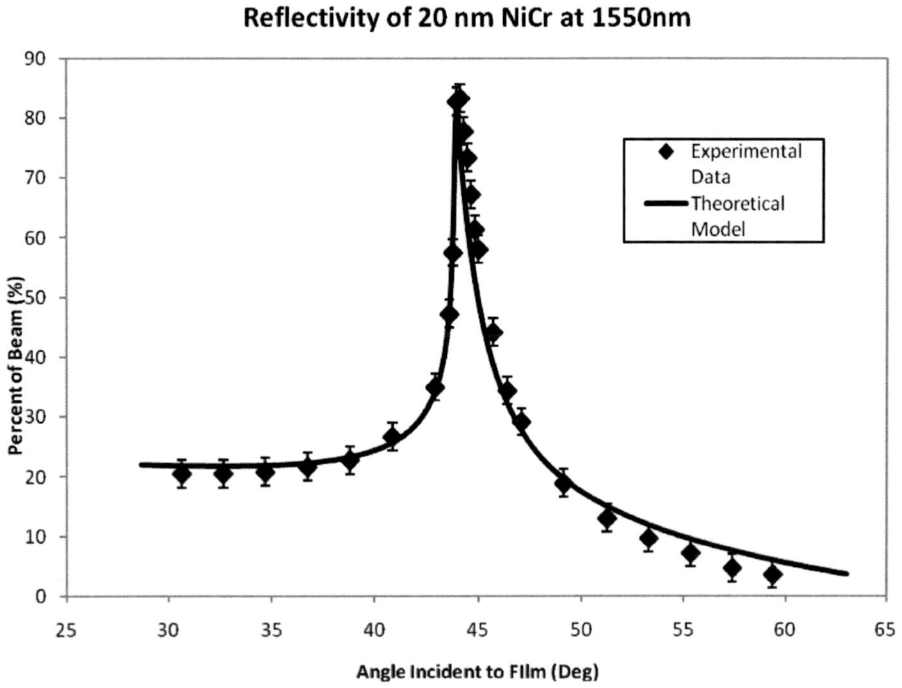

Figure 4. Reflectivity data for 20 nm Nichrome film at 1550 nm wavelength.

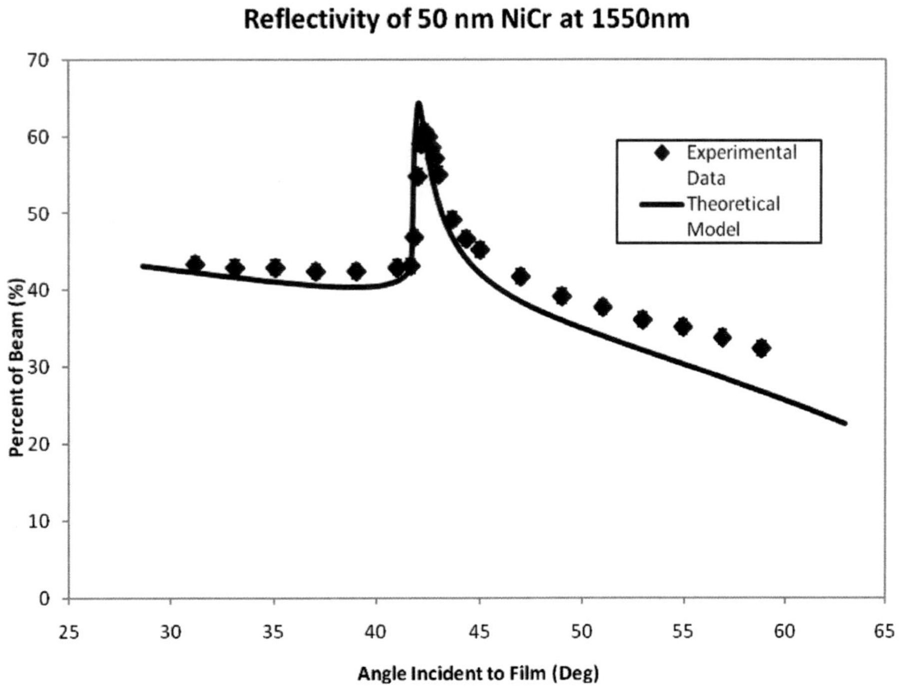

Figure 5. Reflectivity data for 50 nm Nichrome film at 1550 nm wavelength.

CONCLUSIONS

Interesting reflectivity results are obtained for 20 nm and 50 nm nichrome alloy films which show a contrast reversal at 633 nm wavelength. The differences persist at longer wavelengths, 1320 nm and 1550 nm, although both films show reflection peaks at these wavelengths. While the results are modeled as SPR, interesting questions remain regarding the exact nature of the mode responsible for these peaks. Future work will focus on refining the model and investigating the transmittance data to gain insight regarding the nature of the mode in these films.

REFERENCES

1. K. Kurihara and K. Suzuki, "Theoretical Understanding of an absorption-based surface plasmon sensor based on Kretchmann's theory," *Anal Chem.,* 74, 696 -701, 2002.
2. E. Kretschmann & H. Raether, "Radiative Decay of nonradiative surface plasmons excited by light," *Z. Naturforsch*, A23, 2135-2136, 1968.
3. H. J. Simon, D. E. Mitchell, J. G. Watson, "Surface Plasmons in silver films," *Am. J. Phys.*, vol. 43, (7), 630-636, 1975.

Proceedings of the SEM Annual Conference
June 7-10, 2010 Indianapolis, Indiana USA

Effect of Accelerated Ultra Violet and Thermal Exposure on Nano Scale Mechanical Properties of Nylon Fibers

Nandula Wanasekara[1], Vijaya Chalivendra[2*] and Paul Calvert[1]
[1]Materials & Textiles Department
[2]Department of Mechanical Engineering
University of Massachusetts Dartmouth, MA 02747
*Corresponding author, vchalivendra@umassd.edu, 508-910-6572

ABSTRACT

Nano-characterization studies using atomic force microscopy (AFM) was conducted to study the effect of accelerated ultra violet (UV) and thermal degradation on mechanical properties of Nylon textile fibers. Nylon fibers were exposed to UV radiation for six different hours using Q-UV Panel Weatherometer. The exposed fibers were molded in an epoxy plug for nanoindentation using AFM. Progressive nanoindentation from surface to the center of the fiber was used to investigate the effect of degradation on gradation of Young's modulus across fiber cross-section. It was identified that UV degradation decreases the Young's modulus from center to surface of the fibers up to 144 hours of exposure. Reduction of Young's modulus at surface was greater than the center implying more deterioration at the surface. To investigate thermal degradation effect on Nylon fibers, the fibers were exposed to 175°C for four weeks. Initial experimentation indicates that the thermal exposure causes gradual reduction of Young's modulus. Wide angle x-ray spectroscopy (WAXS) and Fourier Transform Infra Red Spectroscopy (FTIR) are used to correlate the fiber chemical and micro-structural changes to the variation of nano-mechanical properties.

1. INTRODUCTION

Nylon is one of the commonly used polymers in commercial applications. It has been known that it undergoes significant deterioration of mechanical properties due to photo-oxidation. Initial investigation of degradation mechanisms on Nylon began as early as 1940 by Fuller et. al. [1] who described that photo-oxidation of nylon 66 proceeds through the removal of the hydrogen on the α-carbon to the amide NH group by a free radical which has been activated by photo energy absorption. In the propagation phase, the α-carbon radical readily react with atmospheric oxygen to form a hydroperoxide. Bunn and Garner [2] have done a detailed study of the crystal structure of drawn Nylon 66. The major phase denoted 'α' was shown to consist of extended molecular chains arranged in triclinic unit cells. A second triclinic unit cell, denoted 'β' was also observed. Y. W. Mai et. al. [3] investigated the degradation of mechanical properties of nylon 6 films due to both oxidative and non-oxidative degradation. Their experiments showed that in a non oxygen environment the rate of degradation is enhanced with rising temperature. Experimental results obtained on samples exposed to natural weathering conditions were found to be difficult to relate to those obtained by accelerated aging. They could not retrieve the samples of nylon sheets exposed at 65°C in nitrogen exceeding 400 hours since upon cutting the sheet it shattered indicating that significant embrittlment had occurred during exposure. It is also possible for nylon 6 to undergo micro-structural changes by cross linking or main chain scissioning by photon energy absorption. It was proposed and proven by the authors' that the specific energy absorption in a tensile test is an effective parameter for characterizing photo degradation in nylon.

Mead JW et. al. [4] conducted accelerated aging of nylon 66 in humid air, ozone and smog to ascertain aging trends in these conditions. Tensile strengths of nylon yarn which were aged in humid environment of 50 % RH at room temperature for six months degraded from 40 % at 110°C to 85 % at 150°C. Based on these data the energy of activation for the nylon degradation is estimated to be is 18 kcal/mol. Nylon exposed to smog for six months lost 63 % of its tensile strength compared to 7% loss in ozone. Krejsa et. al. [5] conducted 1-D and 2-D NMR techniques on nylon 66 to determine the degradation structures resulting from UV exposure of Nylon films and heat aging of Nylon pellets. The dominant degradation

T. Proulx (ed.), *MEMS and Nanotechnology, Volume 2*, Conference Proceedings of the Society for Experimental Mechanics Series 2,
DOI 10.1007/978-1-4419-8825-6_33, © The Society for Experimental Mechanics, Inc. 2011

structure in all cases was found to be the terminal methyl group which was attributed to polymer chain scission reactions. Further evidence for chain scission was found through the presence of aldehyde and formamide structures upon prolonged UV exposure and heat aging in the presence of atmospheric air and terminal vinyl structures upon UV exposure. The substitution of OH adjacent to the amide group at the end of a cleaved alkyl chain may be caused by the presence of oxygen during UV exposure. They also observed the Terminal amide groups in the starting material and same concentration upon heat aging or UV exposure. In heat aged samples, no evidence was found for olefin or hydroxyl-containing structures in nitrogen, but hydroxyl structures were observed in the sample. A degradation mechanism was proposed to account for the structures observed in their research work.

Auerbach [6] studied the tensile strength degradation of nylon 66 yarns at elevated temperatures and over a broad range of relative humidities. It is observed that degradation rates for nylon was initially slow, but increased swiftly suggesting the depletion of an inhibitor. It is concluded that degradation is governed by thermal-oxidative and moisture induced mechanisms. Rate relationships were developed in which contribution from each mechanism was considered. Calculated degradation from these relationships agreed well with observed degradation over a broad range of temperatures and humidities. The goals of the present study were aimed at identifying chemical, micro-structural changes caused by degradation, to evaluate these results in heat-aged and UV-exposed systems, and to use this information on these structures to understand the underlying degradation mechanisms.

2. EXPERIMENTAL DETAILS

2.1. Sample Preparation

Nylon commercial fibers for the analysis were obtained by Providence Yarn Company, Inc., RI, USA. An unknown stabilizer had initially been used in the spinning process of fibers. The typical diameter of the fiber was 30 μm. The fibers were molded in an epoxy plug for AFM characterization. A two-part thermoset epoxy (Epothin, supplied by Buehler Ltd., IL, USA) was specifically chosen since it had the properties of room temperature curing with minimum heat generation during the curing process.

2.2. AFM Characterization

In this study, an atomic force microscope (XE-100, Park Systems Inc, CA, USA) was used in the topographic imaging and nanoindentation of nylon fibers. A contact mode scan was made prior to the indentation of the sample. A typical scanned image with indentation pattern is shown in Figure 1. This facilitates the user to locate the specific location for indentation. Nanoindentaions were performed on fiber cross sections in the radial direction. For each duration of UV and thermal exposure, more than 15 radiuses were selected for indentation and for each radius; 6 points were chosen as shown in Figure 1. The force-indentation depth diagrams were further processed using a previously established methodology to determine Young's modulus [7].

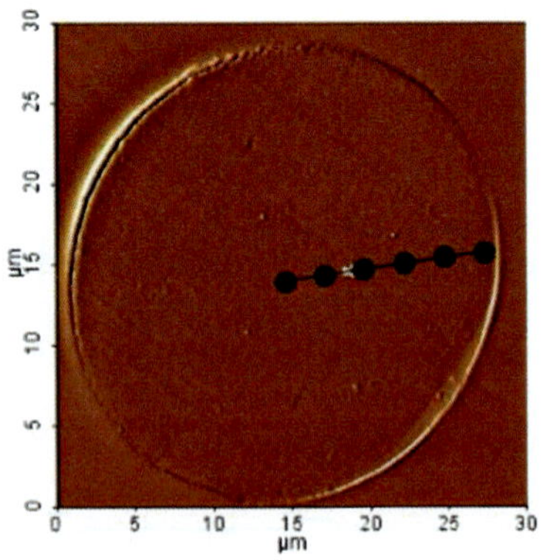

Figure 1: Indentation pattern across the radius of the fiber

2.3. UV and Thermal Exposure of Fibers

The nylon fibers were subjected to Ultraviolet (UV) exposure using a Q-panel UV (QUV) weatherometer (QUV, Q-Lab Corp. OH, USA). Fluorescent UV lamps generated radiation in the QUV chamber. The UV wavelength of 340nm was chosen as per recommendation stated in ASTM G 154 standard for simulation of direct solar radiation, which allows enhanced correlation with actual outdoor weathering. The cyclic exposure conditions were eight hours of UV exposure at 50°C and four hours of condensation at 48°C. The nylon fibers were exposed in the QUV weatherometer for a maximum period of 144 hours in increments of 24 hours. Thermal exposure of nylon fibers was carried out using a controlled temperature/humidity test chamber (AH-202XCC, Bryant Manufacturing Associates, MA, USA). The samples were exposed to a constant temperature of 175°C for a maximum duration of four weeks in increments of one week. A relative humidity level closer to zero was maintained throughout the test.

2.4. Tensile Testing

The specimen for tensile testing was prepared as shown in figure 2. The fiber was held by two epoxy dots at the ends. The uniaxial tensile testing of nylon fibers was carried out using the INSTRON 5569 materials testing system. A load cell having a maximum value of 2.5 N was used in the experiments. The single fiber tensile testing method described in ASTM G3822 was followed. A constant extension rate of 2.54 mm/min was used. The cross-head displacement was used to determine axial strain of the fiber and recorded force values were used to determine the axial stress. The tests were continued until there were failure of fibers to determine tensile strength and percentage elongation at break.

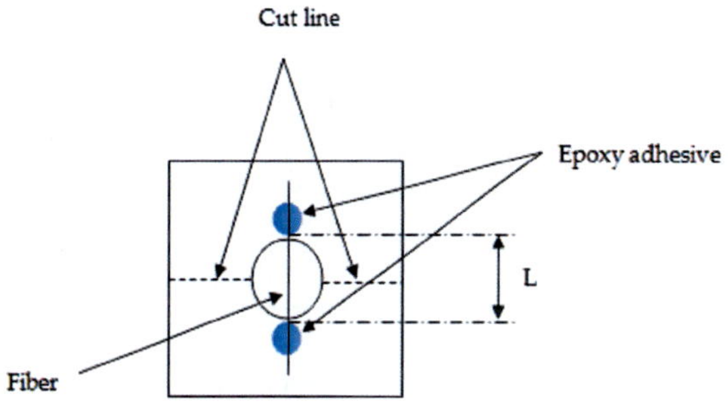

Figure 2: The schematic of tensile testing specimen. The blue dots represent the epoxy adhesive (L: Gage Length = 16mm).

2.5. Wide Angle X-ray and Fourier Transform Infrared Spectroscopy Analysis

WAXS was done using a set-up from Molecular Metrology, Inc. housed at University of Massachusetts, Amherst. It uses a 30w micro-source (Bede) with a 30x30 μm^2 spot size matched to a Maxflux^R optical system (osmic) leading to a low divergence beam of monochromatic CuK_α radiation (wavelength, λ = 0.1542). WAXS is performed using an image plate positioned in the sample chamber at a distance of 39 mm. The image plate (maximum resolution 50 μm) has a hole in its center. The whole system is evacuated while the test is conducted. Fourier transform infrared spectroscopy microscopy (Varian FTIR Bench and Microscope, CA, USA) was used to identify the chemical groups present in the fiber samples. The experiments were carried out in the reflectance mode. The percentages crystallinity of the fibers was calculated using the 'Polar' software.

3. RESULTS AND DISCUSSION

3.1. Nanoindentation of UV degraded nylon fibers

The typical force deflection diagrams of unexposed and UV exposed nylon fibers obtained for the indentation point close to the surface are shown in figure 3. The initial slope of the unloading portion of the curve is calculation of Young's modulus. These curves testify the reduction of Young's modulus for fibers with longer UV exposure time. Variation of Young's modulus of ultraviolet exposed nylon fibers from the center to surface is shown in figure 4. Young's modulus profile of the unexposed fibers remained almost the same from center to surface of the fiber. This proves that the initial stabilizes added to nylon melt during the melt spinning stage have worked well since the fiber has maintained the mechanical properties without deterioration. When the exposure hours are increased by increments of 24 hours up to 144 hours, the Young's modulus was reduced. The decrease of modulus values at the surface is higher than the center of the fiber. This clearly shows that the mechanical deterioration of the nylon fibers happens more severely close to the surface than the center. The Young's modulus values at the center also reduce down to ~2 GPa at 144 hours from ~3 GPa at unexposed level. This suggests a 33% reduction of modulus values at the center, while the reduction of modulus close to the surface is around 65%. The decrease of Young's modulus values can be attributed to the decrease in crystallinity as discussed in section 3.2. This leads to the observed reduction of mechanical properties which is evident from the results of percentage elongation to break and tensile strength which report lowest values as discussed in section 3.3.

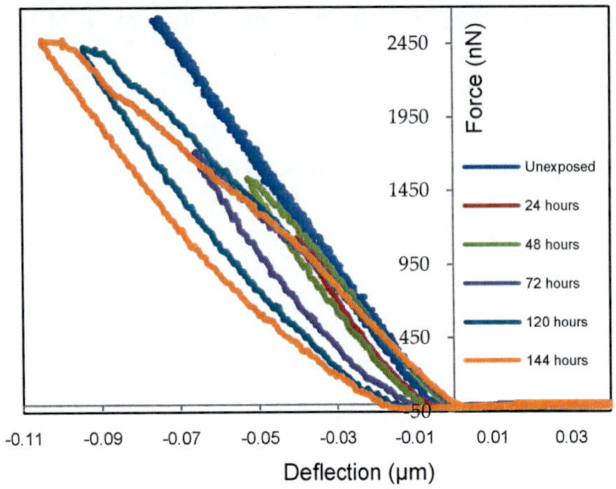

Figure 3: Typical force deflection diagrams of nanoindentation for nylon fibers of different UV exposure hours

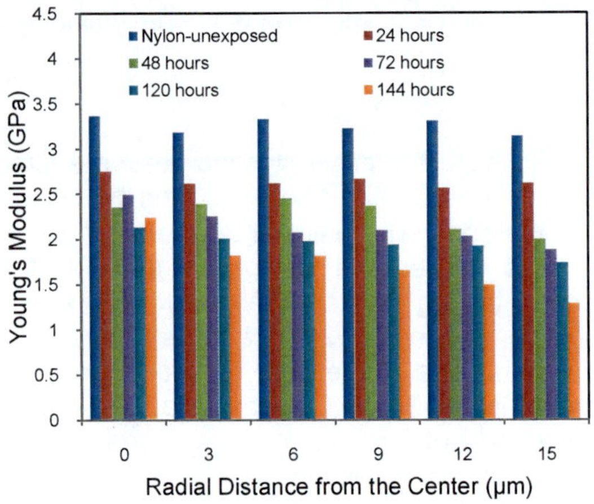

Figure 4: Variation of Young's modulus values across the radius of nylon fibers for different durations of UV exposure

3.2. *Wide Angle X-ray and Infrared Analysis*

WAXS was performed on whole fiber to understand the changes of the fiber crystallinity. Figure 6 shows the WAXS images of UV exposed nylon fibers. Unexposed nylon diffractions are arced with some spots indicating the crystallinity and significant molecular orientation. Even at an ultraviolet exposure of 144 hours, the changes of diffraction arcs showed little noticeable difference. But the calculated percentage crystallinities using the 'Polar' software showed lower values for fibers with increased exposure times. The unexposed fibers were 70 % crystalline, which suggests that the fiber has more crystalline regions with high molecular orientation. So it can be reasonable assumed that the degradation may spread quickly to the crystalline areas making the deterioration of mechanical properties. When the UV exposure is further increased to 144 hours the crystallinity was reduced to 53 %.

Fourier transform infrared spectroscopy was performed on fibers to understand the new formations of chemical bonds upon UV exposure. Figure 5 shows the infrared spectra obtained for fibers with different exposure duration. A broad range of peaks were observed corresponding to amide bonds. The significant observation is the formation of hydroxyl bonds (O-H) corresponding to wavelengths of 3300 – 3400 cm^{-1} with increase of exposure hours. The area of the hydroxyl peak was observed to increase through 144 hours of exposure.

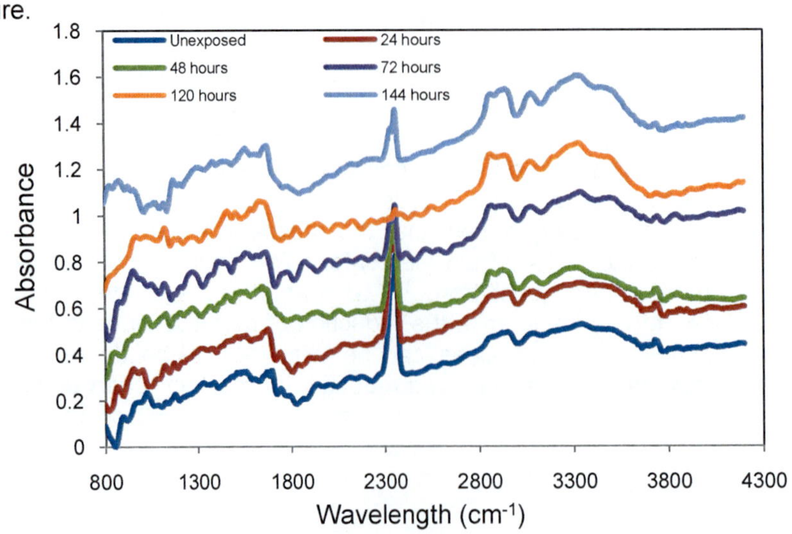

Figure 5: FTIR spectra of UV exposed nylon fibers with UV exposure

3.3. *Tensile Testing Results*

Tensile testing was carried out on UV exposed nylon fibers to determine the tensile strength and elongation at break. Nanoindentation is not capable of measuring those two property values. Using the stress-strain diagrams, the percentage elongation at break and tensile strength of UV exposed fibers as a function of exposure hours is obtained and shown in Figure 7. The unexposed nylon fibers had around 40% percentage elongation at break. A significant reduction could be observed in both tensile strength and percentage elongation at break with increase of number of UV exposure hours. This loss of toughness on photo-oxidation would be crucial for many applications of nylon fibers in industrial textiles.

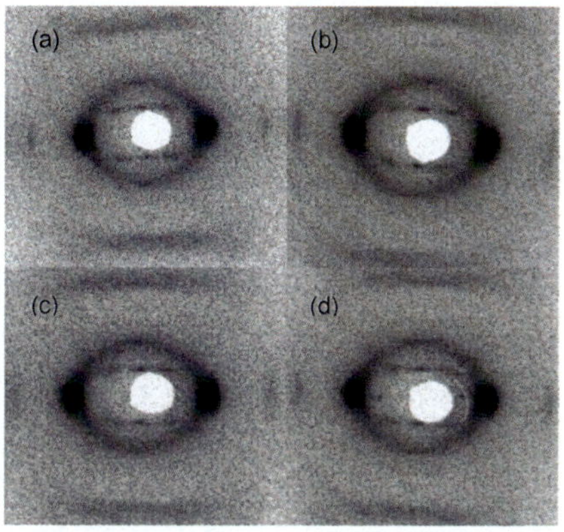

Figure 6: WAXS Images of UV exposed nylon fibers, (a) unexposed (b) 48 hours (c) 96 hours (d) 144 hours.

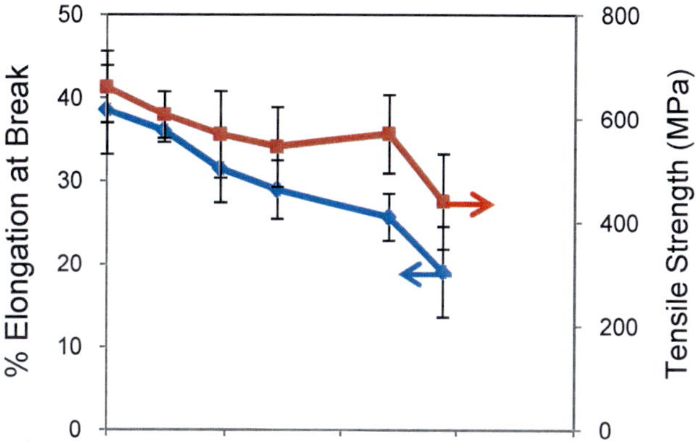

Figure 7: Variation of tensile strength and percentage elongation at break of nylon fibers with number of UV exposure hours.

3.4. *Thermal Degradation*

3.4.1 *Nanoindentation and Tensile Testing Results*

Variation of Young's modulus of thermally exposed fibers from the center to surface is shown in Figure 8. A trend of decreasing Young's modulus can be noticed on the surface at the beginning and later on reduction tends to spread to the interior after three and four weeks of exposure,. Especially at three weeks of exposure, Young's moduli were noticed to be prominently lower in the first three indentation points closer to surface. A significant decrease of Young's modulus was observed throughout the fiber except the center at four weeks of exposure.

Engineering stress strain graphs for the thermally exposed nylon is shown in Figure 9. The slope of the curve at 1% strain was observed to be approximately same for all the fibers. This suggests that the deterioration of mechanical properties measured by elastic modulus cannot be determined using tensile testing. Exposed fibers were tested until they failed and percentage elongation at break and tensile strength was measured. Both tensile strength and percentage elongation at break tend to decrease with increase of thermal exposure time.

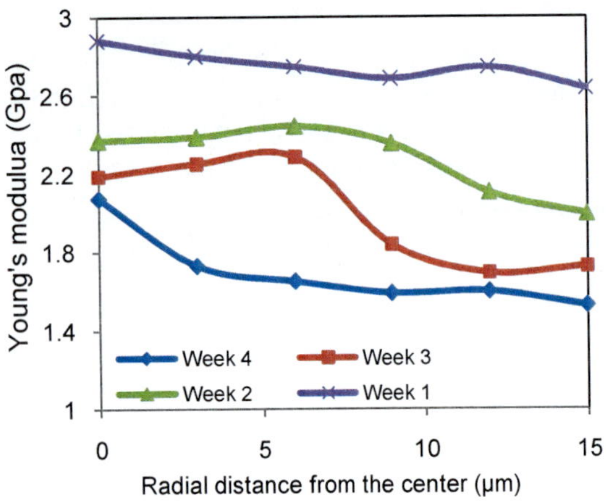

Figure 8: Variation of Young's modulus values across the radius of nylon fibers for different durations of thermal exposure

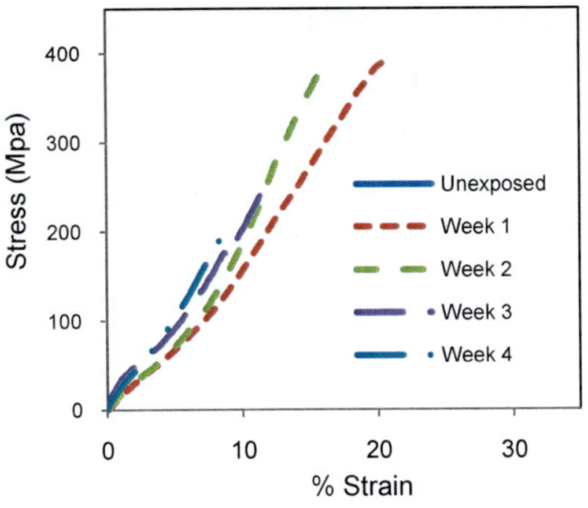

Figure 9: Stress-strain curves for thermal exposed nylon fibers

3. CONCLUSION

Photo-oxidation during the manufacturing process has no significant effect on sub-micron scale measurement of Young's modulus across the nylon fibers. Nanoindentation provided gradation of Young's modulus values across the fiber cross section; however the macro-tensile testing provided no significant change of Young's modulus as a function of UV and thermal exposure hours. It was evident that there is a significant decrease of Young's modulus values near the surface than at the center as the UV exposure duration increases to 144 hours. FTIR spectra proved the formation of hydroxyl bonds with UV exposure. WAXS studies confirmed the destruction of crystalline structure of the nylon fibers when the exposure time increased. Nanoindentation on nylon fibers subjected to thermal exposure showed a similar pattern of decreasing Young's modulus with exposure time increased. A gradual reduction of Young's modulus from surface to the center as the thermal exposure time increased was also observed.

Acknowledgements

The authors greatly appreciate the financial support from the National Science Foundation through Grant CMS0618119 for acquiring an atomic force microscope. Also we would like to acknowledge the financial support from the National Textile Center through project number M08-MD13.

References

1. Fuller CS, Baker NO, Paper NR, "Crystalline Behavior of Linear Polyamides. Effect of Heat Treatment", *Journal of American Chemical Society* 62, 12, 3275, 1940
2. Bunn CW, Garner EV, "The Crystal Structures of Two Polyamides ('Nylons')", *Proceedings of Royal Society* A189, 39, 1947
3. Mai YW, Head DR, Cotterell B, Roberts BW, "Mechanical properties of nylon 6 subject to photo-degradation", *Journal of Material Science* 15, 3057 – 3065, 1980
4. Mead JW, Mead KE, Auerbach I, Ericksen RE, "Accelerated aging of nylon 66 and Kevlar 29 in elevated temperature, elevated humidity, smog, and ozone", *Industrial and Engineering Chemistry, Product Research and Development* 21, 2, 158–163, 1982
5. Krejsa MR, Udipi K, Middleton JC, "NMR analysis of UV and heat aged Nylon 66", *Macromolecules*; 30, 16, 4695-4703, 1997
6. Auerbach I, "Kinetics for Tensile Strength Degradation for Nylon and Kevlar Yarns", *Journal of Applied Polymer Science* 37, 2213-2227, 1959

7. Reynaud C, Sommer F, Quet C, Bounia El N and Duc TM., “Quantitative determination of Young’s modulus on a biphase polymer system using Atomic force microscopy”, *Surface Interface Analysis* 30, 185-189, 2000

Proceedings of the SEM Annual Conference
June 7-10, 2010 Indianapolis, Indiana USA

High Thermal Conductivity Polyurethane-Boron Nitride Nanocomposite Encapsulants

John V. Costa[1], Thomas Ramotowski[2], Steven Warner[1], Vijaya B. Chalivendra[1*]
[1]University of Massachusetts, North Dartmouth, MA 02747, USA
[2]Naval Undersea Warfare Center, Newport, RI, 02841, USA
*Corresponding author, vchalivendra@umassd.edu, 508-910-6572

ABSTRACT

An experimental investigation has been conducted to determine the effect of adding boron nitride (BN) nanoparticles on the thermal conductivity of polyurethane nanocomposites. BN nanoparticles (average size of 70 nm) were functionalized using three different materials: acetone, nitric acid, and an alkoxysilane. Various weight fractions of the alkoxysilane-coated particles were incorporated into a polyurethane via shear mixing and ultrasonication. The dispersion of nanoparticles within the fabricated nanocomposites was verified using field-emission scanning electron microscopy (FE-SEM). Analysis of the FE-SEM images indicated that the particles were well dispersed in the polyurethane matrix, and functionalization with alkoxysilane improved BN adhesion to the polyurethane, as verified by water drop contact angle measurements. Thermal conductivity measurements, made using a thermal conductivity analyzer, indicated that non-functionalized BN particles (5% weight fraction) increased the thermal conductivity of the resulting nanocomposite by 50%, while the increase in thermal conductivity realized by the addition of the silane-treated BN particles was only about 20%. Additional testing is underway to determine why better thermal conductivity results were not obtained from the silane-treated BN-polyurethane nanocomposites.

1. Introduction

Within the U.S. Navy today, there is a growing need for acoustically transparent polymers that exhibit high thermal conductivity but low electrical conductivity. These materials are needed to encapsulate "next-generation" high power/high duty cycle acoustic sources that generate considerable amounts of heat while in operation. Polymers are typically poor heat conductors; their thermal conductivity ranges from 0.1 to 0.6 W/m K [1]. Piezoelectric ceramics such as lead-zirconate-titanate (PZT) that frequently serve as the active elements in acoustic sources typically exhibit thermal conductivities of about 2.0 W/m K [2]. This heat conduction mismatch between the active element and its encapsulant in high power/high duty cycle acoustic sources can result in excessive heating or even thermal runaway that can degrade critical components and/or depole the ceramics. While higher thermal conductivity is a major requirement for encapsulants, they are also required to be acoustically transparent and to exhibit low water permeability to protect underlying electronic components.

In keeping with the requirements of high thermal conductivity and low electrical conductivity, Xu and Chung [3] prepared boron nitride and aluminum nitride micro particle-reinforced epoxy matrix composites and discussed the effects of acetone, nitric acid, sulfuric acid and silane based particle surface treatments on the thermal conductivity of the resulting composites. Later Xu et al.[4] prepared aluminum nitride and silicon carbide whisker/particle reinforced polyvinylidene fluoride (PVdF) and epoxy composites by employing hot pressing techniques. They reported that the addition of a combination of whiskers and particles resulted in higher thermal conductivity than the addition of whiskers or particles alone. Young and Liem [5] fabricated composites using epoxy and both hexagonal and cubic boron nitride particles, and they realized a 271% increase in thermal conductivity through the use of silane surface treatments and multi-modal particle size mixing (i.e. two different h-BN particle sizes and one c-BN particle size). Zhou et al.[6] investigated the effect of boron nitride (BN) content, particle size, and temperature on the thermal conductivity of BN particle-reinforced high density polyethylene (HDPE) composites. Their results indicated that the thermal conductivity of such composites was higher for the larger size HDPE than for the smaller size HDPE, and that the combined use of BN particles and short alumina fibers resulted in higher composite thermal conductivity than when BN particles were used alone. Recently, Terao et al., [7] fabricated catechin surface-modified boron nitride nanotube (BNNT)-polyvinyl fluoride (PVF) composite films containing high BNNT fractions (up to 10% weight fractions) and achieved about a 160% increase in thermal

DOI 10.1007/978-1-4419-8825-6_34, © The Society for Experimental Mechanics, Inc. 2011

Table-I: Specifications of BN Particles

Property	Micro Particles	Nano particles
Average particle size (μm)	1	0.07
Specific area (m^2/kg)	5600	450000
Mass density (kg/m^3)	2290	2300
Crystal structure	Hexagonal	Hexagonal

Table-II: Specifications of Polyurethane

Property	Uralite® FH-3140
Viscosity @ 25°C (Pa-S)	3.8
Specific gravity @ 25°C	1.07
Young's modulus (MPa)	20
Elongation-at-break	700
Tensile strength (MPa)	7.07

conductivity with only 1% weight fraction of BNNT added. Catechin molecules have several aromatic rings and these rings effectively interact with the nanotube surfaces via π- π bonding, so that catechin species mediate between BNNTs and polymer improving heat transfer between the matrix and particles. Very recently, Kemaloglu [8] used three micron-sized and two nano-sized BNs with different particle sizes and shapes as reinforcements to make silicone rubber composites. We have studied the effect of particle loading levels on the thermal, mechanical and morphological properties of fabricated composites. With the addition of particles, the tensile strength and percentage elongation-at-break decreased; while the elastic modulus, coefficient of thermal expansion and thermal conductivity increased.

The work discussed above was focused on the effect of BN particle reinforcement in epoxy, HDPE, PVDF, PVF and silicon rubber matrices and there were no studies reported in the literature on thermal conductivity response of BN particle reinforced polyurethane composites. In this paper, we report on our investigation on the effect of both untreated and silane-treated micro & nano BN particle reinforcement on thermal conductivity of polyurethane composites. Composites of various volume fractions were fabricated and using a combination of shear mixing and ultrasonication. Particle dispersion in the polyurethane matrix was investigated using field-emission scanning electron microscopy (FE-SEM). Thermal conductivity measurements were made using a nondestructive, modified transient plane source technique.

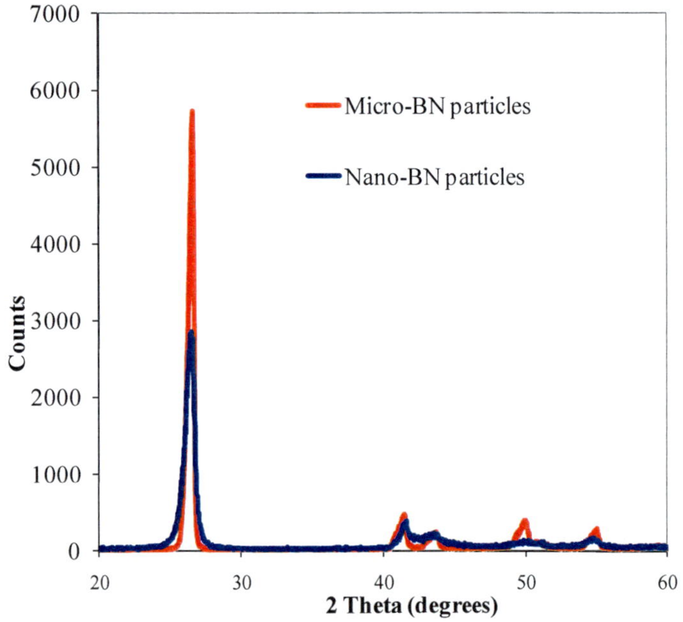

Fig.1. XRD pattern of micro and nano BN particles

2. Experimental Details

In this section experimental details, such as the physical characteristics of the BN particles and the polyurethane are provided. In addition, details on the methods used for functionalization of the BN particles, composite fabrication, FE-SEM sample preparation, and thermal conductivity measurements are discussed.

2.1. Materials and Specifications

Both micro-BN and nano-BN particles were used in fabricating polyurethane composites. The specifications of the procured particles are provided in the Table-I. A thermosetting two-part polyurethane with a trade name of Uralite® FH-3140, was used in this study and its properties are provided in Table-II. In order to determine the quality of BN particles, X-ray diffraction was performed on both micro and nano-particles and the patterns are shown in Fig. 1. Transmission electron microscopy (TEM) elemental analysis was also performed as an additional measure to verify the chemical composition of the procured nano particles. As shown in Fig.2, the elemental analysis indicated that boron, nitrogen and copper (from the grid used to hold the BN powder) were present in the samples.

2.2 Functionalization of BN Particles

Surface treatment of BN particles was performed using three solvents: nitric acid, acetone and Dow-Corning silane Z-6040®. The coating process utilized by Xu and Chung [3] was modified and used for the surface treatment of both micron-sized and nano-sized BN particles. A typical example of the procedure utilized in our work consisted of adding 1.0 gram of h-BN powder to 100 ml of a 2.0% aqueous Z-6040 silane in distilled and de-ionized water. The solution pH was adjusted so that it was between 5.5 and 6.0. This room temperature mixture was stirred and ultrasonicated while being heated between 60°C and 70°C for 20 minutes. After cooling, the mixture was filtered by a Millipore vacuum pump assisted filtration system containing a sheet of ash-less grade, no. 610 qualitative filter paper and an ultrasonic transducer. The ultrasonic transducer was used to break-up any agglomeration of material that occurred during filtration. After initial an filtration, the particles were triple rinsed with distilled water of total volume at least three times greater than that of the initial aqueous solution. After a final filtration, the damp particles were placed in a covered petri dish and dried/cured at 110°C for 12 hours in an oven.

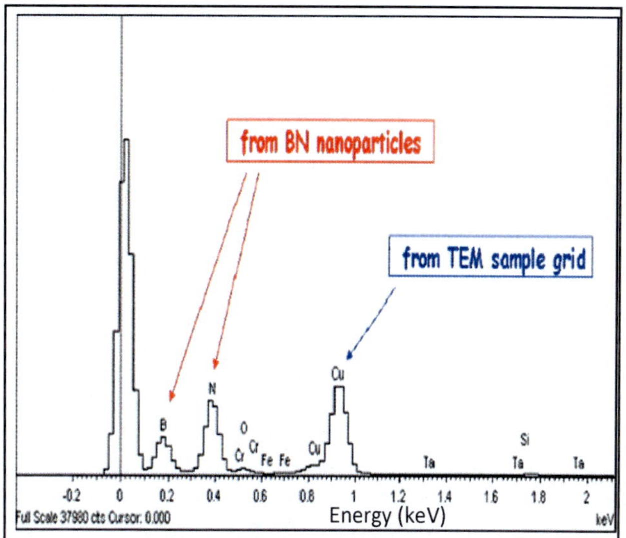

Fig. 2. TEM-EDS analysis of BN particles

Analysis of the coated particles indicated that micron-sized particles absorbed four times more silane than nano-sized particles from the coating solution when simple stirring only (no ultrasonication) was used. However, with ultrasonication, the amount of absorbed silane increased by 100% on micron-sized particles, while for nano-sized particles, the amount of absorbed silane increased by 800%.

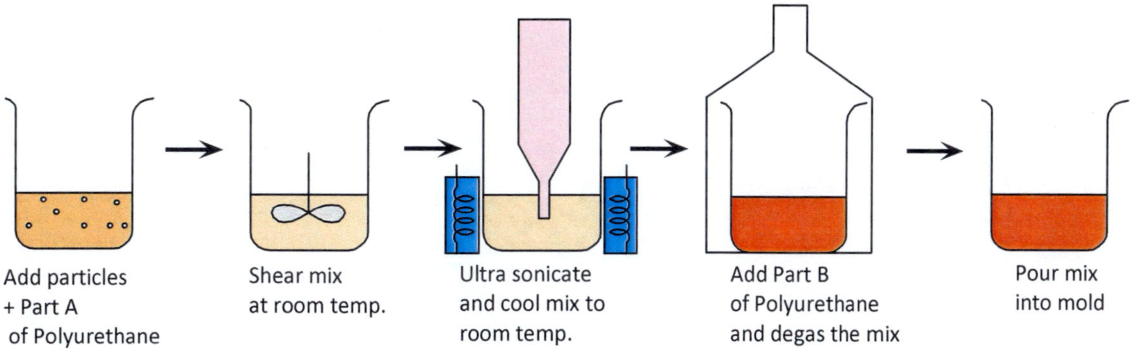

Fig. 3 Fabrication methodology used to create micro/nanoparticle filled polyurethane composites.

2.3 Fabrication Process

Fig. 3 provides a schematic of the fabrication methodology used to make our composites. In the first step, either micro- or nano-particles were hand-stirred into the resin component of the polyurethane for 5 minutes, after which the mixture was shear mixed for 15 minutes at 40% of the machine capacity (200 Watts). In the next step, the mixture was ultrasonicated for 45 minutes at 20,000 Hz at 150 Watts input energy with a 50% duty cycle. Ultrasonication generates heat, and the resin component of the polyurethane can be thermally degraded by such heat. Hence, a nitrogen-gas based cooling system, as shown in Fig. 4, was designed and built to remove the excess heat from the mixtures being sonicated. This system directs cold nitrogen gas around the mixing chamber during sonication while the temperature is monitored by a sensor, connected by controlled feedback, to the ultrasonic mixer. This method also avoids the problem of moisture infiltration into the hygroscopic boron nitride (as would happen with the commonly used ice-water bath), while keeping the mixture at or very near its optimal temperature throughout the mixing process. After completion of ultrasonication, the mixture was degassed for 15 minutes in a vacuum chamber. In the final step, the polyurethane curing agent was added to the degassed polyurethane resin/BN particle mixture and shear-mixed for a few minutes. Then the mixture was degassed a final time for 10 minutes before being poured into various molds.

Fig. 4, N_2-gas cooling system used during ultrasonication.

2.4 FE-SEM sample preparation and imaging

Having identified problems associated with specimen preparation for TEM, we chose field-emission scanning electron microscopy (FE-SEM) to visualize the dispersion of the BN particles within the composites. FE-SEM requires special specimen preparation methodology. Following is the description of the specimen preparation methodology for FE-SEM imaging:

(i) A sample of nanocomposite was trimmed to 3.0 mm square cross section and a length of about 10 mm. The sample was mounted snuggly in a micro-magnetic vice and which was placed in a central location on the magnetic chuck of the cryo-ultramicrotome.

(ii) A glass knife was prepared using a knife maker by cutting a 25 mm square piece from 8 mm thick by 25 mm wide glass bar. The glass square was then cut by the knife maker on a diagonal fashion to create two glass knives, each with a cutting edge 8.0mm long at 40° angle.

(iii) On the cryo-ultramicrotome, the Dewar flask was filled with liquid N_2, the knife was placed in its holder and properly aligned, the cryo-chamber was put in place, and liquid N_2 was allowed to flow into the chamber so that the pre-set thermostat allowed the temperature to drop to -100°C. After 20 minutes at -100°C, the cryo-enclosed knife was used to cut thinner and thinner sections until a smooth-as-glass surface was observed through the 100x viewing scope on the ultramicrotome.

(iv) The smooth surface was used for FE-SEM. This surface was removed from the sample using a razor blade and the 2.0 mm thick specimen was placed on electrically conductive epoxy which had been applied to the top of a 5 mm square silicon wafer. The epoxy was cured for 48 hours and later, the assembly was put inside a metal sputtering chamber where a 5 nm thick layer of gold was sputtered onto the top of the entire specimen.

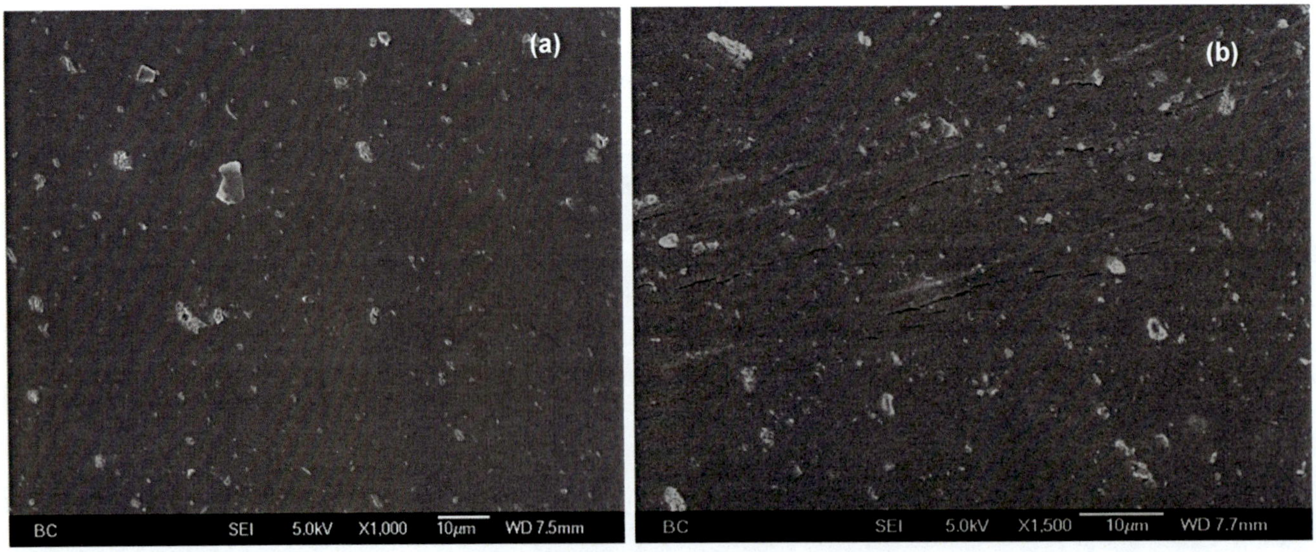

Fig. 5. FE-SEM images of nano-composites of (a) 2 and (b) 3 weight percentage BN nano-particles in the polyurethane matrix

Fig. 5 shows the images of nano-composites containing different weight percentages of BN: (a) 2%, and (b) 3%. A higher maginification photo of the 2% nanocomposite is shown in Fig. 6. Evident are particles of all sizes are present, and many are less than one micrometer. FE-SEM images such as this demonstrated that the ultrasonication process used in the fabrication of these nano-composites generates adequate dispersion of BN particles.

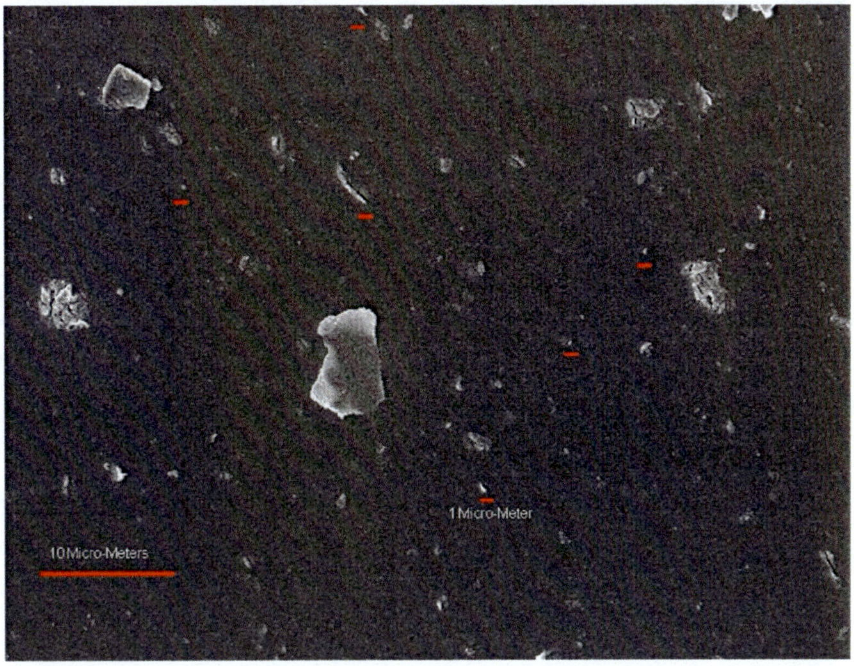

Fig. 6. Detailed FE-SEM image of a nano-composite with 2% by weight BN nano-particles in the polyurethane matrix.

2.5 Thermal Conductivity Measurements

A thermal conductivity analyzer from Setaram/C-Therm was used to measure thermal conductivity of the polyurethane composites. This instrument operates using a modified transient plane source principle. It required a one-sided, interfacial, heat reflectance sensor to apply a momentary heat source to the sample, as shown in Fig. 7. The heat results in a temperature rise at the interface between the sensor and the sample - typically less than 2°C, which induces a change in the voltage drop of the sensor element. The rate of increase in the sensor voltage is used to determine the thermo-physical properties of the sample material. Pyrex glass (1.1 W/m K) was used as a standard for thermal conductivity measurements. Measurements were made on several samples of each composite, treated, untreated micro- and nano-particle composites in sufficient number to generate statistically meaningful thermal conductivity for the polyurethane BN composites.

Fig. 7. Cartoon of the thermal conductivity analyzer (Courtesy: Setaram/C-Therm)

3. Experimental Results

Although measurements on both micro- and nano-particle composites were completed, only results related to BN nanoparticle-filled composites are discussed in this section and they will be presented at conference. Fig. 8 shows the variation of thermal conductivity as a function of weight fraction of both untreated and silane-treated BN nanoparticle-reinforced composites. The polyurethane without BN particles had a thermal conductivity of 0.248 W/m-K. The composite's thermal conductivity value increased to 0.280 W/m-K with the addition of 1% weight fraction of untreated BN particles, and reached a maximum value of 0.367 W/m-K at 5% BN weight fraction. The best percentage increase in thermal conductivity realized by the addition of untreated BN particles was about 50%.

As shown in Fig. 8, changing to silane-treated BN particles produced an overall improvement of about 20% in the thermal conductivity of the resulting nanocomposites. It seems the treatment did not provide better thermal conductivity. Moreover, the silane treatment might have increased the agglomeration of nano-particles in the polyurethane matrix. Careful FE-SEM examination is in progress to identify the reasons for the poor performance of the silane-treated BN particles. The thermal conductivity measurements of acetone- and nitric acid-treated BN particles (currently in progress) may provide additional insight into the unexpectedly low thermal conductivity values obtained from the composites made using silane-coated BN particles.

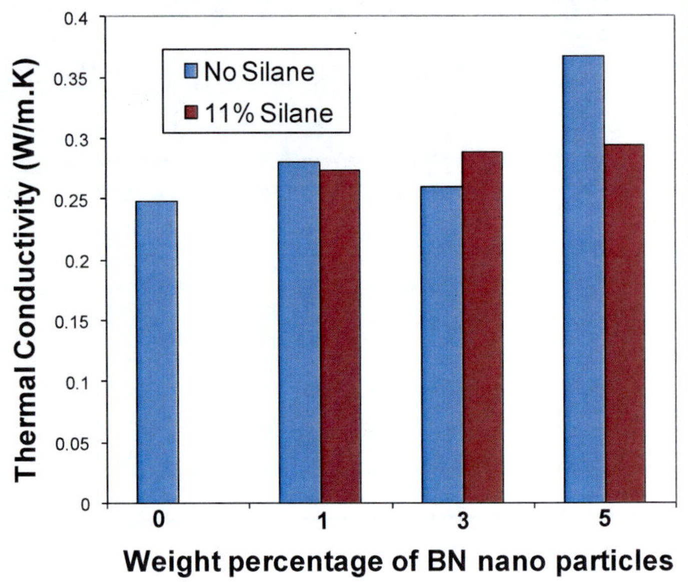

Fig. 8. Thermal conductivity measurements of untreated and treated BN nano-particles reinforced composites

4. Conclusions

A comprehensive study was conducted to fabricate BN-polyurethane nanocomposites and to investigate the effect of adding BN particles to the polyurethane on the resulting nanocomposite's thermal conductivity. The preliminary results indicated that at 5.0% weight fraction of untreated BN nano-particles, the thermal conductivity of the nanocomposite increased by about 50%. However, silane-treated BN particles did not improve the thermal conductivity of their nanocomposites by more than 20%, a result that was unexpected, because the silane treatment should have improved adhesion between the BN and the polyurethane matrix (as shown by decreases in the contact angle of water drops on silane-treated BN surfaces, as compared to untreated BN surfaces). At the present time, we do not have an explanation for these results, though the silane-treated BN particles exhibited a greater tendency to clump together/agglomerate than did untreated BN particles. Additional research is underway to determine why the treated BN particles failed to significantly increase the thermal conductivity of the nanocomposites in which they were contained.

Acknowledgements

The authors greatly appreciate the financial support from the Office of Naval Research's program of University Laboratory Initiative (ULI) to both the University of Massachusetts at Dartmouth, and to the Naval Undersea Warfare Center, Division Newport. The authors thank Program Officer Dr. David M. Drumheller and Program Director, Maria Medeiros for their support in conducting this project.

References

1. Mark, J.E., Physical Properties of Polymers Handbook, Springer, 2nd Edition, 2006.
2. Jiehui, Z., Sadayuki, T., Shoko, Y., Kenji, U. and De Vries, J.W.C., "Heat generation in multilayer piezoelectric actuators," *Journal of American Ceramic Society*, **79(12)**, 3193-3198,1996.
3. Xu, Y. and Chung, D.D.L., "Increasing the thermal conductivity of boron nitride and aluminum nitride particle epoxy-matrix composites by particle surface treatments," *Composite Interfaces*, **7(4)**, 243-256, 2000.
4. Xu, Y., Chung, D.D.L. and Mroz, C., "Thermally Conducting Aluminum Nitride Polymer-matrix Composites," *Composites: Part A*, **32**, 1749-1757, 2001.
5. Yung, K.C. and Liem, H., "Enhanced Thermal Conductivity of Boron-Nitride Epoxy-Matrix Composite Through Multi-Modal Particle Size Mixing," *Journal of Applied Polymer Science*, **106**, 3587-3591, 2007.
6. Zhou, W., Qi, S., An, Q., Zhao, H. and Liu, N., "Thermal Conductivity of Boron Nitride Reinforced Polyethylene Composites," *Materials Research Bulletin*, **42**, 1863-1873, 2007.
7. Terao, T., Bando, Y., Mitome, M., Zhi, C., Tang, C. and Golberg, D., "Thermal Conductivity Improvement of Polymer Films by Catechin-Modified Boron Nitride Nanotubes," *Journal of Physical Chemistry C*, **113**, 13605-13609, 2009.
8. Kemaloglu, S., Ozkoc, G. and Aytac, A., "Properties of thermally conductive micro and nano size boron nitride reinforced silicon rubber composites," *Thermochimica Acta*, **499**, 40-47, 2010.

Proceedings of the SEM Annual Conference
June 7-10, 2010 Indianapolis, Indiana USA

Advanced hard mask approach of ICs copper interconnects processes integration

Chun-Jen Weng
cjweng825@gmail.com
Department of Logistic and Technology Management,
Leader University,
No.188, Sec. 5, An-Chung Road, Tainan, 709, Taiwan, R.O.C.

ABSTRACT

When scaling the critical dimensions into nanotechnology, the impact of layout and line edge becomes important. Implementation of Cu and low dielectric constant (low-k) materials in the manufacturing process requires a complete understanding of these process characteristics and the challenges that appear during the hard mask based dual damascene approach. To create highly reliable electrical interconnects, the interfaces between the Cu metal and low-k must be optimized during the lithography, etching, ashing and copper processes. For higher aspect ratios interconnect profiles however this approach leads to increased sidewall roughness and undercut. To suppress problems in the photolithography and etching processes, the balance of the processes integration should be quantitatively and instantaneously controlled to the optimum manufacturing technologies. For copper filling engineering, this study also clearly demonstrated that the influence of liner barrier Ta and Cu seed performance of dual damascene manufacturing processes integration. Processes parameters that needed to be tuned are gas flow ratio, pressure and bias of the etch process. These process characteristics and manufacturing mechanism optimization will also be discussed.

Keywords: Hard mask, High-aspect-ratio interconnects, Fabrication, Processes integration.

1. INTRODUCTION

Scaling critical dimensions (CD) of CMOS devices to nanotechnology poses great challenges for development of robust lithography and etching processes. Copper (Cu) has been applied in sub 130nm semiconductor metallization processes because of its low resistivity and better reliability. As the semiconductor electronics industry continues its efforts in miniaturizing the integrated circuit (IC), an IC chip with copper interconnect back end of line (BEOL) structures has been developed for reducing RC delay in order to obtain high-speed signal communication. It is important to identify the fundamental processes limits of scaled copper/low-k interconnect schemes and correctly describe the active degradation mechanisms. In BEOL processing, both via hole and trench patterns formation are key processing steps that need to be accurately profiled using integrated module process technologies. The higher pattern density and high aspect ratio always result in pattern abnormal and collapse phenomena, because of the not optimal parameter for fabrication process feasibility. Therefore, for nanotechnology, high-pattern-density technology is desired for achieving a high-performance device.

During BEOL schematic formation, via hole etching, and trench etching are the dominant etching steps to form Cu interconnect layers; these steps are widely applied to manufacture the 90 nm node and beyond. To increase the density of microelectronic devices, the interconnect shrinkage is needed as technology advances. Reliability issue of void opening at the high stress areas as a result of via over etch was reported [1]. Process integration issues of the 0.13 μm Cu/Low K dual damascene integration were discussed [2]. Solutions to the issues were explored and reported. Resist poisoning issue was solved by modifying photoresist and planarizing bottom-anti-reflective-coating (BARC) scheme. The impacts of dual damascene cleaning on the reliability of the 0.13 μm Cu/Low K dual damascene interconnection was studied with splits

T. Proulx (ed.), *MEMS and Nanotechnology, Volume 2*, Conference Proceedings of the Society for Experimental Mechanics Series 2,
DOI 10.1007/978-1-4419-8825-6_35, © The Society for Experimental Mechanics, Inc. 2011

between batch process and single wafer cleaning. The critical applications of isotropic etching, and plasma ashing/cleaning to form precisely controlled profiles of high-aspect-ratio form precise via holes and trenches used in reliable Cu/low-k interconnects in the back end of line (BEOL) are described in detail [3].

Moreover, the investigation of resist pattern collapse with top rounding resist profile was shown [4]. They found that the pattern size is reduced as the device is more integrated. The resist deformation phenomenon has been a serious problem under 100 nm line width patterns. The simulated results show that the pattern collapse phenomenon is reduced for a rounded top resist profile rather than for a flat top profile. Hybrid dual damascenes interconnect approach with organic ultra-low-k for gap filling process was demonstrated [5]. The traditional patter developing approach with via-first process for dual damascene suffers from ashing damage for chemical vapor deposition (CVD) ultralow-k value material. The copper surface condition of via bottom is a big concern of trench etching process as well. The electrical parameter issues that occurred at 65nm trench etch process such as Rs; via resistance (Rc) and VBD were presented [6]. The feasible solutions and related etching mechanisms are also addressed for the above issues from the point view of the improvement of line-edge roughness (LER), within wafer AEI CDU (critical dimension uniformity) and interface conditions of via-bottom.

BEOL process defects of wafer fabrication (e.g., high-resistance via or interconnect defects) have increasing occurrence probability than front-end-of-line (FEOL) defects in nanometer technologies. A defect reduction methodology for advanced copper dual damascene oxide etching was proposed [7]. They discuss the methodology used in determining the source of the defects, correcting marginal hardware, verifying the defect improvement, and using monitor wafers to help separate the tools in question from the rest of the manufacturing line. A simple percolation model shows that if there are existing defects at the trench interface, then percolation requires fewer additional defects [8]. The final failure occurs at the top horizontal interface with delamination and copper penetration.

The reliability of dual damascene Cu / low–k interconnects is limited by electromigration – induced void formation at vias. In [9], they investigated via void morphologies and associated failure distributions at the low percentiles typical of industry reliability requirements. They show that Cu / low–k reliability is fundamentally limited by the formation of slit - voids under vias. The mechanism of copper (Cu) voids formation from electro-chemical plating (ECP) followed by Cu chemical mechanical polishing (CMP) are studied in Cu dual-damascene interconnection. Several challenges regarding diffusion barrier deposition have been identified, primarily (a) ability of barrier to prevent diffusion of Cu into low k material, (b) adherence of Cu to barrier and barrier to low k material, and (c) integrity of barrier conditions [10].

To suppress problems in the fabrication processes, the balance of each process module should be quantitatively and instantaneously controlled to the optimum for each process step. Implementation of Cu and low dielectric constant (low-k) materials in the manufacturing environment requires a complete understanding of these new material characteristics and the challenges that appear during their processes integration. By study and optimization of hard mask stacks and etching fabrication; [11-12] developed a production worthy processes for fabrication of hard mask used for etching in CMOS technology. Dual damascene methodology has dramatically reduced complexity and shortened real process times over traditional metal etch and fill methods. The absorber and buffer etching is a crucial step in the manufacture of EUV (Extreme ultraviolet lithography) masks due to the stringent CD and reflectance requirements. A method of forming optimal dual damascene process for the BEOL process was proposed [13]. To overcome these challenges, accurate and repeatable depth assessment of damascene structures requires the ability to resolve high-aspect-ratio structures in both a high density and isolated structures in the manufacturing process by improving yield.

2. EXPERIMENTS

As technology nodes advance to nanotechnology and beyond, IC companies are investigating the use of a metal hard mask such as TiN in order to gain better etching selectivity to the film materials during the dry etching process. Hard mask was commonly used as etch mask due to the thermal instability of photo-resist in the etch process. Hard mask etching processes have been developed which allowed us to obtain tight profile and CD control. The nanotechnology copper dual damascene architecture was fabricated according to the scheme shown in Fig. 1. Via-first-trench-later approach was used for the formation of dual damascene structure. After the first single damascene process, the wafer was deposited with dielectric film and followed for via patterning process, and again for the trench forming process. In the via-first approach the vias are defined first in the ILD (inter-layer dielectric), followed by patterning the trenches. The inter-metal dielectric layer (IMD1) consisted of a SiN layer and SiO_2 layer. The M1 (metal 1) trench was formed by dry etching after IMD1 deposition. A stack of physical vapor deposition (PVD) barrier layer, Cu seed layer, and electroplating Cu layer was then deposited. Layers of SiN and SiO_2 were deposited as IMD2. M2 trench and via were then formed by a via-first dual-damascene process. The M2 metallization stack was the same as that of M1. In a dual-damascene (DD) structure, only a single metal deposition step is used to simultaneously form the main metal lines and the metal in the vias. Dual-damascene technology in the fabrication of advanced interconnects presents itself as an integration and reliability challenge. The detail physical schematic of the Cu interconnects is shown in Table 1. Moreover, patterns etching development becomes more crucial mainly because of the decayed ion flux down to the bottom of the trench. When the aspect ratio was increased to 5, considering the etching selectivity and loading effects, the top of the photoresist profile collapsed easily after the lithography development process. High aspect ratio dual damascene copper structure fabrication is a complicate lithography patterning process.

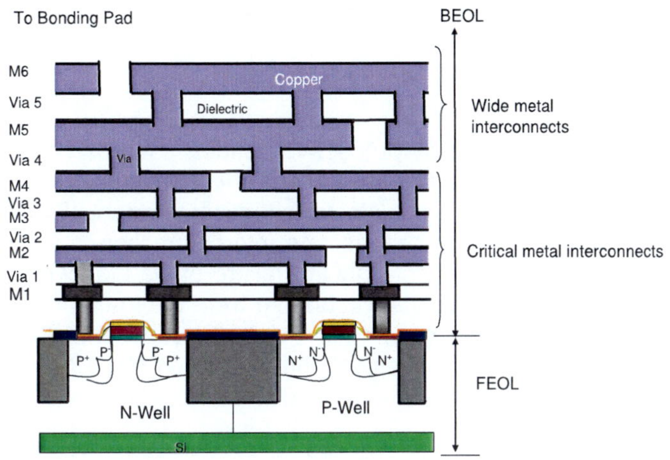

Fig. 1. Semiconductor copper multi-metal interconnect scheme.

Table 1 Dual damascene physical dimension

Interconnect Via/Line width	M1/V1 to V3/M4	V4/M5 to V5/M6
	120 nm	240 nm
Via film deposition	~600nm	~800nm
Via aspect ratio (deposition)	~5	~3.3
Photoresist coating thickness (Via)	~600nm	~800nm
Film process buffer (Etching/ Polish)	~200nm	~200nm
Via film thickness (After Via Processing)	~400nm	~600nm
Trench depth (design)	~200nm	~300nm

The major reliability challenges that must be overcome to achieve the scaling targets for future integrated circuits (ICs) are related to the BEOL process. Etching processes are commonly used to fabricate vertical sidewall trenches and vias for copper (Cu)/low-k dual damascene devices. However, applying layer on top of hard mask, tight thickness control of hard mask becomes a major issue in order to maintain the substrate reflectivity for good for providing tight CD (Critical Dimension) and profile control. Multi-layer stack application will be established at manufacturing nanotechnology process node. In Fig.2, stacked structures of hard mask material and interconnects were demonstrated. Moreover, etching of hard mask presents a challenge, as etching process should find a balance between contradicting requirements of providing sufficient selectivity to photoresist and avoiding formation of excessive sidewall polymer resulting in CD gain.

Via etch and trench etch were needed to be optimized for the profile, electrical and reliability improvement of dense via chain and comb structure. A serious problem in wafer fabrication is the defects issue during the pattern development process, because it decreases the yield of wafer production. Abnormal patterning phenomena lead to yield loss during the electrical device test and productivity yield losses. Dual damascene processes demand cleanliness since defects generated during trough etching result in abnormal data of the electrical device, and can also cause problems in later process steps of next interconnect layer. To evaluate the damascene architecture processes, physical and electrical characterizations were carried out. Therefore, in the nanotechnology, high-pattern-density technology is desired for achieving a high-performance device.

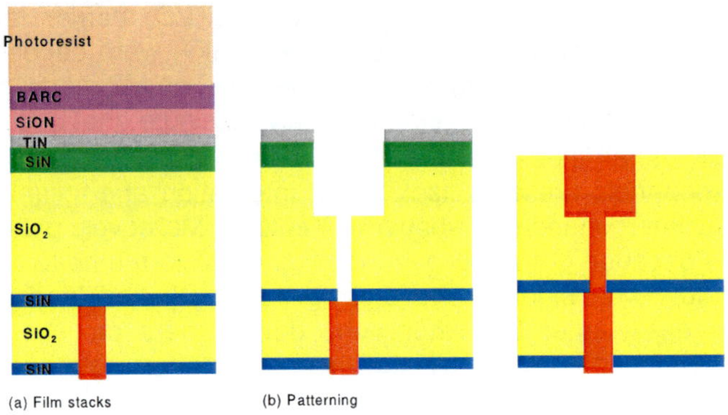

Fig. 2. Top hard mask film stacks and interconnect processes scheme.

3. RESULTS AND DISCUSSION

Copper voiding within vias is strongly process dependent, and can be suppressed by optimization of via barrier deposition, cleaning and etch processes. The mechanisms of copper voids formation from etching pattern profile, linear barrier/seed layer, electro-chemical plating (ECP) followed by copper chemical mechanical polishing (CMP) of Cu dual-damascene interconnection. The formation copper voids at metal lines are the main problem that causes not only the failure of via-induced metal corrosion but also yield loss and abnormal electrical device. Cu voids at via bottom were the major factors resulting in the failure of stress migration and electrical migration testing. In this Fig. 3, the mechanism of the formation of Cu voids of the fabrication processes was investigated. Besides the pattern profile, a possible cause of the void formation at the via bottom is the overhang barrier, Cu metal filling characteristics.

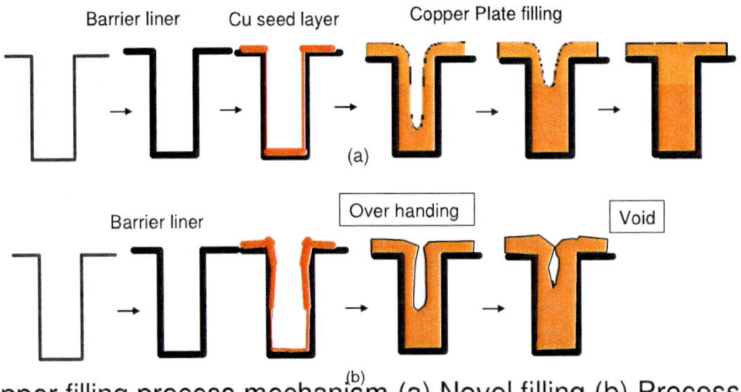

Fig. 3. Interconnect copper filling process mechanism (a) Novel filling (b) Processes induced void.

Processed induced voiding leads to electric circuit open or high electrical resistance. Via structure is very critical for the electrical performance of via chains with dual damascene process. From failure analysis, void always found on metal surface below and/or around via bottom. However, in the dual damascene processes integration, the most difficult and important issues to be solved are not only formation of dual damascene profile, dry process tuning and void-less Cu filling, but also suppressing the pattern dependence of via characteristics, which tend to fail in dense via pattern. It is important to notice that even though defective etch and consequently Cu filling processes are observed process defect review images. Failure analysis (FA) is necessary in order to understand what caused the failure and how it can be prevented in the future. Once the failure mechanism has been determined, the process owner or expert can work with the failure analyst to determine the root cause of the problem. The impact of the non-optimized process on the Cu filling can be observed in Fig. 4, as voids in the center of the trenches are present. The dark regions imply failure mode by comparison of the same pattern. The formation of Cu voids at metal lines is the main problem that causes not only the failure of via-induced metal-island corrosion but also yield loss. The cross-sectional TEM image shows that the defect is a Cu void at via bottom at the failure site, as shown in Fig. 4. The defect map of the Cu void is nearly the same as the yield-loss map. Therefore, it is suspected that the yield loss results from the Cu-void defects. A non-optimized etched process affects the Cu filling of the trenches: voids are present due to the undercut occurring during the trench etch. The review of defect map of the Cu void is nearly the same as the yield-loss map. Therefore, it is suspected that the yield loss results from the Cu-void defects. The defect map of the Cu void is nearly the same as the yield-loss map. Therefore, it is suspected that the yield loss results from the Cu-void defects. It is important to validate the failure of a sample prior to failure analysis in order to conserve valuable FA resources.

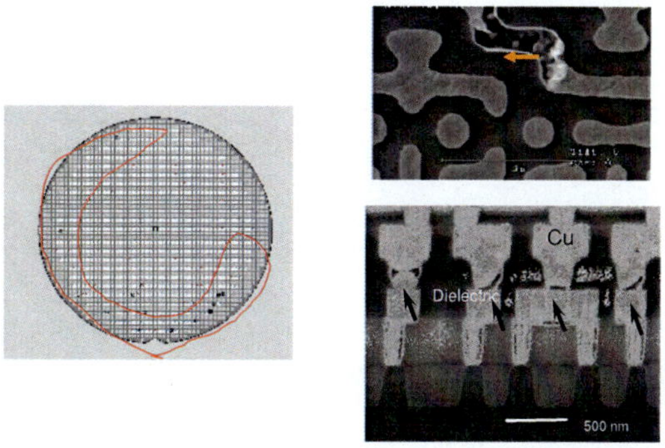

Fig. 4. Comparison images of showing abnormal pattern on metal interconnect.

Dual damascene processes demand cleanliness since defects generated during trough etching result in abnormal data of the electrical device, and can also cause problems in later process steps of next interconnect layer. Therefore, in the deep submicron era, high-pattern-density technology is desired for achieving a high-performance device. The abnormal pattern issue results process issues include pattern missing, peeling, dielectric breakdown, and serious defects on process development control strategies. However, in the dual damascene processes integration, the most difficult and important issues to be solved are not only formation of dual damascene profile, dry process tuning and void-less Cu filling, but also suppressing the pattern dependence of via characteristics, which tend to fail in dense via pattern. Fig. 5 shows the comparisons of the dense abnormal via/line pattern after trench etch on wafer center and edge, respectively. Furthermore, that could cause missing pattern after etch in the serious case as show especially in wafer edge where the trench lines were not defined after trench etch. Systematic yield losses are process-related problems that can affect all die chips on a wafer. It is important to produce better die chip per wafer by minimizing the cycle time to detect and fix yield problems associated with the advanced process module technologies.

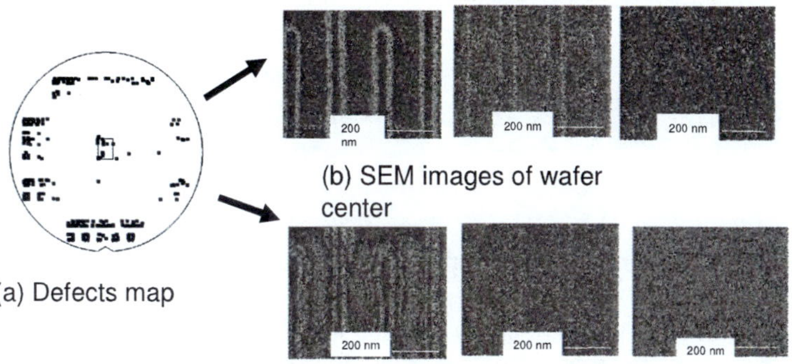

(c) SEM images of wafer edge

Fig.5. SEM images showing abnormal pattern structure after trench pattern etching.

For the copper dual damascene process, the condition of via profile is always affected by the via etching, via cleaning, trench patterning, trench etching, trench cleaning, and dual damascene cleaning process. Both the abrupt and progressive resistance increase failure modes are observed irrespective of the details of the fabrication process associated with the Cu trench or via or inter metal level dielectric. The compare of electrical resistance is also included in the present study. This test pattern is always used in semiconductor process in-line electrical and physical checking. Electrical resistance the box plot distribution for a 120×120 nm^2 copper metal trace were compared and shown in Fig. 6. The distribution of electrical resistance is narrow in the case of the optimal integrated process, and there are no open electrical test data for the optimal integration process. High electrical resistance could still happen if the fabrication processes were not optimized. Void resulted in resistances high also shown in comparison of electronic device verification. Copper voiding within via is strongly process dependent, and can be suppressed by optimization of via barrier deposition, cleaning and etch processes.

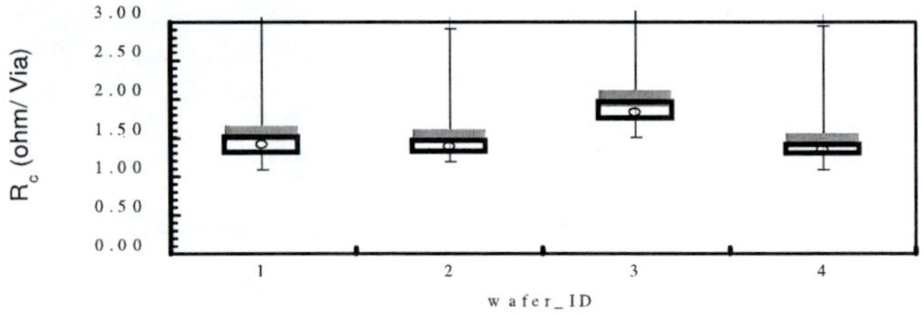

Fig.6. Via resistance distribution of dense via chain subjected to the general process.

As Fig. 7 shown, the via-first approach the vias are defined first in the inter layer dielectric films, followed by patterning the trenches. This is currently the most common method of fabricating dual-damascene interconnects. The sequence of forming the damascene recesses in this approach begins by exposing via patterns with the first mask. After etching the vias completely through the entire dielectric stack and stripping the photoresist, a second mask is used to pattern the trenches. An improved resist etch-back technique using three resist layers also has been developed which offers enhanced planarity over the dielectric film layers process. Between the photoresist treatment and back antireflection coating (BARC), plasma etching is performed to remove the excess gap-fill material above the substrate surface and the dielectric film top surface before photoresist coating and pattern development. A planarizing dielectric film and gap-fill material are used to ensure that the lithography process produces the best profiles and enables critical dimension control. Now, an improved shallow trench isolation process has been developed, whereby high aspect ratio trenches filled with dielectric material exhibit a high degree of integrity and planarity. The processing sequence after via/trench opening consist of barrier metal deposition, copper seed deposition and electrochemical plating of copper and the subsequent processes of Cu chemical mechanical planarization (Cu CMP). The thickness and deposition condition of barrier metal affects directly the result of resistance in interconnection and productivity yield.

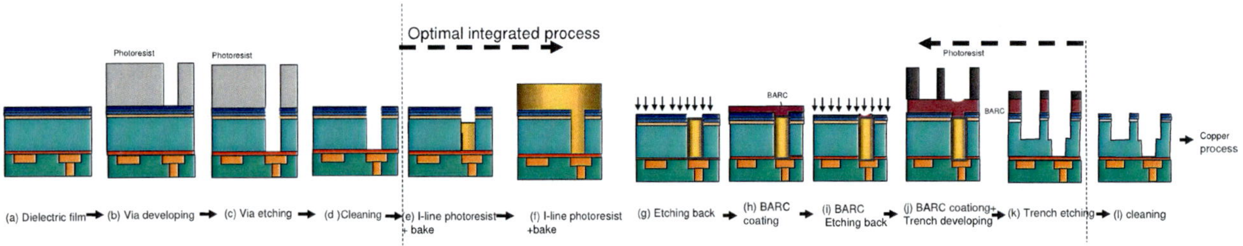

Fig.7. Proposed optimal integrated copper dual damascene interconnects processes.

The profile inspection after step-by-step checking reveals that fabrication processes integration feasibility. The impacts of high via patterns on the electrical and reliability performance of different via chain structure were investigated. For high aspect ratio via etching selectivity, the photoresist remaining after the etching procedure is an important factor in estimating the feasibility of the lithography and etching processes. In the comparisons of top SEM view images of Fig.8 illustrates the different critical via chain structure and critical pattern of the improvement for different locations of wafer. Moreover, Fig.9 also indicates the cross-section view of dense via chain pattern. After the integrated optimal processes methodology was adopted, no abnormal collapse and enough lithography photoresist remain even in relation to the dense via chain patterns after the via etching process. No abnormal pattern of dense and isolated pattern was found for different locations of wafer. Voiding within copper via and abnormal pattern are strongly processes dependent, and can be suppressed by optimization of via barrier deposition, cleaning and etch processes.

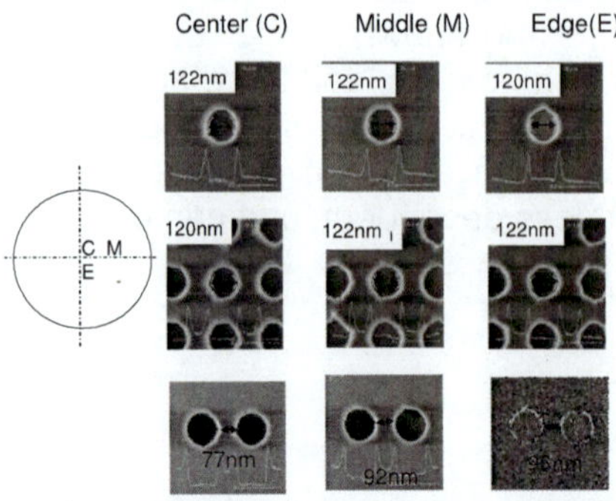

Fig. 8. Top view inspections of critical via chain patterns on different locations of wafer.

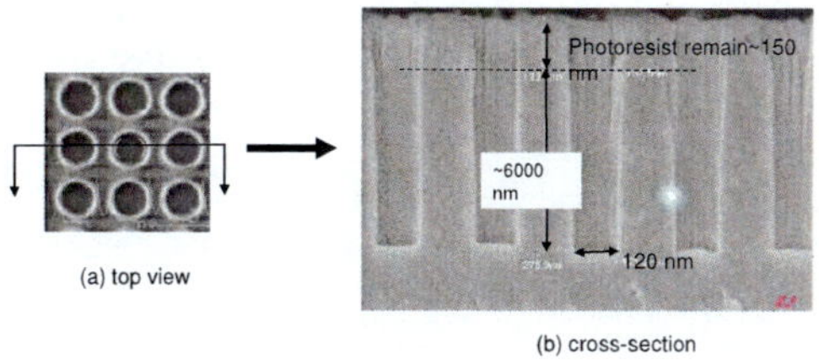

Fig. 9. Via top view inspections and cross-section for dense vias pattern after via etching

The wafer manufacturing process characteristics depend on these basic mechanisms, which form the basis of applications to device fabrication. Consequently, any void in via fill material can lead from deformation of via shape during the trench etch process uniformity. After the new gap-fill process and etching back integration methodology were adopted, no abnormal pattern was observed, and no damage pattern for different critical dense patterns around the wafer is indicated in Fig. 10. In BEOL dual damascene processing, both via hole and trench formation are key processing steps that need to be accurately profiled using optimal integrated processes. The lithography and etching processes are the most critical step in the fabrication of nanostructures for integrated circuit manufacturing. It is important to notice that even though defective etch processes are observed in cross-section images, the electrical results of the same structures shown in the images can be considered normal. When the etch process is optimized, the trench profile does not show any undercutting and damage profile is minimized, which results in good copper filling process. For etching selectivity, the photoresist remaining after the etching procedure is an important factor in estimating the feasibility of the lithography and etching processes. The remaining photoresist thickness affects interconnect physical profile of the pattern after the etching procedure. Figure 11 indicates that the cross section of the lithography photoresist thickness remains in relation to the isolation and dense patterns after the trench etching process. As a sufficient photoresist thickness remains, there will be no top rounding on the pattern after the trench pattern etching process. The electrical device performance will be good.

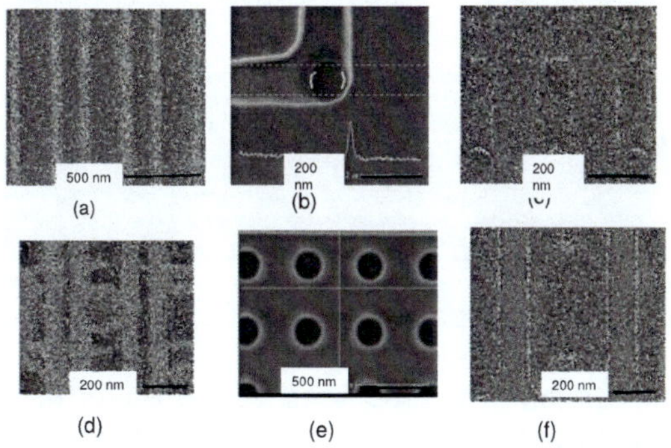

Fig. 10. In-line process SEM inspection on different critical via chain patterns.

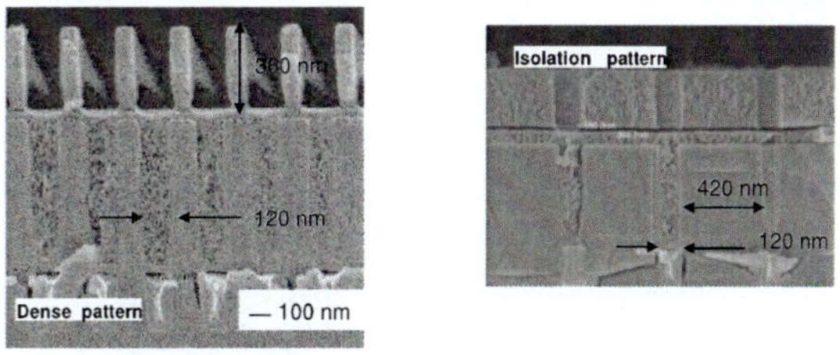

(a) Dense pattern profile (b) Isolation pattern profile

Fig. 11. Inspections of dense/isolation via chain patterns after trench pattern lithography process (Via first/ trench after process).

The key processes for the via first process integration scheme are the protective via fill process and the subsequent trench etch process. Each dual-damascene strategy presents specific challenges from lithography to etch. As Fig.12 proposed integrated processes, the photoresist filling quality of via is important for the subsequent proper shape of the dual damascene structure. Figure 12 indicates that the cross section of the dense patterns step by step verifications after via pattern etching process for wafer center and edge. However, the local via hole density and size have a dramatic effect on the resist thickness on the dielectric surface. Use of a via-fill can eliminate (with etching back) the impact of local via hole density and via hole size on lithography. Photoresist remaining in the via chains after development also varies with via hole density, from being completely developed out in dense structures to essentially undeveloped in isolated structures, thus complicating the etch/strip processes. For etch process, sensitivity of the resist, selectivity of the bottom barrier and profile in the bottom layer are most critical. The control of dense interconnects of lithography and etch processes are also demonstrated.

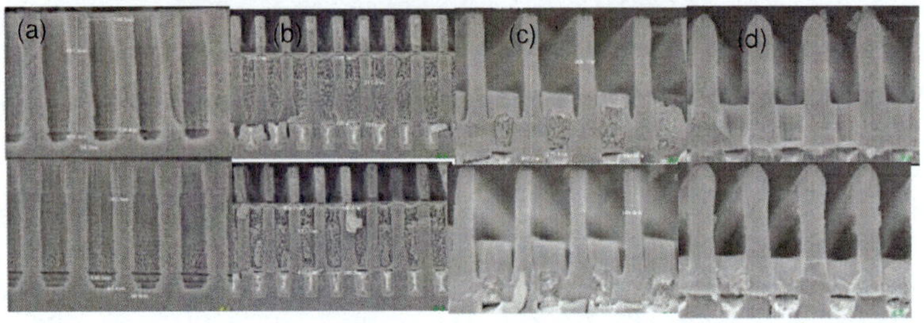

Fig. 12. Step by step profiles of via/trench chain with optimal integrated processes on wafer center/edge (a) Post via etch, (b) Post trench line lithography, (c) Post trench line etch (d) Post trench line etch clean.

The damascene process consists of etching via and trench with high aspect ratio approaching, electrochemical plating (ECP) copper and polishing off the copper overfill with chemical mechanical polishing (CMP). Copper metallization, widely used in logic devices, has led to interconnect geometries that are rapidly approaching nanoscale dimensions. The killer void that causes interconnect failure decreases as the interconnect dimension shrinks. In this study, different types of barrier/ Cu seed schemes for the Cu dual damascene fabrication were evaluated. Fig. 13 shows copper seed effect for copper filling process. It was observed that the void can be minimized by changing the seed layer thickness. The barrier metal profiles with different thickness were identified. It was found in this study that low Cu barrier deposition generally resulted in poor Cu/barrier interface and via void formation. The higher failure open rates were observed from wafer with below 30nm copper seed thickness splits. As each process always trade-off, the thickness of barrier metal was need to be optimized at plasma barrier and copper seed deposition. In this study, it is concluded that wafers processed with higher than 30nm high-density plasma copper seed deposition achieved the best electrical and reliability performance of 120 nm width dual damascene interconnection.

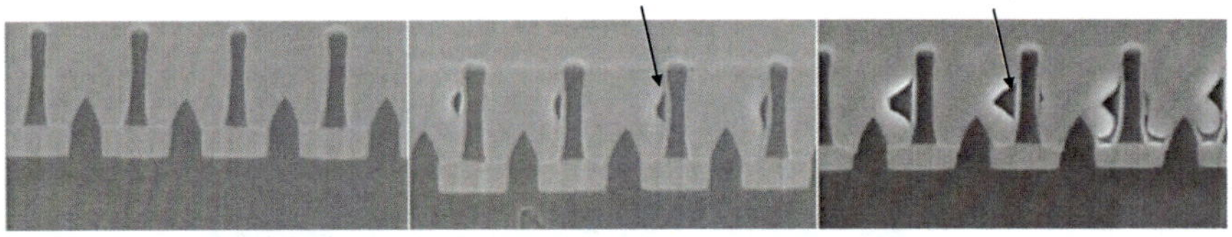

Fig.13 Comparison of copper seed effects of copper dual damascene process.

Fig. 14 shows the cross-section of interspersed dense pattern structures which wafer was fabricated after dual damascene patterning processes. Fig. 14(c) and (d) also illustrates the cross-section of interspersed dense pattern structures which wafer was fabricated after copper deposition and polish processes, high density plasma barrier metal deposition are also indicated. The images for various view shows no damage patterns were found to have formed directly for the dense via/trench chains. Therefore, a planarizing dielectric film and via gap filler material are used to ensure a trench lithography process produces the best profiles and critical dimension control. Comparing results from different critical dense via chain schemes shows that this present optimal integrated processes can also produce effective pattern as demonstrated in copper chemical mechanical polishing process of dual damascene process.

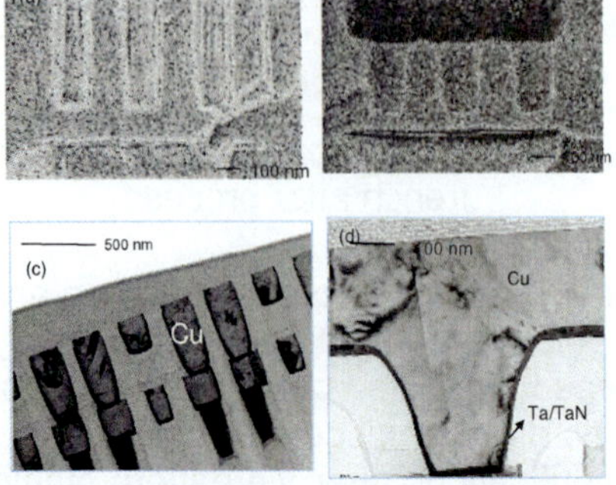

Fig.14 Images showing profiles of via chain with optimal integrated processes: (a) after dual damascene pattern etching (X-X section) (b) after dual damascene pattern etching (Y-Y section) (c) after copper filling and polishing (d) barrier seed verification

As indicated in Fig.15 the test pattern is always to check the semiconductor process in-line electrical WAT

D

voltage (V_S=V_{SUB}=0V),

Measure electrical current (I_d), the Electrical resistance Rc=V_D/I_d

As indicated in Fig.16, the leakage current checking, drain voltage (V_D=1.2V), source voltage (V_S=0), Measure current I_D to verify the interconnect leakage current.

The electrical resistances were also included in present study. This test pattern is always to check the semiconductor process in-line electrical and physical checking. Fig.15 indicate electrical device test on open/short for 120nmx120nm copper metal trace. A comparison of electronic properties between the optimal integration process and the general process was also included. The data distribute tight and no open electrical test data for optimal integration process. From Fig. 15 to Fig.16 show the electrical resistance verification and current leakage checking for copper metal trace, respectively. The distribution of electrical resistance is tight for optimal integrated process and there are no open electrical test data for optimal integration process. There is also no significant difference in the leakage currents for the present optimal integrated dual processes.

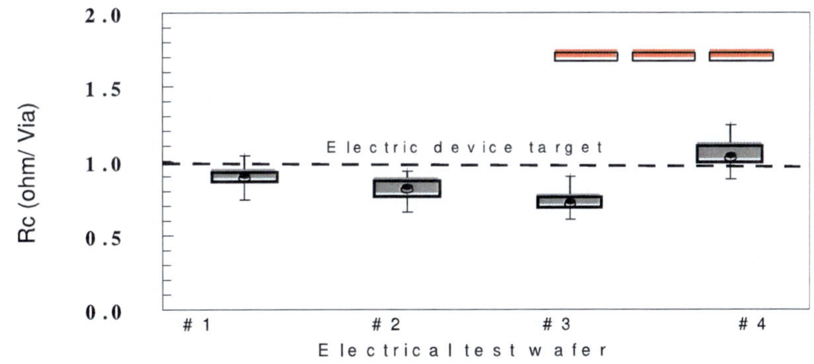

Fig.15 Contact resistance of electrical device test of optimal integration process.

Fig.16 Electrical leakage current test of optimal integration process.

4. CONCLUSIONS

The improvement of high aspect ratio nanotechnology hard mask dual damascene by integrated manufacturing processes was reported. The effects of integrated fabrication process on lithography, etching, barrier and seed in dual damascene copper interconnects manufacturing processes have been investigated. In general, novel integrated copper dual damascene interconnects offer better reliability than traditional interconnects for nanotechnology fabrication processes. However, each dual-damascene process is always trade-off and complicate. With the understanding of the complexities involved in copper interconnects and the associated integration processes, a robust reliability is achievable for this relatively new copper technology.

It is concluded that wafers processed the improvement of dual damascene interconnect process achieved the best electrical and reliability performance.

ACKNOWLEDGMENT

The author gratefully acknowledges the support and assistance of NSC 98-2221-E-426-002.

REFERENCES

[1] Tsang CF, Bliznetsov VN, Su YJ, Study and improvement of electrical performance of 130 nm Cu/CVD low k SiOCH interconnect related to via etch process. Microelectron J. 2003; 34 :1051–8.

[2] Li HY, Su Y.J., Tsang CF, et al. Process improvement of 0.13 μm Cu/Low K dual damascene interconnection, Microelectronics Reliability. 2005; 45: 1134–43.

[3] Haruhiko ABE, Masahiro Y, Nobuo F, Developments of Plasma Etching Technology for Fabricating Semiconductor Devices. Japanese Journal of Applied Physics. 2008; 47: 1435–55.

[4] Lee HJ, Park JT, Yoo JY, et al. Resist Pattern Collapse with Top Rounding Resist Profile, Jpn. J. Appl. Phys. 2003; 42:3922.

[5] Su YN, Shieh JH, Hsu P. P., Low k damage control & its reliability for organic hybrid dual damascene, Proc. of the Int. Symp. Physical and Failure Analysis of Integrated Circuits. 2004; 69-70.

[6] Zhao LL, Shen MH, Han QH, et al. A study of 65nm BEOL trench etch issues, Int. Conf. Solid State Integr. Circuits Technol. Proc. ICSICT, 2008; 1173-76.

[7] Biolsi P, Ellinger S, Morvay D, Defect reduction methodology for advanced copper dual damascene oxide etch , IEEE Int. Symp. Conf. 2000; 312-20.

[8] Borthakur S, Satyanarayana S, Knorr A, et al. Reduction of the initial defect density and improvement of the reliability of Cu/low- k structures by a methylating treatment. J Vac Sci Technol B Microelectron Nanometer Struct 2005; 23:2222-5.

[9] Oates AS and Lee SC, Electromigration failure distributions of dual damascene Cu / low – k interconnects. Microelectronics Reliability 2006; 46: 1581-6.

[10] Mosig K, Cox H, Klawuhn E. et al. Integration of porous ultra low-k dielectric with CVD barriers, IEDM Tech. Dig. 2001; 88–91.

[11] Letzkus F, Butschke J, Irmscher M. et al. Dry etch processes for the fabrication of EUV masks, Microelectronic Engineering 2004 73–74: 282–88.

[12] Tetsuya T, Control of reactive plasmas for low-k/Cu integration, Applied Surface Science 2007; 253:6716–37.

[13] Weng CJ, Lin YS, Chen CY, Method of forming damascene structure, USA Patent, No. US 7,189,640 B2 2007.

Proceedings of the SEM Annual Conference
June 7-10, 2010 Indianapolis, Indiana USA

Advances in Thin Film Nanoindentation

B. Zhou, K. Schwieker, and B.C. Prorok
Auburn University, Department of Mechanical Engineering
275 Wilmore Engineering Labs, Auburn, AL 36849, Email: Prorok@auburn.edu

ABSTRACT: A new model of thin film indentation that accounted for an apparent discontinuity in elastic strain transfer at the film/substrate interface was developed. Finite element analysis suggested that numerical values of strain were not directly continuous across the interface; the values in the film were higher when a soft film was deposited on a hard substrate. The new model was constructed based on this discontinuity; whereby, separate weighting factors were applied to account for the influence of the substrate in strain developed in the film and vice-versa. By comparing the model to experimental data from thirteen different amorphous thin film materials on a silicon substrate, constants in each weighting factor were found to have physical significance in being numerically similar to the bulk scale Poisson's ratios of the materials involved. When employing these material properties in the new model it was found to provide an improved match to the experimental data over the existing Doerner and Nix and Gao models. Finally, the model was found to be capable of assessing the Young's modulus of thin films that do not exhibit a flat region as long as the bulk Poisson's ratio is known.

INTRODCUTION: Nanoindentation plays a very important role in the field of thin film mechanics [1-22]. Nanoindentation on films a few microns or less is usually complicated by substrate effects as the deformation field caused by the indent makes contact with and propagates into the substrate. The effect intensifies when the material properties of the film and substrate become increasingly different. Eliminating or accounting for the substrate effect has been a topic of study for some time and researchers have used both experimental and theoretical methods to investigate this issue. As a result many useful models and insights have been developed [1-9]. There is still much to be learned though as most of the experimental-based studies have been performed on crystalline films where microstructural aspects such as grain boundaries, size, orientation and texture can influence results. To mitigate the microstructural issues, this work investigated the manifestation of substrate effects using thirteen different amorphous thin film materials deposited onto silicon substrates. The films had a range of elastic moduli such that the cases of soft film on hard substrate and hard film on soft substrate were covered. A new model was developed that leverages the empirical data and results from finite element analysis to better explain how substrates effects manifest in thin film indentation.

THEORETICAL BACKGROUND: Substrate effects arise in thin film indentation as the elastic strain field developed in the film, from an indenter punching into its surface, makes contact with and penetrates the substrate. In most cases, the film and substrate have significantly different mechanical properties such that they absorb the strain differently. An example of this effect on measuring the Young's modulus is given in Fig. 1 (a) for an amorphous carbon film with a thickness of approximately 1125 nm on a silicon substrate. At small displacement into the surface the modulus value appears relatively constant, but as the indenter penetrates further to the critical indent depth, h_{cr}, it begins to increase due to the influence of the substrate. It finally reaches a plateau where the mechanical properties of the substrate dominate. In this particular film a flat region can be identified where it is assumed that substrate effects are negligible. However, not all materials exhibit a definitive flat region, complicating determination of film properties.

The first protocol for dealing with the effect of the substrate in thin film indentation came from Bückle [23], who suggested to limit the indentation depth within 10% of the film thickness so that the substrate could be neglected. As the desire to indent ever thinner films increased, limitations on indenter size and sharpness warranted development of a method that could account for substrate effects. A model developed by Doerner and Nix [5] was the first to do so and proposed an empirical function based on the contact depth to describe the film and substrate elastic contributions to the overall composite compliance. A rearranged form of the function is given as,

$$\frac{1}{E'} = \frac{1}{E'_f} + (\frac{1}{E'_s} - \frac{1}{E'_f})\Phi_{D-N}, \quad (1)$$

where,

DOI 10.1007/978-1-4419-8825-6_36, © The Society for Experimental Mechanics, Inc. 2011

$$\Phi_{D-N} = e^{-\alpha(t/h_{eff})}. \tag{2}$$

Here E' is the composite modulus, $E_f' = E_f/(1-\upsilon_f^2)$, $E_s' = E_s/(1-\upsilon_s^2)$, t is the film thickness, h_{eff} is the effective indentation depth, and α is an empirically determined constant, which was suggested as 0.25 for most materials. The function Φ_{D-N} is essentially a weighting factor that accounts for the continuously changing contribution of the film and substrate as the indenter penetrates. Although, it does not have a physical basis, the function works relatively well for most thin film materials. Numerical work by King [24] suggested replacing h_{eff} in the original function by the square root of the projected contact area. Leveraging this insight, Saha and Nix [20] introduced the effective film thickness "*t-h*" to replace *t* in the original function.

Another important contribution came from Gao and co-workers [21] in the form of an analytical model that focused on the total energy transformation from the indenter into a thin film/substrate system during indentation. They derived an analytical solution beginning with a rigid cylindrical stick punching into a semi-infinite elastic body with a different surface layer. Their function is given as,

$$E' = E_f' + (E_s' - E_f')\Phi_{Gao}, \tag{3}$$

where,

$$\Phi_{Gao} = \frac{2}{\pi}\arctan(x) + \frac{1}{2\pi(1-\upsilon)}\left[(1-\upsilon)x\ln\frac{1+x^2}{x^2} - \frac{x}{1+x^2}\right]. \tag{4}$$

Here υ is the Poisson's ratio (assuming the film and substrate have the same υ) and $x = t/a$ where a is the radius of the contact area. In general it has an analogous weighting factor similar to the Doerner and Nix function to account for the substrate. Although approached from different directions, both models have merit and continue to be the standard methods employed by most researchers to account for substrate effects in the thin film indentation community.

EXPERIMENTAL PROCEDURE: The thin films investigated in this work were all deposited onto single crystal (100) silicon wafers. Materials were selected in a wide range of elastic modulus values such that several were softer than the silicon substrate while others were harder. Furthermore, in order to remove microstructural effects such as grain boundaries and grain orientations, only materials capable of forming amorphous structures were considered. These materials included near-frictionless carbon (NFC), chromium oxide (CrO_x), silicon dioxide (SiO_2), iron oxide (FeOx), titanium oxide (TiO_x), amorphous Si (a-Si), aluminum oxide (AlO_x), aluminum nitride (AlN), iron boron (FeB), nickel titanium (NiTi), METGLAS 2826B ($Fe_{40}Ni_{38}Mo_4B_{18}$), silicon nitride (Si_3N_4), silicon carbide (SiC), and tetrahedral hydrogenated carbon (ta-C:H). A list of their properties and deposition techniques is given in Table 1. Thin films were deposited using sputtering and various chemical vapor deposition techniques. X-ray diffraction confirmed the amorphous structure of the films.

Table 1: Deposition parameters for the thin films

Thin Film Materials	Elastic Modulus (GPa)	Deposition Technique	Target materials & Source Gases	Film Thickness (nm)
NFC	52	PECVD	CH_4+H_2	150, 380, 1125
CrO_x	60	Sputtering	Cr+O_2	125, 2000
SiO_2	77	Wet Oxidation	Si+H_2+O_2	395
FeO_x	90	Sputtering	Fe+O_2	125
TiO_x	103	Sputtering	Ti+O_2	120
a-Si	108	Sputtering	Si	450
AlO_x	115	Sputtering	Al+O_2	120
AlN	133	Sputtering	Al+N_2	1100
FeB	143	Sputtering	Fe+B	1010
Metglas	180	Sputtering	METGLAS 2826MB	278
Si_3N_4	208	LPCVD	Si+SiH_2Cl_2+NH_3	300
SiC	212	Sputtering	SiC	500
ta-C:H	230	PECVD	C_2H_2+H_2	868

Finite element analysis was performed using ANSYS 11. The model was constructed using 2-D elements, as shown in Fig. 2. Film surface roughness and residual stress state were ignored to simplify simulations. The element type for the film, substrate, and indenter were set to PLANE183, while TARGE169 and CONTA172 were selected for the target-contact pair surfaces. The material properties were assumed to be linear and elastically isotropic for all materials. The film material chosen was amorphous carbon with a Young's modulus of 52 GPa and a Poisson's ratio of 0.33 [25]. The values used for the silicon substrate [26, 27] and diamond indenter [28] were 170 GPa and 0.27 and 1141 GPa and 0.07 respectively. A film thickness of 1125 nm was assumed on a silicon substrate. To further simplify the simulation, the tip was designed as a half shape Berkovich tip. The right triangle had an angle of 65.3°, which is half of the Berkovich total angle (from one edge to the opposite face). The element size of the film in the local contact area was 10 nm, where a high stress/strain concentration was expected.

The nanoindentation tests were performed with an MTS Nanoindenter XP system operating in continuous stiffness mode. The indenter was a Berkovich type diamond pyramid. For the continuous stiffness measurement, the allowable thermal drift rate was limited to 0.05 nm/s. The Poisson's ratio of the film was assumed as equal to that of the same bulk material. The Poisson's ratio of the substrate was set as 0.27 for the silicon substrate [27]. The continuous stiffness measurement was depth-controlled and the harmonic displacement target was set at 2 nm. For each thin film, 25 or more CSM tests were run with same final indent depth.

RESULTS AND DISCUSSIONS: *Assessment of Alpha in the Doerner and Nix Model:* One of the more indefinite aspects of the Doerner and Nix model was the empirically determined constant α, see Eq. 2. It has been unclear what this constant physically represents other than it is a component of the weighing factor to account for the substrate effect. To investigate its physical significance an experiment was carried out where a film of known Young's modulus was deposited onto a silicon substrate. The film material was near-frictionless carbon [29], an amorphous carbon film, with a Young's modulus of 40-50 GPa [30] measured on the identical film with the membrane deflection experiment [31-33]. A value of 52 GPa was determined from the flat region in Fig. 1(a) and assumed as the modulus. The value of α was then back calculated for each data point and is shown as a function of indent depth in Fig. 1 (b). The results indicate that α increases rather quickly, reaching a maximum near the critical indent depth, h_{cr}. It then decreases in an exponential–like fashion before becoming relatively constant at large displacements. Unfortunately, α does not appear to be constant in the critical region where substrate effects are generally assumed to begin. However, since it is a component of the substrate weighing factor, it makes sense to conclude that it does not physically operate until h_{cr} is reached.

The physical basis of the Doerner and Nix model still remains somewhat unclear. Their weighting factor, $\exp^{(-\alpha t/h)}$ (Eq. 2), appears to physically suggest the ratio of the elastic strain intensity in the substrate to the total strain intensity in the composite. Assuming the total elastic strain intensity in the specimen is confined in an "effective" volume with uniform density, $\Phi_{D\text{-}N}$ can be seen as the ratio of the segment in the substrate to the total volume. In other words, the effective volume can be considered as a volumetric sphere of strain energy encompassing the film and substrate effects. As the indenter penetrates, the volume increases along with continually increasing influence by the substrate, see Fig. 3. To confirm this, a simple spherical mathematical model was constructed for the situation described in Fig. 3. Here the film has a constant thickness, *t*, and the radius of the sphere, *R*,

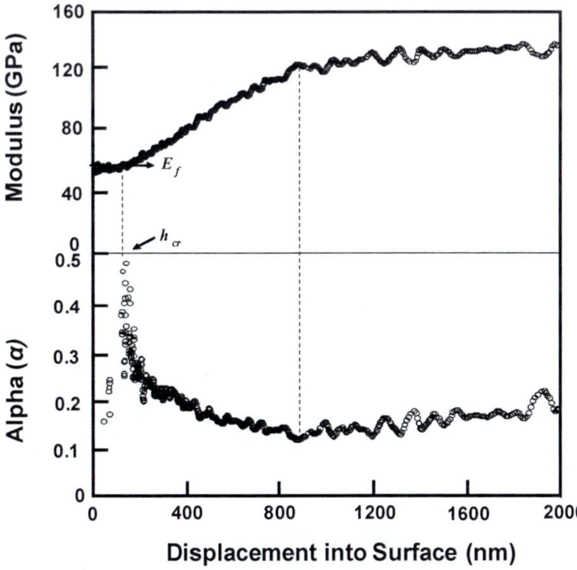

Figure 1: Plot of Young's modulus versus indent depth for an amorphous near-frictionless carbon film 1125 nm thick (a) as compared to the parameter α in the Doerner and Nix model (b).

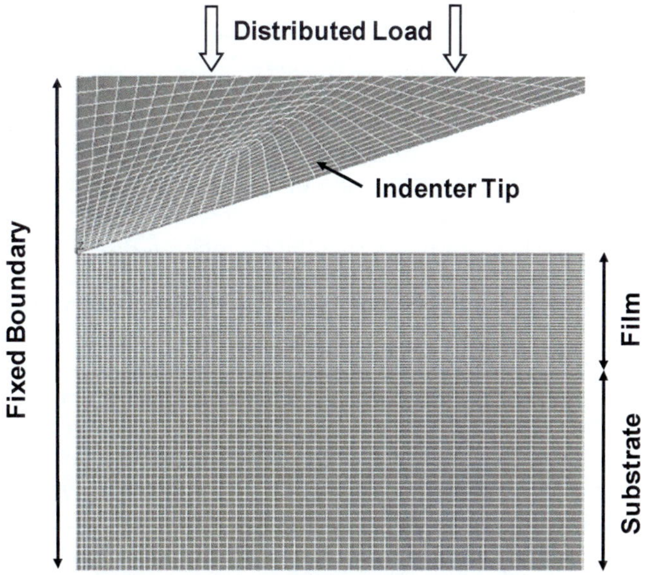

Figure 2: Schematic illustrating the central portion of a 2D axisymmetric mesh for the simulation of thin film nanoindentation.

increases with indent depth, *h*. As the volume of a sphere is equal to (4/3)πR^3, the ratio of volume in the film to total volume can be represented as:

$$\frac{V_{film}}{V_{total}} = \frac{\frac{\pi}{3}t^2(3R-t)}{\frac{4}{3}\pi R^3} = \frac{3}{4}\left(\frac{t}{R}\right)^2 - \frac{1}{4}\left(\frac{t}{R}\right)^3, \quad (5)$$

with the substrate component, or weighting factor, equal to:

$$\Phi_{Sphere} = 1 - \frac{V_{film}}{V_{total}}, \quad (6)$$

Eqs. 5 and 6 are not valid when the radius is less than one half the film thickness as the volume will be contained entirely in the film. Thus, the following linear relationship between radius and indent depth can be suggested:

$$R = \beta h + \frac{t}{2}, \quad (7)$$

where β is the slope. Substituting this into Eq. 6 one can see that a spherical weighting factor depends on β in a manner analogous to α in the Doerner and Nix Function. In fact, values of β can be identified that yield very similar behavior in the spherical weighting function to specific values of α. Fig. 4 is a plot of how the weighting factor changes with indenter depth for both models for different values of α and corresponding β constants. The agreement between the curves suggests that the Doerner and Nix model has a spherical-basis for their weighting factor. Likewise, values of υ in the Gao model can be found that yield similar agreement, but in this case the volume has cylindrical geometry.

Such a straightforward weighting factor approach should be easily replicated in numerical simulations. A finite element model was constructed with an 1125 nm thick carbon film on a silicon substrate, soft film on hard substrate, to match the experimental data in Fig.1. Simulations were performed at indentations depths of 8, 20, 30 and 50 nm to observe the elastic strain field growth in the film and substrate as indentation depth increased, see Fig. 5. At a depth of 8 nm the elastic strain field appeared to be contained entirely in the film. However, at a mere 20 nm of depth the elastic strain field was already propagating in the substrate and at 30 nm had made considerable progress into the substrate. A curious aspect of these results was that elastic strain between the film and the substrate did not appear to be continuous. As the field made contact with the hard substrate a portion of it propagated in the substrate while another portion was forced laterally in the film. Numerical values of strain then are not continuous directly across the interface; the values in the film were higher, resulting in a shear strain at the interface. Both the Doerner and Nix and Gao models assume the interface transfers elastic strain continuously. A modification of these models is necessary to account for the discontinuity of elastic strain across the film/substrate interface.

Discontinuous Elastic Interface Model: To account for the substrate effect Doerner and Nix introduced an

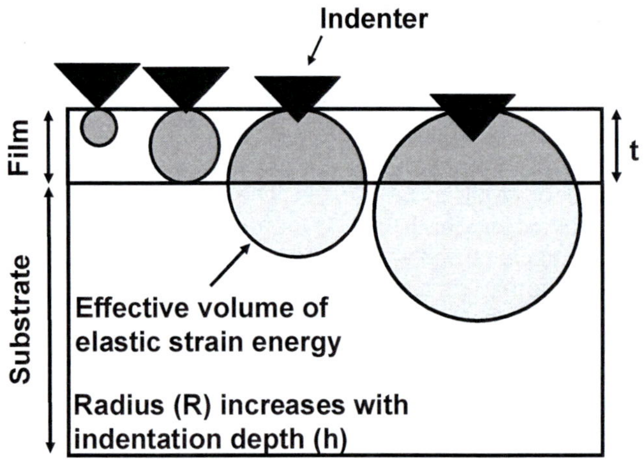

Figure 3: Schematic illustrating the expansion of a continuous elastic strain field in thin film indentation demonstrating how the ratio of film to substrate contribution evolves with indent depth.

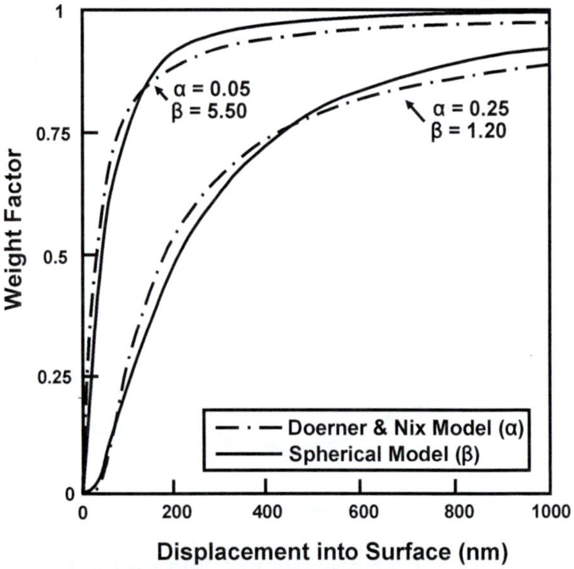

Figure 4: Plot demonstrating how the behavior of the Doerner and Nix model bears a resemblance to a spherical-based model.

empirically-based weighting factor, while the analytical-based Gao model derived a very similar factor. This factor, in both models, appears to be applied to both the film and substrate in the same manner or assuming the interface transfers elastic strain continuously. However, the existence of a discontinuity in elastic strain across the film/substrate interface suggests that the film and substrate should have separate weighting factors as each influences the other in a different manner. To guide the development of these factors a physical model was built to represent the contribution of each component to the composite elastic strain field, see Fig. 6. The top portion represents the elastic strain field as a result of the indent and the bottom portion a decoupling of each, analogous to an identical indent in each as a bulk form. Only the shades regions of each component apply to the actual indent conditions. As the elastic component deals with the indent unloading signature, the contribution to the composite compliance of each component can be modified with a weighting factor. The film and substrate compliance contributions with their weighting factors can be given as,

$$\frac{1}{E'} = \frac{1}{E'_f}\left(1 - \Phi_s\right) + \frac{1}{E'_s}\Phi_f, \tag{8}$$

where Φ_f and Φ_s are the weighting factors to account for the effects of the film on the substrate and substrate on the film respectively and are given as,

$$\Phi_f = e^{-\alpha_f (t/h)} \text{ and } \Phi_s = e^{-\alpha_s (t/h)}. \tag{9}$$

Here, h is the indent depth. Given the geometry of the composite and that the indent starts in the film, the substrate effect on the film compliance is $(1 - \Phi_s)$ as Φ_s represents the portion of the substrate above the film/substrate interface and we are interested in the portion below. If the film/substrate interface transfers strain continuously, then Φ_f and Φ_s are equivalent and one obtains a formulation similar to the Doerner and Nix model.

Experimental Verification: To verify the new function it was applied to the experimental data for all of the amorphous films listed in Table 1. For the discontinuous function to work properly the parameters α_f and α_s should each be a constant related to how the material of each influences the other. Using the amorphous SiO_2 film on a silicon substrate as an example, soft film on a hard substrate, a value of α_s was assumed and varied until α_f was also found to be constant. For the SiO_2 film this occurred when α_s was 0.27 and α_f was then 0.20. Fig. 7 is a plot of the values of α parameter versus indent depth comparing the Discontinuous and Doerner and Nix models. The constant nature of α_f and α_s lends confidence that these values can be employed in the Discontinuous model for SiO_2 to best match the experimental data. Fig. 8 is a plot of Young's modulus versus the normalized displacement (h/t) for the SiO_2 film on a silicon substrate. The data points are the experimental obtained values and the curves represent the various models using the modulus of the flat region as the film modulus. In comparing the models, the Discontinuous curve matches the data well at both small and large indent depths. The Doerner and Nix model appears to match well at low indent depths but not so well at larger depths and vice-versa for the Gao model. The treatment was repeated for the amorphous Si_3N_4 films, a hard film on a soft substrate, see Fig. 9. Here, when α_s was 0.28 and α_f was found to be 0.30. Again, the Discontinuous model appeared to do a better job at matching the experimental data.

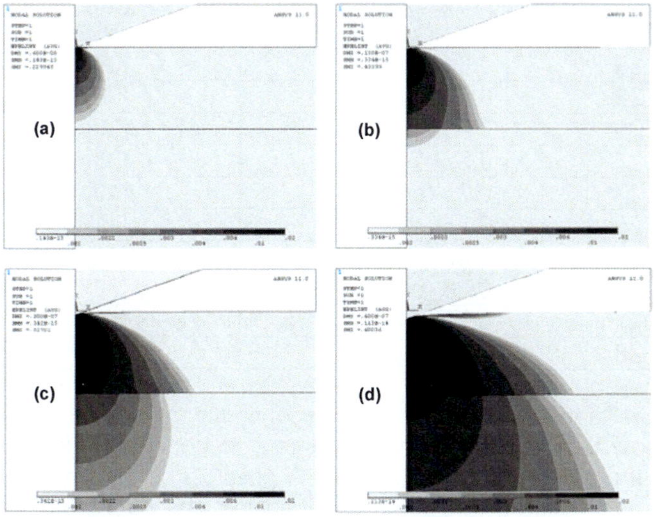

Figure 5: Finite element results of nanoindentation on a near-frictionless carbon film on a silicon substrate for (a) 8, (b) 20, (c) 30, and (d) 50 nm indent depths.

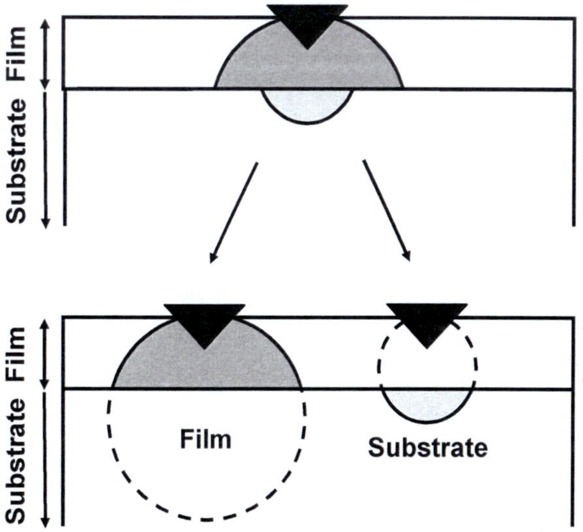

Figure 6: Schematic illustrating the development of a model that accounts for the discontinuity of elastic strain at the film/substrate interface.

This process was repeated for all films and the values of the α parameters are given in Table 2. All curves for the Discontinuous model matched well with the experimental data. However, the most astonishing aspect was that for nearly every film the α_s and α_f appeared to be numerical similar to the bulk Poisson's ration of the substrate and film materials respectively, see Table 2. Some values of Poisson's ration could not be found for the amorphous form and the crystalline value was used for comparison. The value of α_s for the silicon substrate varied somewhat between 0.27 and 0.29 and may reflect that the Berkovich indenter may have different rotations with the (001) crystal structure of silicon substrate. The literature value of the SiC film varied over a wide range, but matched well in the upper range. Here, the deposition technique is known to have a significant influence on film stoichiometry and Poisson's ratio [34].

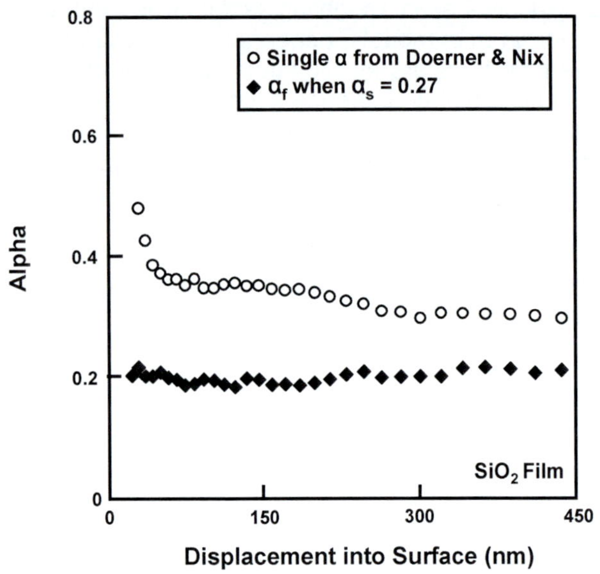

Figure 7: Plot comparing the α parameters in the Doerner and Nix and Discontinuous models as a function of indent depth for an SiO_2 film on a silicon substrate.

The physical connection between the α parameters and Poisson's ratios of the materials involved indicated that the real indents likely possessed a discontinuity in elastic strain at the film/substrate interface. It also revealed that the Poisson effect of each material played an important role in the indent and is likely an important factor causing the discontinuity in elastic strain. By accounting for the discontinuity in Eqs. 8 and 9, the experimental data was well matched and revealed the complex interplay between the film and substrate during indentation. The Discontinuous-based Interface model appeared adept at describing the influence of the substrate effects for both soft on hard and hard on soft situations. More importantly though, films such as Si_3N_4, which do not exhibit a well formed flat region can have their moduli determined when the bulk Poisson's ratio is known. The model should be applicable to different indenter geometries as this factor was not considered in the model development and should work for any film-substrate combination. Finally, the film materials in this work were all amorphous, thus, direct application of the Discontinuous model to crystalline films may yet be complicated by microstructural-based influences.

Table 2: List of the α_s *and* α_f values for each thin film with the film bulk Poisson's ratio.

Thin Film	α_s	α_f	**Poisson's Ratio**	**[REF]**
NFC	0.27	0.33	0.33	[25]
CrO_x	0.28	0.27	0.25	[35]
SiO_2	0.27	0.2	0.17 (quartz)	[36]
FeO_x	0.28	0.35	0.34	[37]
TiO_x	0.27	0.28	0.28	[38]
a-Si	0.28	0.25	0.25	[39]
AlO_x	0.29	0.27	0.27	[38]
AlN	0.28	0.27	0.25 (crystal)	[40]
FeB	0.27	0.25	0.25	[41]
Metglas	0.28	0.31	0.33	[42]
Si_3N_4	0.28	0.3	0.28 (crystal)	[38]
SiC	0.28	0.3	0.19 – 0.29	[34]
ta-C:H	0.28	0.28	0.28	[43]

CONCLUSIONS: A new model of thin film indentation that accounts for the discontinuity of strain at the film/substrate interface has been developed to reveal how the presence of the substrate influences the indent. The model accounts for the discontinuity and matched the experimental data well. A physical connection was made between parameters in the model and the Poisson's ratios of the materials involved and supported the premise that real indents likely possessed a discontinuity in elastic strain at the film/substrate interface. The model revealed the complex interplay between the film and substrate during indentation and can reveal the properties of films that do not exhibit a well formed flat region.

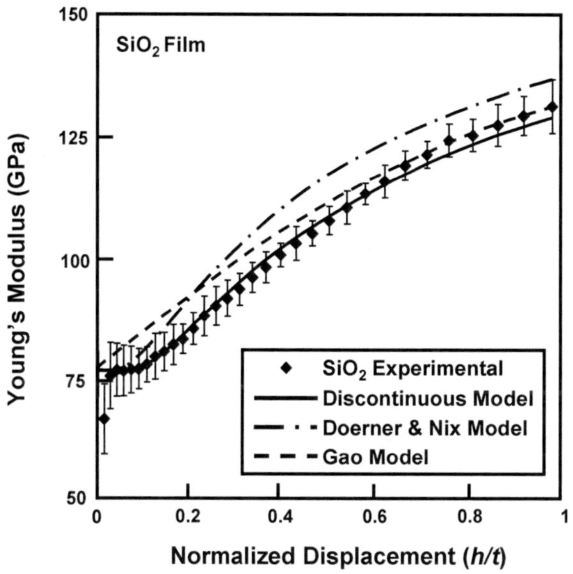

Figure 8: Plot of Young's modulus versus normalized displacement comparing the Gao, Doerner and Nix, and Discontinuous models for an SiO_2 film on a silicon substrate (soft on hard case).

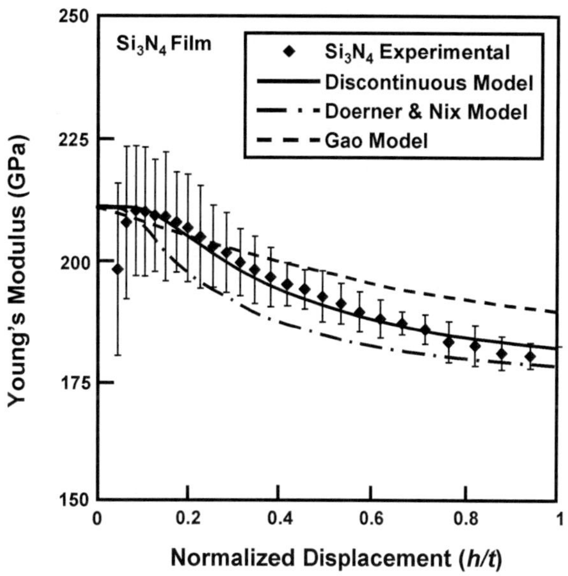

Figure 9: Plot of Young's modulus versus normalized displacement comparing the Gao, Doerner and Nix, and Discontinuous models for an SiN film on a silicon substrate (hard on soft case).

REFERENCES:

1. Ahn JH, Kwon D (2000) Micromechanical estimation of composite hardness using nanoindentation technique for thin-film coated system. Mater Sci Eng A 285:172-179.
2. Bolshakov A, Pharr GM (1998) Influences of pileup on the measurement of mechanical properties by load and depth sensing indentation techniques. J. Mater. Res 13:1049-1058.
3. Chen X, Vlassak JJ (2001) Numerical study on the measurement of thin film mechanical properties by means of nanoindentation. J. Mater. Res. 16:2974-2982.
4. Cho SJ, Lee KR, Eun KY, Hahn JH, Ko DH (1999) Determination of elastic modulus and poisson's ratio of diamond-like carbon films. Thin Solid Films 341:207-210.
5. Doerner MF, Nix WD (1986) A method for interpreting data from depth-sensing indentation instruments. J. Mater. Res. 1:601-609.
6. Frohlich F, Grau P, Grellmann W (1977) Performance and Analysis of Recording Microhardness Tests. Phys. Status Solid a 42:79-89.
7. Huang Y, Aziz MJ, Hutchinson JW, Evans AG, Saha R, Nix WD (2001) Comparison of mechanical properties of Ni3Al thin films in disordered FCC and ordered L12 phases. Acta Mater. 49:2853-2861.
8. Li X, Bhushan B (2002) A review of nanoindentation continuous stiffness measurement technique and its applications. Mater. Char. 48:11-36.
9. Li X, Diao D, Bhushan B (1997) Fracture mechanisms of thin amorphous carbon films in nanoindentation. Acta Mater. 45:4453-4461.
10. Mayo MJ, Siegel RW, Narayanasamy A, Nix WD (1990) Mechanical Properties of Nanophase TiO_2 as Determined by Nanoindentation. J. Mater. Res. 5:1073-1082.
11. McElhaney KW, Vlassak JJ, Nix WD (1998) Determination of indenter tip geometry and indentation contact area for depth-sensing indentation experiments. J. Mater. Res. 13:1300-1306.
12. Mencik J, Munz D, Quandt E, Weppelmann ER, Swain MV (1997) Determination of elastic modulus of thin layers using nanoindentation. J. Mater. Res 1997:2475-2484.
13. Nix WD (1997) Elastic and plastic properties of thin films on substrates: nanoindentation techniques. Mater. Sci Eng. A 234:37-44.
14. Nix WD, Gao H (1998) Indentation size effects in crystalline materials: A law for strain gradient plasticity. J. Mech. Phys. Sol. 46:411-425.
15. Nix WD, Greer JR, Feng G, Lilleodden ET (2007) Deformation at the nanometer and micrometer length scales: Effects of strain gradients and dislocation starvation. Thin Solid Films 515:3152-3157.
16. Oliver WC (1986) Progress in the Development of a Mechanical Properties Microprobe. MRS Bull. 11:15-19.

17. Oliver WC, Pharr GM (1992) Improved Technique for Determining Hardness and Elastic Modulus Using Load and Displacement Sensing Indentation Eexperiments. Journal of Materials Research 7:1564-1583.
18. Pharr GM (1998) Measurement of mechanical properties by ultra-low load indentation. Mat. Sci. & Eng. A:151-159.
19. Pharr GM, Callahan DL, McAdams SD, Tsui TY, Anders S, Anders A, Ager Iii JW, Brown IG, Bhatia CS, Silva SRP (1996) Hardness, elastic modulus, and structure of very hard carbon films produced by cathodic-arc deposition with substrate pulse biasing. Appl. Phys. Lett. 68:779-781.
20. Saha R, Nix WD (2002) Effects of the substrate on the determination of thin film mechanical properties by nanoindentation. Acta Mater. 50:23-38.
21. Gao H, Chiu C-H, Lee J (1992) Elastic Contact versus Indentation Modelling of Multi-Layered Materials. Int. J. Solids Structures 29:2471-2492.
22. King RB (1987) Elastic Analysis of Some Punch Problems for a Layered Medium. International Journal of Solids and Structures 23:1657-1664.
23. Buckle H (1973) The Science of Hardness Testing and its Research Applications. ASM, Metals Park, Ohio.
24. King RB (1987) Elastic Analysis of Some Punch Problems for a Layered Medium. Int. J. Solids Struct. 23:1657-1664.
25. Pastorelli R, Ferrari AC, Beghi MG, Bottani CE, Robertson J (2000) Elastic constants of ultrathin diamond-like carbon films. Diam. Relat. Mater. 9:825-830.
26. Bhushan B, Li X (1997) Micromechanical and tribological characterization of doped single-crystal silicon and polysilicon films for microelectromechanical systems devices. J. Mater. Res. 12:54-63.
27. Wortman JJ, Evans RA (1965) Young's Modulus, Shear Modulus, and Poisson's Ratio in Silicon and Germanium. J. Appl. Phys. 36:153-156.
28. Stillwell NA, Tabor D (1961) Elastic Recovery of Conical Indentations. Proc. R. Soc. 78:169.
29. Erdemir A, Eryilmaz OL, Fenske G (2000) Synthesis of diamondlike carbon films with superlow friction and wear properties. J. Vac. Sci. Tech. A 18:1987-1992.
30. Zhou B, Wang L, Mehta N, Morshed S, Erdemir A, Eryilmaz O, Prorok BC (2006) The mechanical properties of freestanding near-frictionless carbon films relevant to MEMS. J. Micromech. Microeng. 16:1374-1381.
31. Prorok BC, Espinosa HD (2002) Effects of nanometer-thick passivation layers on the mechanical response of thin gold films. J. Nanosci. Nanotech. 2:427-433.
32. Espinosa HD, Prorok BC, Fischer M (2003) A methodology for determining mechanical properties of freestanding thin films and MEMS materials. J. Mech. Phys. Sol. 51:47-67.
33. Espinosa HD, Prorok BC, Peng B (2004) Plasticity size effects in free-standing submicron polycrystalline FCC films subjected to pure tension. J. Mech. Phys. Sol. 52:667-689.
34. El Khakani MA, Chaker M, Jean A, Boily S, Kieffer JC, O'Hern ME, Ravet MF, Rousseaux F (1994) Hardness and Young's Modulus of Amorphous a-SiC Thin Films Determined by Nanoindentation and Bulge Tests. J. Mater. Res. 9:96-103.
35. Pang X, Gao K, Yang H, Qiao L, Wang Y, Volinsky A (2007) Interfacial Microstructure of Chromium Oxide Coatings. Adv. Eng. Mater. 9:594-599.
36. Mizuno S, Verma A, Tran H, Lee P, Nguyen B (1996) Dielectric constant and stability of fluorine doped PECVD silicon oxide thin films. Thin Solid Films 283:30-36.
37. Mober A (2008) Blast Cleaning Technology. Springer, Berlin.
38. Shackelford JF, Alexander W (2001) CRC Materials Science and Engineering Handbook. CRC Press, Boca Ratan, Florida.
39. Hess P (1996) Laser diagnostics of mechanical and elastic properties of silicon and carbon films. Appl. Surf. Sci. 106:429-437.
40. Boch P, Glandus JC, Jarrige J, Lecompte JP, Mexmain J (1982) Sintering, oxidation and mechanical properties of hot pressed aluminum nitride. Ceram. Int. 8:34-40.
41. Engdahl G (2000) Handbook of Giant Magnetostrictive Materials. Academic Press, San Diego, CA.
42. Liang C, Morshed S, Prorok BC (2007) Correction for Longitudinal Mode Vibration in Thin Slender Beams. Appl. Phys. Lett. 90:221912.
43. Ferrari AC (2004) Diamond-like carbon for magnetic storage disks. Surf. Coat. Tech. 180:190-206.

Proceedings of the SEM Annual Conference
June 7-10, 2010 Indianapolis, Indiana USA

Mechanical and Piezoelectric Behavior of Thin Film PZT Composites for MEMS Applications

S. Yagnamurthy, I. Chasiotis

Aerospace Engineering, University of Illinois at Urbana-Champaign, IL 61801, U.S.A.

ABSTRACT

The elastic and failure mechanical properties, the d_{31} piezoelectric coefficient and the effect of applied stress on the hysteresis curves of freestanding PZT composite films, comprised of SiO_2, Pt, PZT and Pt, were measured from microscale tension specimens. The d_{31} coefficient was measured from the out-of-plane deflection of biased PZT specimens with dimensions similar to those of MEMS components. An analytical solution for the bending of a multilayered piezoelectric beam was used to compute a first estimate of d_{31} as 176±27 pm/V. The field induced in-plane stress hysteresis loops were asymmetric at small in-plane stresses becoming of similar magnitude as the applied stress was increased beyond 300 MPa. Similarly, the intersection of the hysteresis loops shifted from negative to positive electric field at stresses larger than 150 MPa. The applied stress resulted in reduction of the hysteresis magnitude due to mechanical constraints imposed on 90° domain switching. The effect of high in-plane stress on domain switching was also the reason for the hysteretic non-linear stress-strain curves that were recorded for unbiased PZT films.

1. INTRODUCTION

MEMS devices have been used for sensor applications because of their small size and the ability to integrate with microelectronics. Lead Zirconium Titanate (PZT) has emerged as a multifunctional material for MEMS, which has piezoelectric properties and brittle mechanical behavior. Utilizing its piezoelectric properties, MEMS devices based on PZT thin films, such as RF switches [1], microphones and microscale robots and flapping wing micro-aerial vehicles have been conceptualized but very few of them have been successful due to limitations arising from their mechanical and piezoelectric reliability. Hence, accurate knowledge of the mechanical and piezoelectric behavior of this material is required to design durable devices with better performance.

Current literature on PZT film mechanics is limited and the elastic modulus and strength values show large variations. Nanoindentation [2,3] and membrane deflection [4-6] are common experimental methods employed to measure the mechanical behavior of PZT films. In previous work by this group [7], uniaxial tension experiments of freestanding composite films were employed to extract from full-field strain measurements the mechanical behavior of PZT. In addition to the mechanical behavior of the PZT stacks, appropriate design of PZT-based MEMS device also requires their piezoelectric behavior, which depends on the fabrication method, the residual stresses, the poling state and the microstructure of PZT. Various experimental techniques have been used to obtain the piezoelectric coefficients but there is a wide range of values for the d_{31} and d_{33} values reported, ranging from 30 pm/V to 220 pm/V and 29 pm/V to 400 pm/V, respectively [8-12]. Most values reported were obtained from experiments performed on PZT films bonded to a substrate and could be unreliable for designing MEMS devices with freestanding PZT structures as the film state of stress determines its deformation evolution and piezomechanical response. In this paper we present a study on mechanical and piezoelectric behavior of PZT composite thin films by performing experiments directly on their freestanding form. The transverse piezoelectric coefficient d_{31} was calculated by using an approximate multilayer beam bending model and the effect of pre-stress on the piezoelectric hysteresis of PZT composite films was investigated.

T. Proulx (ed.), *MEMS and Nanotechnology, Volume 2*, Conference Proceedings of the Society for Experimental Mechanics Series 2,
DOI 10.1007/978-1-4419-8825-6_37, © The Society for Experimental Mechanics, Inc. 2011

2. MATERIALS AND EXPERIMENTAL METHODS

The materials used in this study were thin film PZT composites fabricated by a combination of SiO_2, Ti, Pt, and PZT. Uniaxial tension experiments were conducted on PZT composite specimens with gauge length of 1,000 μm and widths of 50 and 100 μm as shown in Figure 1. The microscale tension specimens were fabricated by a combination of chemical vapor deposition, physical vapor deposition and chemical solution deposition on silicon substrate followed by various etching techniques to obtain freestanding dog bone thin film specimens as shown in Figure 1. Though the nominal thickness of the individual layers could be calculated from the deposition time for each layer, the actual thickness of each layer was measured from SEM micrographs. The PZT composite specimens were made of a 1-μm thick PZT layer sandwiched between 100 nm thick Pt electrodes with a 300 nm thick sacrificial layer of SiO_2 that served as thermal and electrical insulator.

The uniaxial tension tests were conducted by using an in-situ setup that had a PZT actuator to generate loading and a loadcell to measure the load in the sample, identical to one reported before [13-15]. The test apparatus was placed under an optical microscope to capture photos of the specimen gauge section having a fine and dense speckle pattern as shown in Figure 1. These pictures were later analyzed by using Digital Image Correlation (DIC) to compute the full-field strain on the specimen surface and the stress-strain curves were constructed with the use of the loadcell measurements. The mechanical response of PZT was then derived from the stress strain plots of the individual materials comprising the PZT composites and the composite behavior with the aid of basic lamination theory [7].

The PZT composite specimens used to evaluate piezoelectric behavior were identical to the dog bone shaped specimens shown in Figure 1. The SiO_2-TiPt-PZT-Pt composite was used to measure the d_{31} coefficient by biasing the Pt electrodes that were firmly bonded on either side of PZT. Due to their freestanding nature, these specimens underwent out-of-plane bending upon biasing the electrodes with very fine electrical probes. The die containing specimens was mounted vertically under an optical microscope as shown in Figure 2. Optical images of the specimen's sidewalls were recorded to measure the beam deflection upon application of electric field. The optical images of the specimen's sidewall at different bias voltages, ranging between 0-6V (see later Figure 4), were processed to calculate the deflection of the beam and fit it to an analytical model for a piezoelectric beam subjected to bending, which was developed by Balls et al [16], to compute the d_{31} piezoelectric coefficient.

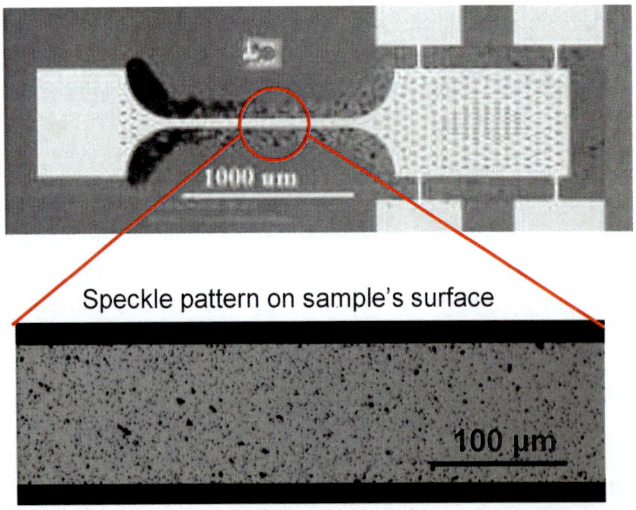

Figure 1. A freestanding SiO_2-TiPt-PZT-Pt specimen with a fine speckle pattern deposited to measure full field mechanical strains.

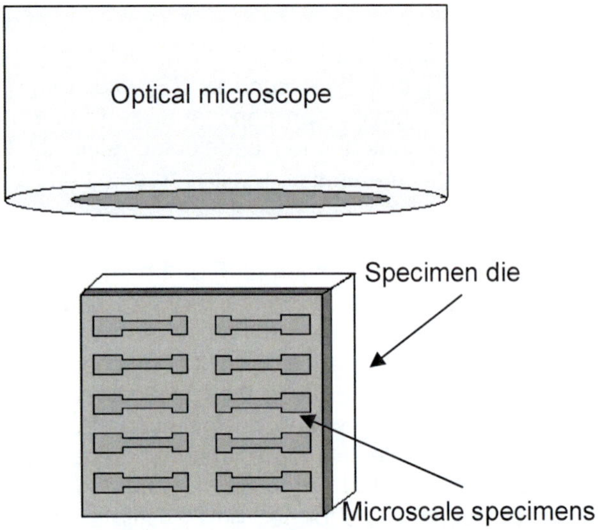

Figure 2. Schematic of a die containing PZT specimens positioned vertically under an optical microscope for measuring out of plane deflection of PZT composite beams under an applied bias.

3. RESULTS AND DISCUSSION

The mechanical behavior of the individual layers SiO_2, SiO_2-TiPt, SiO_2-TiPt-PZT and SiO_2-TiPt-PZT-Pt was reported in our previous work [7]. The stress-strain plot of the PZT thin film stack including an elastic loading-unloading segment and the final reloading to fracture is shown in Figure 3. The composite modulus for SiO_2-TiPt-PZT-Pt was 94.7 GPa with the failure strain being 0.65%. The stress vs. strain curves showed nonlinear behavior after approximately 0.35% strain.

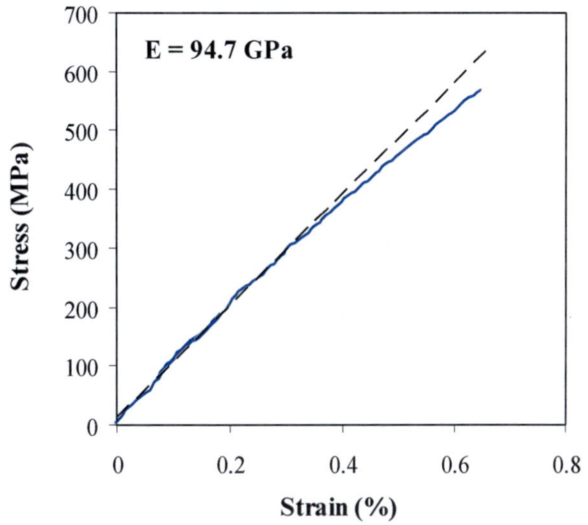

Figure 3. Stress vs. strain of a freestanding PZT composite film including the loading-unloading and final reloading segments.

The piezoelectric behavior of this PZT material was evaluated using the same freestanding composite specimens. Optical images of the beam's profile taken at bias voltages between 1.5V and 6V, shown in Figure 4, were used for this purpose. The stress mismatch between the PZT layer and the attached Pt and SiO_2 layers resulted in composite beam deflection up to 480 µm at bias voltages of 6V. A first estimate for piezoelectric coefficient d_{31} was calculated as 176±27 pm/V by using multilayer beam bending equations developed by Ballas *et al* [16]. This d_{31} value is higher than that reported for bulk PZT [17] (93.5 pm/V) and is in the upper range of d_{31} values reported in literature for sol-gel prepared PZT thin films. This may be attributed to the presence of fewer defects, lack of microcracking, perfect bonding between layers, the influence of fabrication conditions in the formation of PZT with high ferroelectric properties, and the use of freestanding films as opposed to PZT films attached to a rigid substrate that were used in previous studies.

Table 1. Mechanical properties of the PZT composite and the component layers

Material	Modulus (GPa)	Failure Stress (MPa)	Failure Strain (%)
SiO_2	72.3±2	1,060±200	1.4±0.2
SiO_2-TiPt	87.9±1	780±70	0.8±0.1
Pt [13]	173	1,876±10	3.84±0.26
SiO_2-TiPt-PZT	87.9±1	412±50	0.5±.0.05
SiO_2-TiPt-PZT-Pt	93.5±6	511±50	0.6±0.05
PZT (Extracted)	84±2	510±35	0.5±0.1

Further investigation was conducted on the PZT composites to understand their nonlinear and hysteretic behavior. Uniaxial tension experiments were carried out on PZT composite specimens to deduce the origins of the non-linear behavior shown in Figure 3, by performing multiple loading (beyond 0.35% strain) and unloading to zero strain cycles with a final reloading to failure. The resulting multiple stress vs. strain curves of SiO_2-TiPt-PZT-Pt composites did not deviate from a single curve indicating a nonlinear elastic behavior of the PZT composite stacks. In literature, it has been reported that PZT exhibits nonlinear elastic behavior attributed to 90° domain switching upon applying stress [18,19], which is a possible explanation for the non-linear mechanical behavior recorded in the present stress vs. strain curves. The mechanical behavior of the thin films made of SiO_2, SiO_2-TiPt, SiO_2-TiPt-PZT-Pt, and SiO_2-TiPt-PZT reported in our previous work [7] are tabulated in Table 1. Using simple lamination theory, the mechanical behavior of PZT was extracted from the properties of individual layers of SiO_2, Pt and PZT composites.

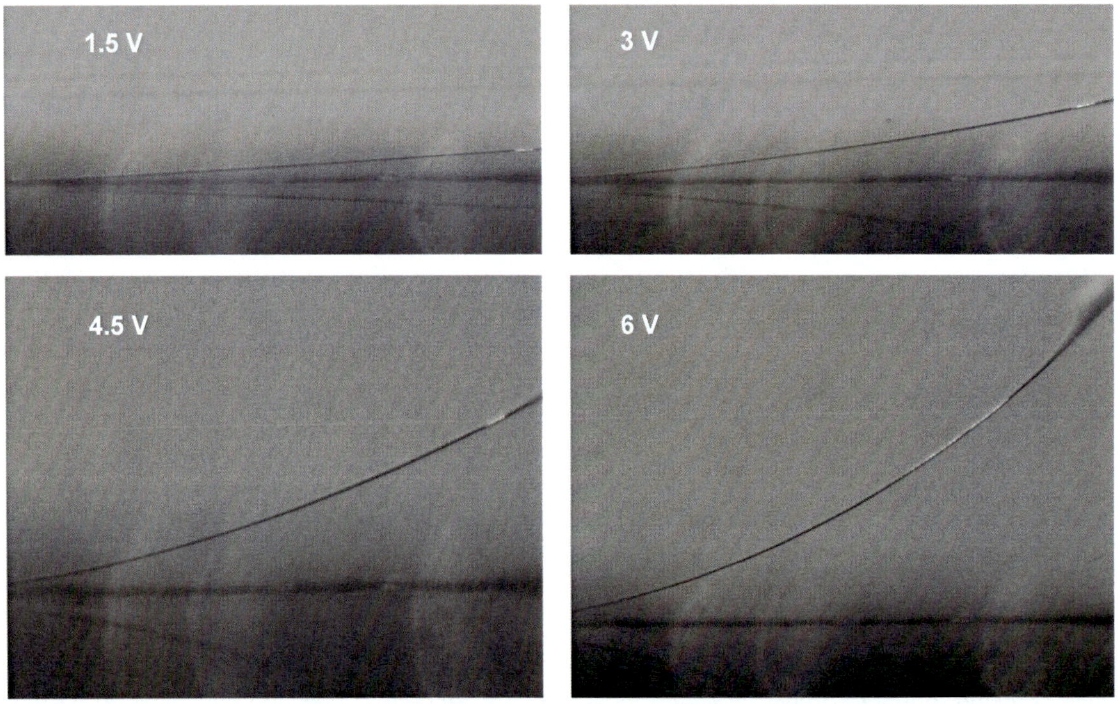

Figure 4. PZT beam deflection at bias voltages between 1.5V and 6V.

Additionally, the effect of applied pre-stress on the electric field induced stress hysteresis of the PZT composite films was measured. Our in-situ microscale tensile testing apparatus was used to preload the specimens that were then biased by DC voltage varying between -10 V and 10 V. The force on the specimens was recorded by the loadcell attached at one end of the specimen while the other end was firmly attached to an actuator. The electric field induced stress is plotted against applied voltage in Figure 5. The field induced stress hysteresis loops were asymmetric at small applied stresses becoming of equal magnitude as the applied stresses larger than ~300 MPa, before accounting for residual stresses in the PZT layer. Residual and mechanical stresses do affect domain switching in piezoelectric ceramics [17]: at higher stresses, the domains that could easily undergo 90° switching in the direction that causes hysteresis on the right half of butterfly loops were mechanically constrained. This led to reduced hysteresis in the right half of the butterfly curve thereby making the hysteresis loops equal in magnitude.

Similarly, the intersection of the hysteresis loops shifted from negative to positive electric field values at stresses larger than 150 MPa. This intersection point compares the relative population of domains that remained polarized in one direction vs. the opposite direction upon removal of the electric field. The remnant polarization exhibited by a piezoelectric crystal is caused mainly due to 90° switching of domains that are in a metastable state [17]. Higher applied stresses impose greater mechanical constraint on domains that could undergo 90° switching thereby lowering the number of domains that cause hysteresis at ~200 MPa when the remnant polarization is identical on both sides of hysteresis loop. Symmetrical butterfly loops would imply equal number of domains able to remain polarized in each direction and depend on the film thickness, residual stresses and the fabrication conditions.

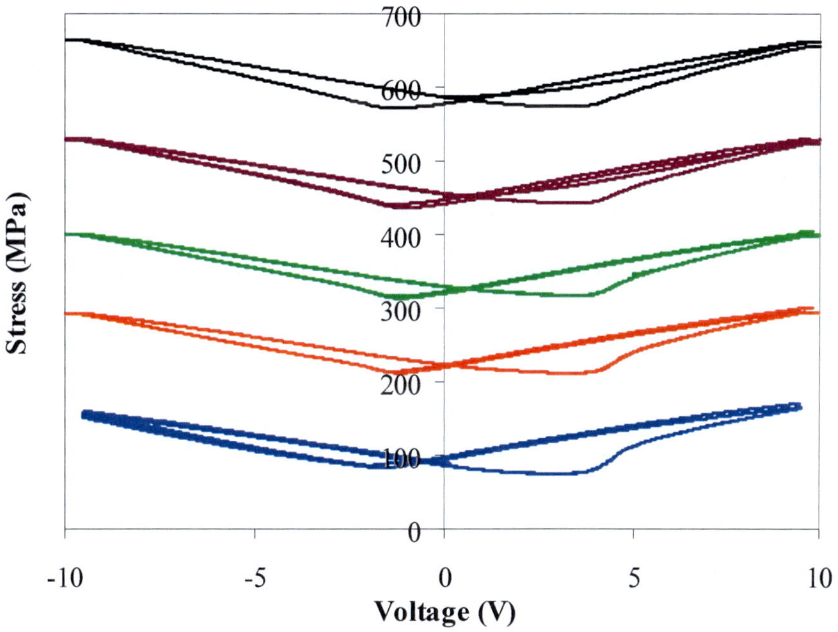

Figure 5. Electric field induced stress hysteresis loops for a PZT stack.

4. CONCLUSIONS

In this paper, the mechanical and piezoelectric behavior of PZT films was investigated. The stress vs. strain curve of the PZT composite was nonlinear due to domain switching at strains larger than 0.35%. The initial modulus of PZT at up to 0.3% strain was extracted from the stress vs. strain curves of SiO_2, SiO_2-TiPt, SiO_2-TiPt-PZT, and SiO_2-TiPt-PZT-Pt freestanding films, by using simple lamination theory, as 84±2 GPa with the failure strength averaging 510±35 MPa, after accounting for the residual stress in the PZT layer

The same PZT stack films were employed to measure the piezoelectric coefficient d_{31} and to quantify their hysteretic response as a function of applied pre-stress. A multilayer beam bending model was used to calculate an estimate for the d_{31} coefficient as 176±27 pm/V. The alternating electric field induced stress hysteresis "butterfly" loops that were asymmetric at smaller applied stresses, became symmetric at applied stresses beyond 300 Mpa which corresponded to the stress at which the PZT films showed elastic softening. Due to a change in the number of domains that switched with applied stress, the intersection of the hysteresis loops shifted from the negative to positive electric field at about 200 MPa of applied pre-stress. This non-linear electromechanical behavior has a significant effect on the performance of MEMS devices when operated at hundreds of MPa applied stresses due to the inaccurate estimate of stress and strain associated with the application of a combination of a voltage bias and a mechanical stress.

5. ACKNOWLEDGMENTS

The authors acknowledge the support by the Army Research Office Grant #W911NF-05-1-0063 with Dr. Bruce LaMattina as the Program Manager and R. Polcawich and J. Pulskamp from the Army Research Laboratory in Adelphi, MD for fabricating the thin film specimens.

6. REFERENCES

[1] Polcawich, R., Judy, D., Pulskamp, J., Trolier-McKinstry, S., Dubey, M., "Advances in Piezoelectrically Actuated RF MEMS Switches and Phase Shifters", Microwave Symposium IEEE/MTT-S International, pp. 2083-2086, 2007

[2] Bhar, D., Robach, J., Wright, J., Francis, L., Gerberich, W., "Mechanical deformation of PZT thin films for MEMS applications", Materials science and Engineering, 259, pp. 126-131, 1999

[3] Fang, T., Jian, S., Chuu, D., "Nanomechanical properties of lead zirconate titanate thin films by nanoindentation", Journal of Physics: Condensed Matter, 15, pp. 5253-5259, 2003

[4] Zhou, J., McMcollough, T., Mantell, S., Zurn, S., "Young's Modulus Measurement of Thin Film PZT", Proceedings of 13 biennial Microelectronics symposium, pp. 153-157, 1999

[5] Hong, S., Kim, M., Lee, S., Lee, C.S., "Characterization of Deformation Behaviors and Elastic Moduli of Multilayered Films in Piezoelectric Inkjet Head", Journal of Microelectro Mechanical Systems, 17 (5), pp. 1155-1163, 2008

[6] Wakabayashi, S., Totani, H., Sakata, M., Ikeda, M., Goto, H., Takeuchi, M., Yada, T., "Study on mechanical characteristics of PZT thin film for sensors and actuators", Proceedings of SPIE - The International Society for Optical Engineering, 2639, pp. 304-314, 1995

[7] Yagnamurthy, S., Chasiotis, I., Lambros, J., Polcawich, R., Pulskamp, J., Dubey, M., "Mechanical Properties of PZT Films and their Composites for RF-MEMS", Proceedings of Society of Experimental Mechanics, Orlando, June 2008

[8] Tagawa, N., Seki, H., Kitamura, K., Mori, A. "Development of novel PZT thin films for active sliders based on head load/unload on demand systems", Microsystem Technologies, 8 (2-3), pp. 133-138, 2002

[9] Kobayashi, T., Tsaur, J., Ichiki, M., Maeda, R., "Fabrication and performance of a flat piezoelectric cantilever obtained using a sol–gel derived PZT thick film deposited on a SOI wafer", Smart Materials and Structures, 15 (1), pp. S137-S140, 2006

[10] Zhao, Z., Luo, Z., Liu, C., Wu, W., Gao, C., Lu, Y., "Quantitative measurement of piezoelectric coefficient of thin film using a scanning evanescent microwave microscope", Review of Scientific Instruments, 79 (6), 064704, 2008

[11] Ong, R.J., Berfield, T.A., Sottos, N.R., Payne, D.A., "Sol-Gel derived Pb(Zr,Ti)O3 thin films: Residual stress and electrical properties", Journal of the European Ceramic Society, 25, pp. 2247-2251, 2005

[12] Berfield, T.A., Ong, R.J., Payne, D.A., Sottos, N.R., "Residual stress effects on piezoelectric response of sol-gel derived lead zirconate titanate thin films", Journal of Applied Physics, 101 (2), 024102, 2007

[13] Jonnalagadda, K., Chasiotis, I., Yagnamurthy, S., Lambros, J., Pulskamp, J., Polcawich, R., Dubey, M., "Experimental Investigation of Strain Rate Dependence of Nanocrystalline Pt Films", Experimental Mechanics, 50 (1), pp. 25-35, 2010

[14] Chasiotis, I., Knauss, W., "A New Microtensile Tester for the Study of MEMS Materials with the aid of Atomic Force Microscopy", Experimental Mechanics, 42 (1), pp. 51-57, 2002

[15] Cho, S., Chasiotis, I., "Elastic Properties and Representative Volume Element of Polycrystalline Silicon for MEMS," Experimental Mechanics, 47 (1), pp. 37-49, 2007

[16] Ballas, R., Schlaak, H., Schmid, A., "The constituent equations of piezoelectric multilayer bending actuators in closed analytical form and experimental results", Sensors and Actuators, A: Physica,l 130-131, pp. 91-98.2006

[17] Jaffe, B., Cook, W.R., Jaffe, H., "Piezoelectric Ceramics", Academic Press London and Newyork, 1971

[18] Fan, J., Stoll, W., Lynch, C. "Nonlinear constitutive behavior of soft and hard PZT: Experiments and modeling", Acta Materialia, 47 (17), pp. 4415-4425, 1999

[19] Xu, F., Trolier-McKinstry, S., Ren, W., Xu, B., Xie, Z., Hemker, K., "Domain wall motion and its contribution to the dielectric and piezoelectric properties of lead zirconate titanate films", Journal of Applied Physics, 89 (2), pp. 1336-1348, 2001

Proceedings of the SEM Annual Conference
June 7-10, 2010 Indianapolis, Indiana USA

FRACTURE BETWEEN SELF-ASSEMBLED MONOLAYERS

S.R. Na[1], B. Doynov[1], A. Hassan[2], K. M. Liechti[1] and M.J. Krische[2]
[1]Aerospace Engineering and Engineering Mechanics
[2]Chemistry and Biochemistry
The University of Texas at Austin
Austin, TX 78712
kml@mail.utexas.edu

While it is intuitive that molecular interaction should correlate to the mechanical properties of a material, it has only recently become possible to make the measurements necessary to trace the effects of molecular interaction across length scales to properties at the level of the material. The utility of "classical" polymeric adhesives is underscored by their widespread use in primary structural applications ranging from aerospace, automotive, and civil structures to biomedical implants and microelectronic devices. To date, the vast majority of the efforts directed toward improving the strength and durability of adhesives have been largely empirical. At the same time, the drive towards miniaturization in MEMS and NEMS devices and nano patterning means that an understanding of adhesion and fracture at smaller and smaller scales needs to be developed. This actually provides an opportunity to decrease the amount of empiricism as the number of variables is essentially decreased.

The paper describes the development of an experiment with associated analysis to determine the toughness of a molecular level adhesive joining two silicon strips. Si (111) surfaces were coated with amine and carboxy-terminated self-assembled monolayers (SAMs). The silicon beams were pressed together to form miniature laminated beam specimens which were then separated using a specially developed high vacuum fracture tester. The deposition of the SAMs and associated spectroscopic, scanning probe and ellipsometric diagnostics are described. The delaminations between the silicon strips and the associated normal crack opening displacements (NCOD) were measured via infra-red crack opening interferometry.

The fracture toughness of the specimens was measured. Traction-separation laws were extracted from measurements of crack opening displacements and an associated fracture analysis. These compared favorably with traction-separation laws based on potentials for ionic bonding.

DOI 10.1007/978-1-4419-8825-6_38, © The Society for Experimental Mechanics, Inc. 2011

Proceedings of the SEM Annual Conference
June 7-10, 2010 Indianapolis, Indiana USA

Control and Quantification of Residual Stresses in Anodically Bonded MEMS Structures

R. Inzinga, T. Lin, M. Yadav, H.T. Johnson, G.P. Horn
Department of Mechanical Science and Engineering, University of Illinois at Urbana-Champaign
1206 W. Green St. Urbana, IL 61801

ABSTRACT

Residual stresses in anodically bonded silicon devices can result in quality control and process control deficits if the stresses are not controlled. At the same time several geometries may benefit from a controlled introduction of residual stresses. For example, long, thin structures may utilize a residual tensile stress to minimize the likelihood of buckling, while etched cavities with sharp corners can benefit from a residual compressive stress to suppress crack initiation and growth.

In the present work, we quantify the residual stress fields present in silicon wafers that are anodically bonded to virgin Pyrex wafers. Anodic bonding is conducted using standard procedures as well as a proposed alternative method that utilizes differential thermal bonding to control the residual stress state. The experimental stress state is compared to theoretical finite element calculations to determine the capability of controlling stresses based on a simple thermal model.

I. INTRODUCTION

Anodic bonding is used in the MEMS industry for the production of microfluidic devices [1], pressure sensors [2], and accelerometers [3,4] among other applications. It is a very useful tool for many applications that require the formation of a reliable hermetic seal between silicon and glass wafers (or bare silicon and glass coated silicon wafers for silicon-on-insulator applications) at relatively low temperatures. However, one noteworthy issue with anodic bonding is the unwanted curvature of the bonded structure at room temperatures that results from residual stresses in the bonded wafer pair. As anodic bonding requires the joining of two dissimilar materials at elevated temperatures, the very slight mismatch in coefficient of thermal expansion between the glass and silicon wafers results in residual stresses in every process currently used in industry and academia. Several types of borosilicate glass wafers are manufactured, including the Corning Pyrex 7740, to have a coefficient of thermal expansion that relatively closely matches that of silicon up to high temperatures. However, at temperatures above about 320 °C, the coefficient of thermal expansion of silicon becomes larger than Pyrex 7740, which typically results in wafer bow with silicon in tension ("convex" in our geometry) [5-7]. Bond strength depends strongly on bond temperature, so in most applications temperatures range from about 350 to 450 °C. At 400 °C, the difference in CTE between the two wafers is only about 7%, but creates a significant wafer bow on the order of tens to hundreds of microns deflection over a 100 mm wafer [6,7]. For equally thick bonded silicon and Pyrex wafers, the induced strain in the silicon changes from being in a compressive state to a tensile state above 315°C [8].

The characterization and manipulation of residual stresses in anodically bonded structures can be useful in the production of MEMS devices. Compressive stresses in a slender beam application or pressure sensor membrane [9] can result in buckling of the structure. In these cases, the typical bonding profile with silicon in tension is often useful to ensure that buckling does not occur. On the other hand, the presence of any residual stresses in resonant sensors can alter the resonant frequency of a device [10]. In many applications, the overall wafer bow must be minimized for post processing reasons, particularly if lithography is required. Finally, in applications where etched features are present at the bond interface, cracks have been found to grow from areas of high stress concentration. Processing to introduce a compressive stress in the silicon wafers with these structures can

T. Proulx (ed.), *MEMS and Nanotechnology, Volume 2*, Conference Proceedings of the Society for Experimental Mechanics Series 2, DOI 10.1007/978-1-4419-8825-6_39, © The Society for Experimental Mechanics, Inc. 2011 provide a measure of protection from crack initiation and growth. Therefore, for different devices, it is important to understand the magnitude of residual stresses near etched features and to be able to control the state of residual stress during processing.

Methods to control residual stresses and wafer bow include utilizing different glass chemistries [5,6] post bond annealing steps [8] and current-limited anodic bonding [10]. However, each of these approaches is restricted by other processing requirements, additional processing times, or a limited effect. As such, no simple approach for residual stress control in anodic bonded devices has been devised.

The purpose of this presentation is to describe a simple modification to the standard bonding process as a method to control the residual stress in anodically bonded samples. Furthermore, post bond wafer bow generated by these residual stresses are compared to numerical predictions from a ABAQUS finite element model accounting for the nonlinear coefficients of thermal expansion of the bonded substrates.

II. EXPERIMENTAL PROCEDURES

The anodic bonding process utilized in these experiments joins a 100 mm single crystal, double-side polished silicon wafer and a 100 mm borosilicate glass wafer, in our case a Pyrex 7740 wafer. For some samples, circular and square features were etched into the silicon wafer to study stress concentrating effects of these geometry, specifically to investigate effects of corner radii as well as feature orientation, size and proximity. Feature in-plane diameters are in the range of 250-1000 μm while the depths were typically near 100 μm. To pattern the interfacial features in virgin silicon wafer a positive photoresist is spun prior to being exposed through the photomask. After developing the photoresist, the silicon etching process is completed in an ICP-DRIE etching system using the Bosch process. The Bosch process combines an etching step using accelerated Ar^+ ions to etch the exposed silicon surface with an inhibiting step that produces vertical sidewalls [11].

Standard RCA-1 and RCA-2 cleans are performed on the patterned silicon wafers to remove any remaining organic materials or heavy metals [11]. To reduce surface roughening, a diluted RCA-1 solution of deionized water, hydrogen peroxide and ammonium hydroxide at a ratio of 5:1:0.1 is prepared. Immediately after the RCA-1 clean, the wafer is placed in an RCA-2 solution of deionized water, hydrogen peroxide and hydrogen chloride at a ratio of 5:1:1. The Pyrex wafer is cleaned with a Piranha solution, with a 5:1 mixture of sulfuric acid and hydrogen peroxide. The wafers are bonded in an Electronic Visions Group (EVG) 501 vacuum bonding system. For these samples, anodic bonding takes place in a vacuum better than $2x10^{-4}$ mbar. For isothermal bonding protocols, the bond chuck temperatures are set to 400°C. As a means of comparison, an anisothermal bonding protocol is used to provide roughly opposite curvature by bonding at silicon temperatures of 388°C and Pyrex at temperatures of 343°C. Upon reaching temperature, pressure is applied to bring the centers of the wafers into contact and a voltage of 400V is maintained for 25 minutes to ensure the bonding process is complete.

To verify the magnitude of bulk stresses in bonded wafer pairs, an Alpha Step IQ Surface Profiler is used to scan 1000 μm sections of the wafer surfaces. The profilometer obtains curvature measurements of the bonded samples at 25 mm from the wafer center. It is assumed that the bulk curvature of the substrate is represented by the local curvature over these sections. The measurements are averaged to estimate the wafer bow after bonding.

Residual stress concentrations at etched features are quantified using an infrared photoelastic stress analysis system known as the IR-GFP [13]. Most of images shown here are taken with a 5X optical magnification that provides a spatial viewing window of approximately 1.78 mm x 1.33 mm with a spatial resolution of 1.3 μm. This magnification allows for a sufficient viewing area for any feature in the silicon. A through thickness spatially resolved value of the shear stress cannot currently be obtained using the IR-GFP with this geometry, since the shear stress measurement is integrated through the entire sample thickness, including both the glass and the silicon wafers. A comprehensive modeling framework is being developed to relate the retardation of the elliptically polarized light to in-plane shear stress. The data presented here are in terms of ellipticity or "intensity" of depolarization, and are directly related to residual shear stress.

III. RESULTS AND DISCUSSION

The surface profile line scans from the isothermal and anisothermal bonded wafers are shown in Figure 1, showing the wafer bow that resulted from each bond recipe. The isothermal wafer protocol resulted in a wafer bow with the silicon concave as is expected due to the mismatch in silicon and Pyrex CTE. The standard isothermal bonding recipe yields a significant wafer bow with a radius of curvature of approximately -8.1 m. This magnitude of wafer bow is similar to that reported in the literature with an isothermal recipe [7-10]. Meanwhile, the wafer pair bonded via the anisothermal recipe shows that a nearly opposite curvature (+6.5 m) can be achieved by reducing the Pyrex wafer temperature by approximately 45°C from the silicon temperature..

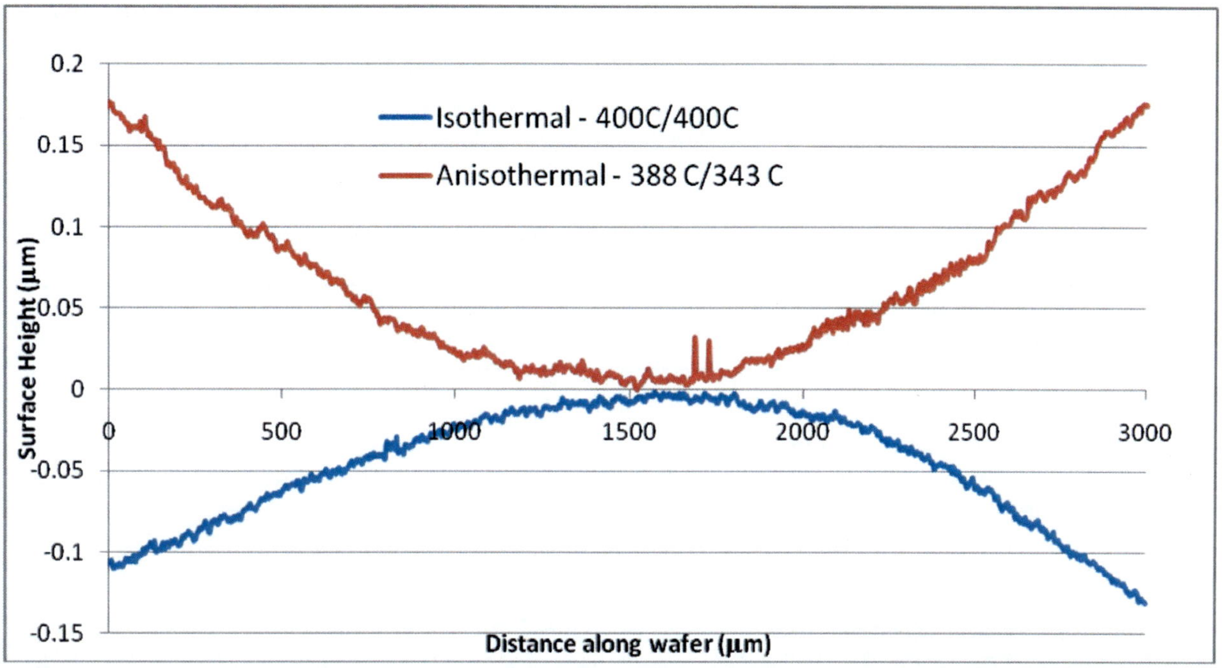

Figure 1: Representative local area line scans depicting wafer curvature for wafers bonded with Pyrex and silicon temperatures of (blue) 400°C, 400°C and (red) 388°C, 343°C respectively.

With no other change in bond process (in terms of time, power and input requirements), it is possible to completely reverse the wafer bow that results from anodic bonding. This simple modification to the bond recipe does not cause any changes to the processing time or require any addition power input. The only additional control parameter for anisothermal bonding is the independent control of platen temperatures, which can be done in a fairly straightforward manner with modern bonding equipment and control capabilities. The 45 °C temperature differential requirement to change the resultant wafer bow is small enough that it can easily be supported within a bond chamber without requiring special feedback or cooling of either chuck, yet it provides a large enough range that it should be possible to easily define a final curvature based on the anisothermal inputs. As such, structures can be designed to have a larger bow than shown here or to have bow as small as the virgin wafers prior to bonding.

As a means of comparison, an ABAQUS finite element model was developed, whereby silicon and Pyrex wafers are joined at their interfacial nodes with each wafer at a uniform, elevated wafer bonding temperature. The wafers then cool to room temperature and the resulting bulk wafer curvature is calculated based on the temperature dependent coefficient of thermal expansion for each wafer. It is assumed that both silicon and Pyrex are at uniform temperature after bonding. As a baseline, an isothermal bond procedure is studied, where both silicon and Pyrex wafers are initially at 400°C. In this case, a resulting wafer radius of curvature is estimated to be roughly -9.1m (silicon "concave" in our geometry). The Pyrex temperature is then reduced until a roughly opposite curvature is predicted by the model. When the silicon wafer is maintained at 400 °C and the Pyrex wafer temperature is reduced to 360 °C, the model predicts a curvature of roughly 8.5 m (silicon "convex" in our geometry).

Table 1 provides a comparison between experimental results and the corresponding numerical predictions roughly match the experimental conditions. Model and experiment agree to within approximately 20%, which is quite good for the simplicity of the model. This suggests that the bulk of the residual stress state can be explained by the mismatch in temperature dependent material thermal expansion coefficients. There are several other mechanisms that are not included in the model, including glass shrinkage and development of a depleted zone or initial wafer bow variations. The initial shape of the silicon and Pyrex wafers prior to bonding affects the overall bonded wafer shape and could be considered to further improve modeling accuracy. Furthermore, inaccuracies in the thermal measurement system in the bond systems can result in different anisothermal conditions than expected. Finally, the model assumes that each wafer is held at a constant and uniform temperature. The anodic bonder used in this study requires that the wafers be maintained in close proximity during the heating phase, separated by an air gap of less than 1 millimeter and with thin steel "flags in contact with the both wafers at three points on the outer periphery. As such, it is expected that the wafers will not be at a uniform temperature prior to bonding. The temperatures reported here are those measured at the bonding chuck in contact with the unbounded surface of both the Pyrex and silicon wafer. However, other bonder designs can provide a large gap between the wafers with no direct contact, reducing this concern.

Table 1: Experimentally measured and numerically predicted wafer pair radius of curvature.

	Experimental Radius of Curvature (m)	ABAQUS Predicted Radius of Curvature (m)
Isothermal (ΔT =~0 °C)	-8.1	-9.1
Anisothermal ((ΔT =~40 °C)	6.5	8.5

Figure 2 displays IR-GFP results from the isothermal and anisothermal bond recipes. It is immediately apparent that significantly different stress states are associated with these features. The residual shear stress patterns at these defects displays an opposite orientation, resulting from the concentration of a tensile stress field in the silicon wafer from isothermal conditions and a compressive residual stress field in the silicon wafer from the anisothermal conditions chosen for this investigation.

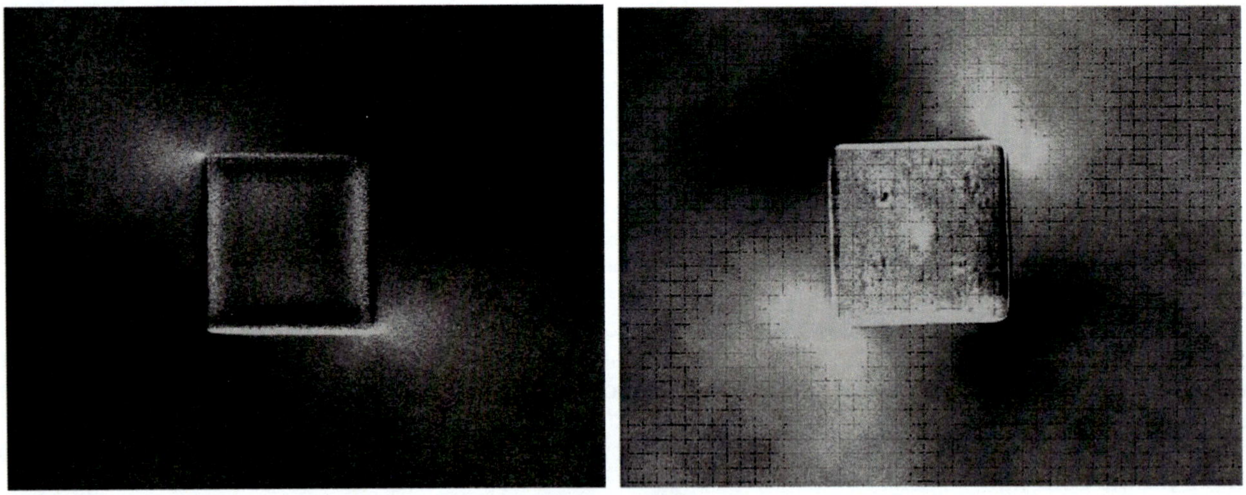

Figure 2: Representative IR-GFP images of an etched cavity in a bonded wafer pair using the isothermal (left) and anisothermal (right) bond recipes. Note the opposite orientation of the local residual shear stress field indicating that the anisothermal recipe results in a concentrated residual maximum normal stress distribution near the cavity corners.

IV. CONCLUSIONS AND FUTURE WORK

This study has provided the foundation for a simple method to control the residual stresses and thus resulting wafer bow in anodically bonded structures by controlling the initial bond temperatures of the silicon and Pyrex wafers independently. The process should be capable of inducing a range of residual stresses states including leaving the silicon devices layer in a stress free state or introduction of a residual compressive stress in the silicon wafer that often contained etched features. Finally, it was shown that the stress concentrations associated with etched features are modified through this process such that a highly tensile stress can be eliminated thus reducing the driving force for crack initiation and even providing a protective residual compressive field at these locations that may retard crack growth.

Further study in this area is required to understand the effects of nonisothermal conditions on bond strength, failure modes of the wafer pair, and to develop models to improve the predictive capability of the resulting wafer curvature based on additional stress generation mechanisms.

ACKNOWLEDGEMENTS

This material is based on work supported by the National Science Foundation under grant # CMS-0700704.

REFERENCES

1. Knowles, K. M., van Helvoort, A. T. J. Anodic Bonding. *Institute of Materials, Minerals and Mining*, **51** (5), 273-311, 2006.
2. Henmi, H., Shoil, S., Shoil, Y., et. al. Vacuum Packaging for Microsensors by Glass-Silicon Anodic Bonding. *Sensors and Actuators A*, **43**, 243-248, 1994.
3. Barth, P., Pourahmadi, F., Mayer, R., Poydock, J., Petersen, K. A Monolithic Silicon Accelerometer with Integral Air Camping and Over-range Protection. *Technology Digest IEEE Solid-State Sensor and Actuator Workshop, Hilton Head, SC*, **35**, 1988.
4. Schmidt, M. Wafer-to-wafer Bonding for microstructure Formation. *Proceedings of the IEEE*, **86** (8), 1575-1585, Aug. 1998.
5. Rogers, T. Considerations of Anodic Bonding for Capacitative Type Silicon/Glass Sensor Fabrication. *Journal of Micromechanics and Microengineering*, **2** (3), 164-166, 1992.
6. Rogers, T., Kowal, J. Selection of Glass, Anodic Bonding Conditions and Material Compatibility for Silicon-Glass Capacitive Sensors. *Sensors and Actuators: Physical A*, **46** (1-3), 113, 120, 1995.
7. Harz, M., Engelke, H. Curvature Changing or Flattening of Anodically Bonded Silicon and Borosilicate Glass. *Sensors and Actuators, A: Physical*, **55** (2-3), 201-209, 1996.
8. Harz, M., Bruckner, W. Stress Reduction in Anodically Bonded Silicon and Borosilicate Glass by Thermal Treatment. *Journal of the Electrochemical Society*, **143** (4), 1409-1414, 1996.
9. Cozma, A., Puers, B. Characterization of the Electrostatic Bonding of Silicon and Pyrex Glass. *Journal of Micromechanics and Microengineering*, **5** (2), 98-102, 1995.
10. Rogers, T., Aitken, N., Stribley, K., Boyd, J. Improvements in MEMS Gyroscope Production as a Result of Using In Situ, Aligned, Current-Limited Anodic Bonding. *Sensors and Actuators, A: Physical*, **123-124**, 106-110, 2005.
11. Madou, M. J., June (2005), *Fundamentals of Microfabrication: the Science of Miniaturization*, CRC Press, Boca Raton.
12. Kern, W. The Evolution of Silicon Wafer Cleaning Technology. *Journal of the Electrochemical Society*. **137** (6), 1887-1892, June 1990.
13. Horn, G., Lesniak, J., Mackin, T., Boyce, B. Infrared Grey-Field Polariscope: A Tool for Rapid Stress Analysis in Microelectronic Materials and Devices. *Review of Scientific Instruments*, **76**, (045180)-1-10, 2005.

Proceedings of the SEM Annual Conference
June 7-10, 2010 Indianapolis, Indiana USA

Sub-micron Scale Mechanical Properties of Polypropylene Fibers Exposed to Ultraviolet and Thermal Degradation

Nandula Wanasekara[1], Vijaya Chalivendra[2*] and Paul Calvert[1]
[1]Materials & Textiles Department
[2]Department of Mechanical Engineering
University of Massachusetts Dartmouth, MA 02747
*Corresponding author, vchalivendra@umassd.edu, 508-910-6572

ABSTRACT

Nanoindentation studies using atomic force microscopy (AFM) was conducted to investigate the effect of accelerated ultra violet (UV) and thermal degradation on mechanical properties of Polypropylene textile fibers. The Polypropylene fibers with initial stabilizers were exposed to UV degradation using Q-UV Panel Weatherometer. The effect of degradation on gradation of Young's modulus values across fiber cross-section was investigated by doing progressive nanoindentation from the surface to the center of the fiber. It was identified that UV degradation initially increases the Young's modulus values from center to surface of the fibers till 120 hours of exposure and the values show decreasing trend at 144 hours of exposure. The Youngs modulus values at 144 hours exposure are less than those of unexposed fibers. To investigate thermal degradation effect on Polypropylene fibers, the fibers were exposed to 125°C as a function of number of weeks in increments of one week for four weeks. Results indicate that the thermal exposure did not have much impact on variation of Young's modulus values for the first three weeks and showed increase in Young's modulus values at the surface when they are exposed to four weeks. The increase in Young's modulus values of Polypropylene fibers exposed to UV will be correlated with oxidation chemistry and micro-structural changes using Wide Angle X-ray spectroscopy and infra red spectroscopy techniques.

1. INTRODUCTION

Photo-oxidative degradation of polypropylene has been a major concern since its invention. This has drawn the attention of many researchers to investigate the mechanisms involved in degradation using various experimental techniques. The polypropylene melt is spun to fibers using a melt spinning process followed by drawing. Initial work was focused on finding out how these process parameters affect the degradation. The work conducted on weatherability of polypropylene involved laboratory extrusion of monofilaments and photo oxidation using a xenon arc lamp weatherometer. Oxidation is more likely to result from free radicals produced by bond rupture on drawing the polymer melt to fibers. It has been found that filament sensitivity to photo-oxidation is significantly affected by the extrusion and draw conditions and photosensitivity increases with increasing draw speed and decreasing draw temperature [1]. Microstructural changes that occur in polypropylene upon photo-oxidation were explained using established models of microstructure. These significant morphological and structural changes take place in crystalline or fibrillar regions showing much more resistance to scission than the amorphous regions [2]. It is well established that oxidation is confined to amorphous regions between crystallites since oxygen is insoluble in the crystal phase of polyolefins [3, 4].

Various types of analytical techniques were also used by researchers to characterize the non-uniformity of oxidation. These techniques include Fourier transform infrared (FTIR) Spectroscopy [5, 6] and photoacoustic spectroscopy [7]. A detailed study has been carried out using infrared absorption spectroscopy on the effect of molecular structure of polypropylene fibers that revealed the extent of changes on molecular structure of fibers under UV radiation depended on initial fiber structure, the added modifiers and macroscopic features [8]. Thermal degradation of polypropylene is also explained by similar mechanisms. The molecular weight distributions of polypropylene samples subjected to various degrees of thermal degradation were experimentally determined and compared with those expected theoretically for random scission of the polymer chains. The comparison demonstrated that the chain scission is completely random [9]. It has been found that the rate of oxidation increases with the increase in the partial pressure of oxygen and also with temperature. The most significant

T. Proulx (ed.), *MEMS and Nanotechnology, Volume 2*, Conference Proceedings of the Society for Experimental Mechanics Series 2,
DOI 10.1007/978-1-4419-8825-6_40, © The Society for Experimental Mechanics, Inc. 2011

observation is that in polypropylene films, the degree of crystallinity based on the infrared data was found to increase with the degree of oxidation of the polymer [10].

In addition to bulk effects, it is expected that fibers will show radial gradients in composition, structure, and properties. These can arise from non-uniform drawing, from surface oxidation during spinning, and evaporative loss of stabilizers during or after spinning. Since thermal and photo-oxidation are autocatalytic processes, these gradients may give rise to subsequent rapid degradation at the surface. Such rapid surface embrittlement would lead to premature failure of fibers or films. Direct demonstration of property gradients on the sub-micron scale was not possible when most work on polymer oxidation was carried out. The major product of the oxidation are carbonyl bonds and tertiary hydroperoxide, which can also become oxidized to give free radicals that cause oxidation chain reactions to propagate further [1, 12]. Thickness profiles of radio chemically aged low density polyethylene LDPE indicated higher concentrations of carbonyl groups closer to the surface [12]. Micro and nano-scale indentation methods have been used to measure modulus profiles after the heterogeneous oxidation of thick nitrile rubbers [13]. This is mainly based on measurement of modulus indirectly using the inverse tensile compliance indentation method. But this technique fails to detect changes in the sub-micron scale. The x-ray photoelectron spectroscopy (XPS) study on photo oxidized polypropylene films has shown enhanced surface oxidation and advance more into the film as the exposure level was increased

2. EXPERIMENTAL DETAILS

2.1. Sample Preparation

Polypropylene commercial fibers for the analysis were obtained by Providence Yarn Company, Inc., RI, USA. An unknown stabilizer had initially been used in the spinning process of fibers. The typical diameter of the fiber was 45 μm. The fibers were molded in an epoxy plug for AFM characterization. A two-part thermoset epoxy (Epothin, supplied by Buehler Ltd., IL, USA) was specifically chosen since it had the properties of room temperature curing with minimum heat generation during the curing process.

2.2. AFM Characterization

In this study, an atomic force microscope (XE-100, Park Systems Inc, CA, USA) was used in the topographic imaging and nanoindentation of polypropylene fibers. A contact mode scan was made prior to the indentation of the sample. A typical scanned image with indentation pattern is shown in Figure 1. This facilitates the user to locate the specific location for indentation. Nanoindentaions were performed on fiber cross sections in the radial direction. For each duration of UV and thermal exposure, more than 15 radiuses were selected for indentation and for each radius; 6 points were chosen as shown in Figure 1. The force-indentation depth diagrams were further processed using a previously established methodology to determine Young's modulus [14].

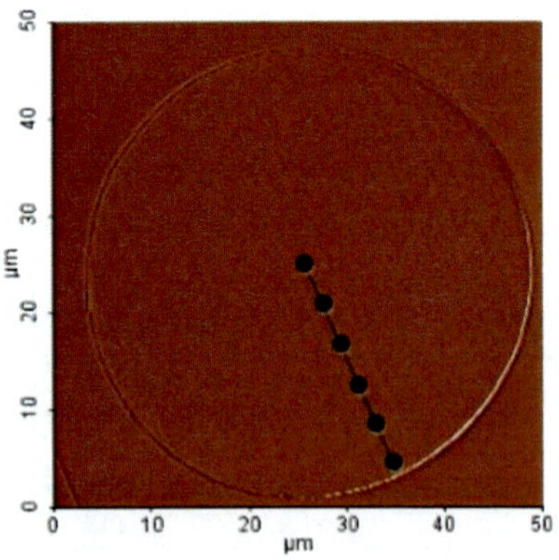

Figure 1: Indentation pattern across the radius of the fiber.

2.3. UV and Thermal Exposure of Fibers

The polypropylene fibers were subjected to Ultraviolet (UV) exposure using a Q-panel UV (QUV) weatherometer (QUV, Q-Lab Corp. OH, USA). Fluorescent UV lamps generated radiation in the QUV chamber. The UV wavelength of 340nm was chosen as per recommendation stated in ASTM G 154 standard for simulation of direct solar radiation, which allows enhanced correlation with actual outdoor weathering. The cyclic exposure conditions were eight hours of UV exposure at 50°C and four hours of condensation at 48°C. The polypropylene fibers were exposed in the QUV weatherometer for a maximum period of 144 hours in increments of 24 hours. Uniformity of exposure was confirmed by the nanoindentations performed on selected points on eight radii taken at 45° to each other. Thermal exposure of polypropylene fibers was carried out using a controlled temperature/humidity test chamber (AH-202XCC, Bryant Manufacturing Associates, MA, USA). The samples were exposed to a constant temperature of 125°C for a maximum duration of four weeks in increments of one week. A relative humidity level closer to zero was maintained throughout the test.

2.4. Tensile Testing

The uniaxial tensile testing of polypropylene fibers was carried out using the INSTRON 5569 materials testing system. A load cell having a maximum value of 2.5 N was used in the experiments. The single fiber tensile testing method described in ASTM G3822 was followed. A constant extension rate of 2.54 mm/min was used. The cross-head displacement was used to determine axial strain of the fiber and recorded force values were used to determine the axial stress. The tests were continued until there were failure of fibers to determine tensile strength and percentage elongation at break. Only tests that had failure within the gage length of the fibers were considered in this study.

2.5. Wide Angle X-ray and Fourier Transform Infrared Spectroscopy Analysis

WAXS was done using a set-up from Molecular Metrology, Inc. housed at University of Massachusetts, Amherst. It uses a 30w micro-source (Bede) with a 30x30 μm^2 spot size matched to a Maxflux^R optical system (osmic) leading to a low divergence beam of monochromatic CuK_α radiation (wavelength, $\lambda = 0.1542$). WAXS is performed using an image plate positioned in the sample chamber at a distance of 39 mm. The image plate (maximum resolution 50 µm) has a hole in its center. The whole system is evacuated while the test is conducted. Fourier transform infrared spectroscopy microscopy (Varian FTIR Bench and Microscope, CA, USA) was used to identify the chemical groups present in the fiber samples. The experiments were carried out in the reflectance mode. The percentages crystallinity of the fibers was calculated using the 'Polar' software.

3. RESULTS AND DISCUSSION

3.1. Nanoindentation of UV degraded polypropylene

The typical force deflection diagrams of unexposed and UV exposed polypropylene fibers obtained for the indentation point close to the surface are shown in Figure 2. The variation of Young's modulus values is shown in Figure 3. The closeness of the indenter to the surface is defined by the square pointer when its right most edge falls on top of the fiber edge. This point is located with an accuracy of $\pm$ 800 nm from the fiber circumference. The error bars shown in this figure (and all other figures of this article) were determined using student t-distribution with a 95% confidence level. The initial gradient of the unloading portion of the curve increases for up to 120 hours exposure and then it drastically reduces by 144 hours. These raw data curves illustrate the rationale behind increasing Young's modulus values of fibers for up to 120 hours of UV exposure and then reducing Young's modulus values for fibers with 144 hours of UV exposure. Also it is clearly evident from the above graph that residual indentation depth is higher for 144 hours of UV exposure. The retraction force or the adhesion force almost remained the same for all fibers.

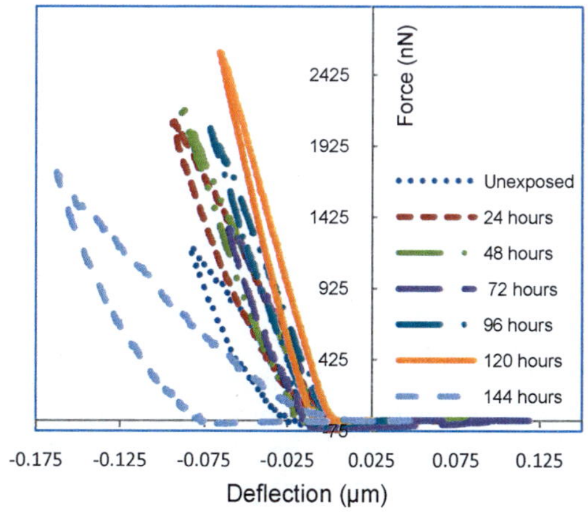

Figure 2. Typical force deflection diagrams of nano-indentation for polypropylene fibers of different UV

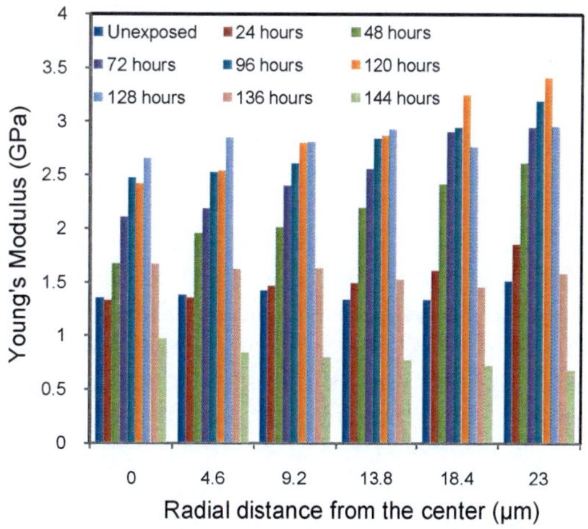

Figure 3. Variation of Young's modulus of UV exposed polypropylene fibers

The Young's Modulus values, which were obtained using nanoindentation of unexposed fibers suggests that the initial degradation in manufacturing and processing stages is insignificant. Nanoindentation on UV exposed fibers proves that UV degradation initially increases the Young's modulus values from center to surface of the fibers until 120 hours of exposure and then reducing when the exposure is further increased. The Young's modulus values at 128 and 136 hours show lower values than that of 120 hours. This implies that fiber microstructural changes causing reduction of mechanical properties start at 120 hours. The modulus values are the lowest at 144 hours of exposure. The Young's modulus values at 144 hours exposure are 50.9% of modulus values of unexposed fibers near the surface. Young's modulus values near to the surface of 120 hours UV exposed fibers showed an increase of 144.84% and at the center showed an increase of 73% compared to unexposed ones.

The increase of Young's modulus values can be attributed to the increase in crystallinity as discussed in section 3.2. The oxidation process causes the scission of tie molecules in the disordered regions, which allows further crystallization [2]. The noticeable reduction of Young's modulus values for fibers with 144 hours of UV exposure can be attributed to complete destruction of fiber crystalline regions. This leads to the observed reduction of mechanical properties which is evident from the results of percentage elongation to break and tensile strength which report lowest values as discussed in section 3.3.

3.2. Wide Angle X-ray and Infrared Analysis

WAXS was performed on whole fiber to understand the changes of the fiber crystallinity. Unexposed polypropylene diffractions are arced with some spots indicating the crystallinity and significant molecular orientation. At an ultraviolet exposure of 120 hours, the diffraction arcs are turned into sharp spots and have been very prominent implying that crystallinity and crystal orientation are very highly developed. The 2D WAXS image of polypropylene exposed to 144 hours showed 'NO' crystallinity at all, only the amorphous halo was seen indicating the breakdown of crystal lamellar structure of polypropylene fibers at 144 hours. These qualitative observations were verified by the calculation of percentage crystallinity. Table 1 shows the percentage crystallinity for each duration of exposure. The unexposed fibers were 48.7 % crystalline while the 120 hours UV exposed fibers were 71.13 % crystalline. When the UV exposure is further increased to 144 hours the crystallinity reduced to 32.16 %.

Table 1: Percentage crystallinity for UV exposed PP fibers

UV Exposure Hours	Percentage Crystallinity (%)
Unexposed	48.70
24	69.68
48	67.71
72	66.58
96	66.70
120	71.13
128	39.62
136	38.13
144	32.16

Fourier transform infrared spectroscopy was performed on fibers to understand the new formations of chemical bonds upon UV exposure. Figure 6 shows the infrared spectra obtained for fibers. The significant observation is the formation of carbonyl bond (C=O) corresponding to wavelengths of 1681 – 1753 cm^{-1} with increase of exposure hours. This observation was in accordance with previous literature [1]. The area of the carbonyl peak was calculated using mathematical software. The carbonyl index shows an increase up to 120 hours of exposure and shows a kind of stabilizing trend towards the end.

3.3. Tensile Testing Results

Figure 4 represents the engineering stress- strain diagrams for UV exposed fibers as a function of exposure duration. Young's modulus values observed from stress-strain graphs for polypropylene fibers with different hours of UV exposure showed no significant change. This observation, together with higher elongation to break of unexposed fibers, suggests that the experimented fibers had medium orientation of molecular chains. Using the stress-strain diagrams, the percentage elongation at break and tensile strength of UV exposed fibers as a function of exposure hours is obtained and shown in Figure 5. The unexposed polypropylene fibers had significant percentage elongation at break of more than 100%. A significant reduction could be observed in percentage elongation at break with increase of number of UV exposure hours. This loss of toughness on photo-oxidation would be crucial for many applications of polypropylene fibers in industrial textiles. The tensile testing results reported in the literature has also shown similar pattern of reduction in elongation to break with increase of exposure times [2].

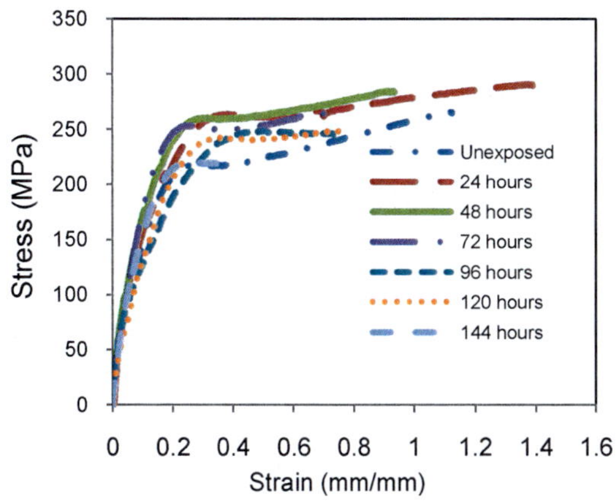

Figure 4. Stress-strain curves for UV exposed polypropylene fibers.

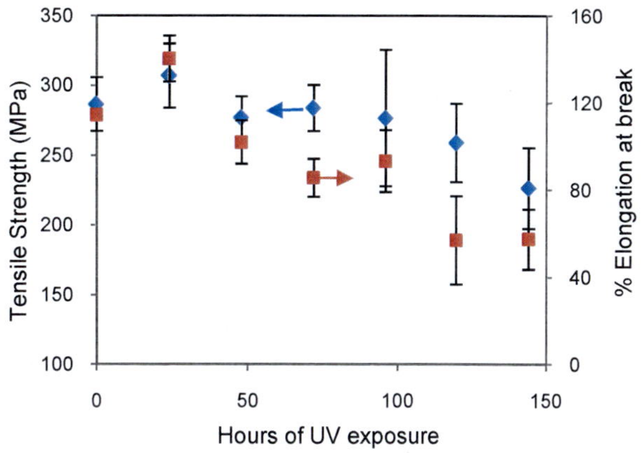

Figure 5. Variation of tensile strength and percentage elongation to break of polypropylene fibers with number of UV exposure hours.

The tensile strength of 24 hours UV exposed fibers are slightly higher than unexposed fiber and 48, 72 and 96 hours exposed fibers maintain almost same tensile strength as unexposed and slight reduction is noticeable at 120 hours. The tensile strength of 144 hours exposed fibers is much less with compared to the unexposed fiber. These results suggest the increase of Young's modulus values up to 120 hours does not necessarily increase the tensile strength of fibers. Our results are worthwhile to note that increase in crystallinity up to 120 hours by chain scission in non-crystalline regions does not cause formation defects that affect the tensile strength. But when chain scission starts at crystalline regions at 144 hours of exposure, the defects start to create and tensile strength is reduced.

3.4. Thermal Degradation

3.4.1. Nanoindentation and Tensile Testing Results

Variation of Young's modulus values of thermally exposed fibers from the center to surface is shown in Figure 6. Young's modulus values were found to be almost the same for fibers up to three weeks of thermal exposure. A significant increase of Young's modulus values were observed closer to the surface in the fourth week, while the modulus values in the center remained the same. This implies that thermal degradation starts to have an effect on fiber microstructure closer to the surface on the fourth week at 125°C temperature and may propagate to fiber interior with an increase of time. The observed increase in Young's modulus values can again be attributed to the increase in crystallinity [8] caused by relaxation of chains scissioned by thermal energy to folded structures within the amorphous regions. Thermal degradation experiments were also carried out at a higher temperature of 150°C. It was found that the polypropylene fibers were degraded to powdery substances at the end of first week. This implied that thermal degradation is accelerated at temperatures closer to the melting temperature (which is around 160°C) of polypropylene.

Variation of tensile strength and percentage elongation to break of polypropylene fibers as a function of weeks of thermal exposure is shown in Figure 7. Percentage elongation at break values showed no significant change for fibers until three weeks of thermal exposure. A slight reduction of elongation at break is noticed at four weeks of exposure. It can be anticipated from these results that the continuous exposure of polypropylene fibers for more than four weeks can lead to significant reduction of elongation to break. Tensile strength values are the same for the first three weeks of thermal exposure and a slight reduction for four weeks of exposure is noticed

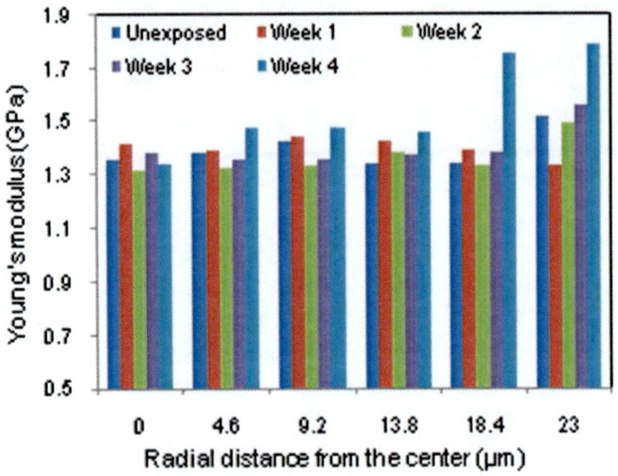

Figure 6: Variation of Young's modulus values across the radius of polypropylene fibers with thermal exposure.

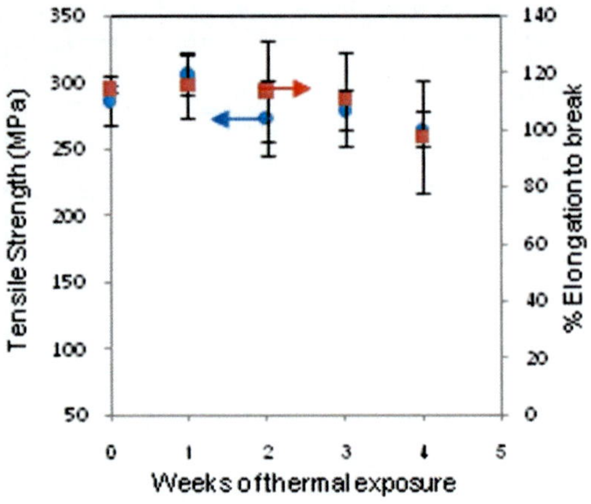

Figure 7: Variation of tensile strength and percentage elongation at break of polypropylene fibers with weeks of thermal exposure.

4. Conclusion

Photo-oxidation during the manufacturing process has no significant effect on sub-micron scale measurement of Young's modulus across the polypropylene fibers. Nanoindentation provided gradation of Young's modulus values across the fiber cross section, however the macro-tensile testing provided no significant change in Young's modulus as a function of UV and thermal exposure hours. It was evident that there is a significant increase of Young's modulus values near the surface than at the center as the UV exposure duration increases to 120 hours. The Young's Modulus values of UV exposed fibers for 144 hours showed a remarkable decrease of Young's Modulus values even less than the unexposed ones. WAXS images proved the increase of crystallinity up to 120 hours of exposure and total destruction of crystallinity at 144 hours. FTIR spectra proved the formation of carbonyl bonds with UV exposure. Nanoindentation on polypropylene fibers subjected to thermal exposure showed no significant change. A slight increase of Young's modulus values can be noticed near to the surface of the fibers for four weeks of thermal exposure. Macro tensile testing results verified the sub-micron scale nanoindentation results.

Acknowledgements

The authors greatly appreciate the financial support from the National Science Foundation through Grant CMS0618119 for acquiring an atomic force microscope. Also we would like to acknowledge the financial support from the National Textile Center through project number M08-MD13. And finally, we will like to acknowledge Mr. Naresh Mehta from Cooley Group, RI, USA for letting us use their weatherometer and Providence Yarn Company, Inc., RI, USA for providing free samples of polypropylene fibers.

References

1. Carlsson DJ, Garton A, and Wiles DM. "The weatherability of polypropylene monofilaments", *Journal of Applied Polymer Science* 21: 2963-2978, 1977.
2. Blais P, Carlsson DJ, Clark FRS, Sturgeon PZ, and Wiles DM. "The photo-oxidation of polypropylene monofilaments Part 2: Physical Changes and Microstructure", *Textile Research Journal* 46: 641-648, 1976.
3. Billingham NC, Calvert PD. *Development of Polymer Stability* 3: 139, 1980.
4. Allen NS. In: Allen NS, editors, "Degradation and Stabilization of Polyolefins", London: *Applied Science Publishes* 1-28, 1983.
5. Adam C, Lacoste J, Lemaire J, "Photo-oxidation of elastomeric materials: Part 3—Photo-oxidation of acrylonitrile-butadiene copolymer", *Polymer Degradation and Stability* 24: 185, 1989.
6. Jouan X, Adam C, Fromageot D, Gardette JL and Lemaire J. *Polymer Degradation and Stability*, 25: 247-265, 1989.

7. Delprat P, and Gardette JL. “Analysis of photo-oxidation of poymer materials by photoacoustic Fourier Transform Infrared spectroscopy”, *Polymer* 34: 933-937, 1993.
8. Lipp-Symonowicz B, Sztajnowski S, Kardas I. “Influence of UV radiation on the mechanical properties of polyamide and polypropylene fibers in aspect of their restructuring”, *AUTEX Research Journal* 6: 196-203, 2006.
9. Davis TE, Tobias RL, Peterli EB. “Thermal Degradation of Polypropylene”, *Journal of Polymer Science* 56: 485-499, 2003.
10. Ismat Abu-Isa. “Thermal degradation of thin films of isotactic polypropylene and polypropylene with ketonic additives”, *Journal of Polymer Science. Part A-1 Polymer Chemistry* 8: 961 – 972; 2003.
11. Carlsson DJ, Clark FRS, Wiles DM. “The photooxidation of Polypropylene monofilaments, Part I, Chemical Changes in Mechanical Deterioration”, *Textile. Research Journal* 46: 590-599, 1976
12. Audouin L, Langlois V, Verdu J, Bruijn de JCM. “Role of oxygen diffusion in polymer ageing: kinetic and mechanical aspects”, *Journal of Material Science* 29: 569-583, 1994.
13. Gillen KT, Terrill ER and Winter RM. “Modulus Mapping of Rubbers using micro and nano indentation techniques”, *Rubber Chemistry and Technology* 74:3: 428-450, 2001.
14. Reynaud C, Sommer F, Quet C, Bounia El N and Duc TM. “Quantitative determination of Young’s modulus on a biphase polymer system using Atomic force microscopy”, *Surface Interface Analysis* 30**:** 185-189, 2000.